Volume 7 · 1987 · Number 1

Acta Biotechnologica

Journal of microbial, biochemical and bioanalogous technology

Akademie-Verlag
Berlin

ISSN 0138-4988
Acta Biotechnol., Berlin 7 (1987) 1, 1–100

Instructions to Authors

1. Only original papers that have not been published previously will be accepted.
Manuscripts may be submitted in English, German or Russian (in dublicate). The name of the institute (with the full address) from which the manuscript originates should be stated below the name(s) of the author(s). The authors are responsible for the content of their contributions.

2. Original papers should not consist of more than 20 typewritten pages (two-lined spacing), including references, tables and figures; short original communications may have a maximum of six typewritten pages.

3. Each paper should be preceded by a summary in English and also by an English translation of the title.

4. Latin names of species as well as passages to be printed in italics for greater emphasis should be denoted by a waving line. Please use only units and symbols of the SI-system.

5. Tables may be used to shorten the text or to make understanding easier. They should be numbered consecutively throughout and supplied with a brief heading. They should not appear in the text, but should be written on separate sheets.

6. The numbers and sizes of illustrations should be limited to the essential minimum, they should be numbered consecutively and be supplied on separate sheets.
Line drawings, including graphs and diagrams, should be drawn in black ink. Half-tone illustrations should be supplied as white glossy prints. Figure legends are to be typed in sequence on a separate sheet.
The name(s) of the author(s) should be written on the back of each sheet.

7. References listed at the end of the contribution should contain only works quoted in the text. They should be numbered in the order in which they are first mentioned in the text.
Please give surnames and initials of all authors, the name of the journal abbreviated according to "Chemical Abstracts — List of Periodicals", volume number, year of publication, issue number or month, first page number. Books are to be cited with full title, edition, volume number, page number, place of publication, publisher and year of publication.

8. Notes to the text may be given as footnotes on the same page.

9. 50 printed copies are free of charge.
Additional reprints may be ordered (at full cost).

10. The author will receive two galley proofs for correction. They are to be returned to the managing editor (Permoserstr. 15, DDR - 7050 Leipzig, Dr. L. Dimter) as soon as possible.

Acta Biotechnologica

Journal of microbial, biochemical and bioanalogous technology

Edited at the
Institute of Biotechnology of the Academy of Sciences of the G.D.R., Leipzig
and at the
Kombinat of Chemical Plant Construction Leipzig—Grimma
by
M. Ringpfeil, Leipzig and G. Vetterlein, Leipzig

Volume 7

1987

Number 1

AKADEMIE-VERLAG · BERLIN

"Acta Biotechnologica" publishes original papers, short communications, reports and reviews from the whole field of biotechnology.
The journal is to promote the establishment of biotechnology as a new and integrated scientific field.
The field of biotechnology covers microbial technology, biochemical technology and the technology of synthesizing and applying bioanalogous reaction systems.
The technological character of the journal is guaranteed by the fact that papers on microbiology, biochemistry, chemistry and physics must clearly have technological relevance.

Terms of subscription for the journal "Acta Biotechnologica"

Orders can be sent

- **in the GDR:** to Postzeitungsvertrieb or to the Akademie-Verlag Berlin, Leipziger Str. 3–4, PF 1233, DDR - 1086 Berlin;
- **in the other socialist countries:** to a book-shop for foreign languages literature or to the competent news-distributing agency;
- **in the FRG and Berlin (West):** to a book-shop or to the wholesale distributing agency Kunst und Wissen, Erich Bieber, Wilhelmstr. 4–6, D-7000 Stuttgart 1;
- **in the other Western European countries:** to Kunst und Wissen, Erich Bieber GmbH, General Wille-Str. 4, CH-8002 Zürich;
- **in other countries:** to the international book- and journal-selling trade, to Buchexport, Volkseigener Außenhandelsbetrieb der DDR, PF 160, DDR - 7010 Leipzig, or to the Akademie-Verlag Berlin, Leipziger Str. 3–4, PF 1233, DDR - 1086 Berlin.

Acta Biotechnologica

Herausgeber: Institut für Biotechnologie der AdW,
Permoserstr. 15, DDR - 7010 Leipzig
(Prof. Dr. Manfred Ringpfeil) und

VEB Chemieanlagenbaukombinat Leipzig–Grimma,
Bahnhofstr. 3–5, DDR - 7240 Grimma
(Dipl.-Ing. G. Vetterlein).

Verlag: Akademie-Verlag Berlin, Leipziger Straße 3–4, PF 1233, DDR - 1086 Berlin;
Fernruf: 2236201 und 2236229; Telex-Nr.: 114420; Bank: Staatsbank der DDR, Berlin,
Konto-Nr.: 6836-26-20712.

Redaktion: Dr. Lothar Dimter (Chefredakteur),
Käthe Geyler (Redakteur),
Permoserstr. 15, DDR - 7010 Leipzig; Tel.: 2392255.

Veröffentlicht unter der Lizenznummer 1671 des Presseamtes beim Vorsitzenden des Ministerrates der Deutschen Demokratischen Republik.

Gesamtherstellung: VEB Druckhaus „Maxim Gorki", DDR - 7400 Altenburg.

Erscheinungsweise: Die Zeitschrift „Acta Biotechnologica" erscheint jährlich in einem Band mit 6 Heften. Bezugspreis eines Bandes 180,— DM zuzüglich Versandspesen; Preis je Heft 30,— DM. Der gültige Jahresbezugspreis für die DDR ist der Postzeitungsliste zu entnehmen.
Bestellnummer dieses Heftes: 1094/7/1.

AN (EDV) 18520

03000

Acta Biotechnol. 7 (1987) 1, 3—8

Oil Feeding during the Oxytetracycline Biosynthesis

Ettler, P.

Research Institute for Antibiotics and Biotransformations
252 63 Roztoky u Prahy, CSSR

Summary

The ability of utilization of vegetal oils belongs to common properties of *Streptomycetes*. During the cultivation of *Streptomyces rimosus* and biosynthesis of oxytetracycline in 15 m^3 fermenters, a fixed regime of soya bean oil feeds as antifoam agent and partial carbon source according to subjectively designed mechanism was investigated and further also feeding regime after exhausting starch as main carbon source and finally his dosing with ammonia sulphate in accordance with physiological activity of the culture. In the second case with semicontinuous supplements of soya bean oil it was possible to prevent premature autolysis of mycelia and to extend the biosynthetic process with keeping the oxytetracycline production rate. The third approach, the most progressive one, represents besides the automatisation of control loop the necessity of assurance continuous pH adjustment for elevation of the process production rate.

Introduction

Metabolic activity of producing microorganism can be influenced by the adjustment of chemical potential of substrates, intermediates and products by changing the cell permeability, by the change of activity of one or more enzymes together with a multitude of genetical interventions to the functional activity of cells [1]. The lack of sterilizable sensors for the specific determination of adequate substrates in media is commonly known [2]. Therefore, for the technological practice are more advantageous indirects methods of substrates dosing [3].

Nowadays practically not one process for antibiotics biosynthesis can be considered as a typical batch process. Different ways of the feeding performance were summarized by Pirt [4], for several of them mathematical apparatus for optimization of substrate feeding velocity were developed [5—7].

For antibiotic production feed -on- demand control can be solved by dosing the programme so that the accumulation of the base amount of biomass can be followed by shift to a lower value of specific growth rate assuring high antibiotic formation rate [8]. Several autors [8] regulate the nutrient feed according to D.O.C., pH, RQ, OUR, whilst Küenzi [9] considers as simplest one the dosing on a basis of a constant designed in advance.

In this work, finding of optimal monitoring system for dosing of soya-oil as antifoam agent and partial carbon source during the cultivation of *Streptomyces rimosus* and biosynthesis of oxytetracycline has been studied.

Material and Methods

Experimental studies were performed in 15 m^3 fermenters whose geometry was described earlier [10]. Complex nutrient medium contained starch, dry yeast and ammonia sulphate as main carbon and nitrogen sources. As chemical antifoaming agent was used vegetal oil, mainly soya bean oil. The inoculation material of industrial strain *Streptomyces rimosus* was prepared from the 2nd vegetative generation. Exit gas analysis of CO_2 was performed by Uras 2 analyser (Hartmann-Braun, FRH). Spectrum of analytical methods citated in work [10] was extended by simplified extraction of lipids into ether [11].

Results and Discussion

The optimization of the OTC biosynthesis can be seen in these spheres:

- the manner of production media preparation with the exploitation of knowledge about hydrolysis of polysaccharides;
- the manner of inoculation material preparation with the induction of hydrolytic activity in steps of inoculation;
- the elevation of initial concentration of substrates in agreement with historical trend of enrichment of media;
- the development of fed-batch technology.

After the technological balance about the mode of feed dosing the experiments were focused on soya-bean oil [12]. In laboratory scale, maximal concentration of soya-bean oil utilizable either for the growth or for product formation was checked. It was established that to the concentration of 6% soya-bean oil the oxytetracycline formation rate is not depressed. These laboratory experiments were performed with the knowledge that this scale is not optimal for complete evaluation of the influence of surface-active agents on the biosynthesis of OTC.
The continuous slow feed of animal or vegetable oils is frequently found to be an economically attractive method for maintaining antibiotic fermentations at a highly productive rate [13, 14]. In order to maintain a culture in the stationary phase and cover the maintenance energy needs, sugars or oil feeds are used. The general reactions for the oxidation show that oil contains 1.9 times the amount of carbon compared to glucose on a weight basis. However, on a per 10^4 kcal ($4186 \cdot 10^4$ J) oil price corresponds to 40% of the sugars cost and is roughly equivalent to in cost to starches and molasses in world. The mechanism of the simultaneous utilization of both C- sources is unknown. Hypothetically the ability of oil to regulate the pH of the broth and to elevate the "metabolic pool" of acetyl Co A [14] was published. The consumption of fatty acids depends on the chain length and the degree of their unsaturation [15].
For oil feeds in 15 m^3 fermenters following strategy was chosen:

a) the evaluation of fixed doses, prevention of broth foaming, the possibility of alternative use of soya-bean oil as a carbon source,
b) dosing of the soya-bean oil after the exhaustion of the starch as main carbon source,
c) dosing of the emulsion of soya-bean oil with nitrogen source.

For correct selection of feeding strategy the demands of *Streptomyces rimosus* by calculation important metabolic parameters as specific rate of antibiotic synthesis (q_{OTC}), specific rate of carbohydrate combustion, specific uptake rate of ammonia nitrogen, and specific oxygen uptake rate were determined. The last two coefficients are difficult to evaluate for the whole process, while the level of ammonium ions is controlled and the elevation of dissolved oxygen concentration from the limitation level can be hardly classified as the consumption of oxygen in a given period of time according to the principles of quantitative physiology [16, 17]. In Table 1 it can be seen that the maximal elevation of specific rate of antibiotic synthesis can be localized between 30 and 40 h of cultivation

Table 1. The determination of demands of *Streptomyces rimosus* in 15 m^3 fermenter during the OTC biosynthesis

time period Δt [h]	0–10	10–20	20–30	30–40	40–50	50–60	60–70
q_{OTC} [μg OTC/g wet biomass $\times \Delta t$]		2.6		45.8	60	100	150
$q_{C/X}$ [g of starch/ g of wet biomass $\cdot \Delta t$]	0.38	0.38	0.58	0.66	0.3	0.25	0.25

time period Δt [h]	70–80	80–90	90–100	100–110	110–120	120–130	130–140
q_{OTC} [μg OTC/g wet biomass $\cdot \Delta t$]	183	166	300	233	150	150	100
$q_{C/X}$ [g of starch/ g of wet biomass $\cdot \Delta t$]	0.66	1.4	2.5	5.0	6.0	8.0	10.5

and it is in relation with the end of secondary mycelium formation and exhaustion of inorganic phosphate. More illustrative is Fig. 1., from which it follows that the specific rate of product formation expressed in µg of active substance on the unit of wet biomass (sediment) in time elevates to 60 h, afterwards the increase is insignificant. Therefore, the aim of feeding technology should be to shift this coefficient to a quantitative higher level using higher metabolic activity of cells and amelioration of their physiological

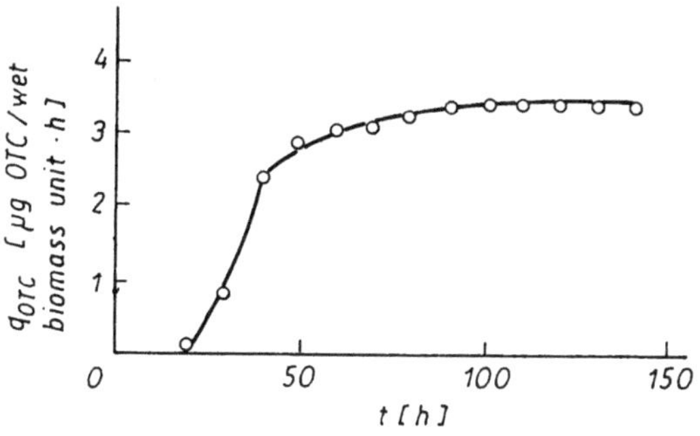

Fig. 1. The profile of the specific production rate (q_{OTC}) during the biosynthesis of oxytetracycline.

state to prolong the period of production phase. Calculating the specific uptake rate of carbon source for fixed time period, two maxima can be determined; the first one between 30 and 40 h corresponds to maximal utilization of carbon substrate, maximal growth of the culture and maximal release of CO_2 [10]. The elevation of calculated coefficient at the end of the biosynthetic process corresponds to stationary growth of the culture.

Fixed regime of oil additions was choosen so, that maximal doses of soya-bean oil correspond to coeficient q_{OTC} determined and introduced in Table 1 and to maximal enzymatic activity of α-amylase (α-1,4-glucan-4-glucanohydrolase) and anhydrotetracyclineoxidase [16].

From the 112 h of cultivation a partial autolysis of mycelia becomes visible, a drop in amylolytical activity with a stagnation of OTC production was found. Therefore, the process could not be prolonged.

Table 2. Regime of fixed doses of soya-bean oil during the biosynthesis of OTC

cultivation time [h]	48	60	68	84	90	96	108	114	120	126	130	138
concentration of the oil in batch [vol%]	0.1	0.1	0.2	0.2	0.3	0.3	0.3	0.2	0.2	0.2	0.2	0.2

The total amount of the oil in feeds was 2.4% of batch volume. This part of experimental study can be closed with the statement that antropomorfical access to the determination of the extent of soya-bean oil doses was without a positive point of incidence on the process productivity elevation. We confirmed the results of Bajpai and Reuss [7] which consider fixed regime of feeds as a naive one.

Results of dosing the soya-bean oil after the exhaustion of starch as a main carbon source are shown in Table 3.

Table 3. Feeds time regime of soya-bean oil doses after the exhaustion of the main carbon source

cultivation time [h]	114	117	120	123	126	129	132	135	138	141	148
concentration of the oil in batch [vol%]	0.39	0.35	0.37	0.33	0.39	0.38	0.30	0.30	0.34	0.35	0.44

Dosing of the soya-bean oil was started at the starch concentration $\sim$ 10 $g \cdot l^{-1}$ with the knowledge of the fact that in the liquid phase there is still a multitude of lower grafts of polysaccharide. The total amount of the oil in feeds was 1.3% of the batch volume in this case. The magnitude and the frequency of soya-bean oil doses was determined from the slope of carbohydrate uptake rate at various times. For the maintenance of conformable oxygen-transfer rate, the aeration rate was elevated after the initiation of oil feeding from 0.33 VVM to 0.7 VVM. The elevation of aeration is described as a mean for raising of secondary metabolite production with the use of oily substrates and also helps for sampling of homogeneous aliquots from the bottom of the fermentation vessel [21, 22].

Fig. 2 shows differences in course of the culture growth determined by the sediment content in fed-batch culture and the check variant without any additions of oil. From the results it is obvious that semicontinuous addition of soya-bean oil can prevent from

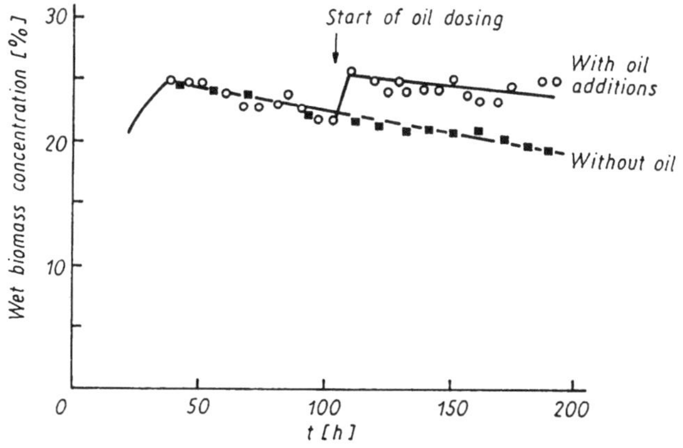

Fig. 2. Growth patterns of *Streptomyces rimosus* cultivation in 15 m³ fermenter.

partial autolysis of mycelia and makes possible to lengthen the process at preservation of the same production rate.
The stimulating effect of the oil was connected with the accumulation of large amount of biomass rather than with its specific effect on the biosynthesis. In the penicillin biosynthesis [14] oil accelerates the culture growth in the presence of lactose.

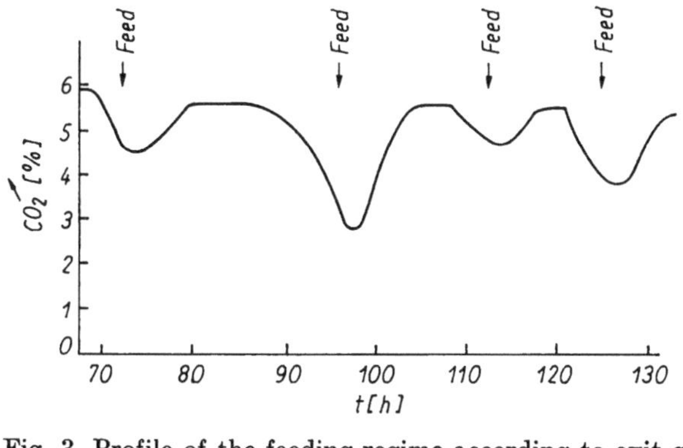

Fig. 3. Profile of the feeding regime according to exit gas analysis during OTC biosynthesis

As the most perspective from the point of view of the production elevation appeared complex nutrient feed with C- and N-sources. New findings from the rationalization of antibiotics biosynthesis [7, 8, 20] are dealing with the necessity to form the optimal environmental conditions for producing microorganisms excluding the shocks in more intense changes of pH, gas hold-up, osmotic pressure that is influenced by the concentrations of the nutrients and the surface tension. All these realities lead us to subordination of all the control loops in the biosynthesis of OTC in 15 m³ vessels to parameter or complex of several parameters that are reflecting the physiological state of the culture in relation to hydrodynamics of the process. Analyzing the concentration of CO_2 in the exhaust air it was established that after the peak between 30 and 40 h when respiration reaches till 6% of CO_2 appears permanent decrease until the end of the biosynthesis [10]. Fig. 3 demonstrates the mode of dosing C- and N-feed. The dosing of feed was performed from 60 h of cultivation with the aim to interact flexibly on the decrease of physiological activity with the nutrient addition at simultaneous buffering of secondary influences. Continuous pH adjustment to physiological level belongs to one of them The increased concentration of carbon source followed by acceleration of catabolic process produce namely the pH decrease.

As an impulse for switching the feed dose a register slope was chosen after three hours decrease without its interruption minimally of 1%. The production rate can be elevated by this dosing method. For total automatization of feeding regime and elimination of the subjectivity of human attendance factor, it is possible to connect the phase current entering into a register of the gas analyser and to derivative regulator that binds (according to the velocity of decrease of the respiration) the feed pump with the time pause for filtration of fluctuations and process disturbances.

Received October 23, 1985

References

[1] Edwards, V. H.: Biotechnol. Bioeng. **12** (1970), 679.

[2] Sikyta, B.: Bioinženýrské metody ve fermentační technologii. Kurs der ČSVTS (Kammer der Technik) MBÚ ČSAV, Praha, 1983.

[3] Adámek, Z.: II. Konferenz: Racionalisace kvasných výrob, Zvikov, ČSSR, 1975.

[4] Pirt, J.: J. Appl. Chem. Biotechnol. **24** (1974), 415.

[5] Yamane, T., Kume, T., Sada, E., Takamatsu, T.: J. Ferment. Technol. **55** (1977) 6, 587.

[6] Kao, E. I.: Biotechnol. Bioeng. **18** (1976), 1493.

[7] Bajpai, R. K., Reuss, M.: Biotechnol. Bioeng. **23** (1981), 717.

[8] Icygin, C. B., Biriukov, V. V.: Chim. farm. J. **10** (1976), 82.

[9] Küenzi, M. T.: Process design and control of antibiotic fermentations. FEMS Basel, 1977.

[10] Ettler, P., Cechner, V.: Konferenz: Entwicklung von Laborfermentoren. Sonderband der AdW der DDR. Reinhardsbrunn, DDR, 1978. 129.

[11] Páca, J., Ettler, P., Grégr, V.: J. Ferment. Technol. **56** (1978) 2, 147.

[12] Rattray, J. B. M.: JAOCS **61** (1984) 11, 1701.

[13] Bader, F. G., Boekeloo, M. K., Graham, H. E., Cagle, J. W.: Biotechnol. Bioeng. **26** (1984), 848.

[14] Lurje, L. M., Stepanova, N. E., Bartosevitch, J. E., Levitov, M. M.: Antibiotiki **2** (1979), 86.

[15] Řezanka, T., Vanek, Z., Klanova, K., Podojil, M.: Folia Microbiol. **29** (1984), 306.

[16] Ryu, D. D. Y., Hospodka, J.: Biotechnol. Bioeng. **22** (1980), 289.

[17] Queener, S., Swartz, R.: Secondary Products of Metabolism. Ed.: Rose, A. H. San London, New York, San Francisko: Academic Press, 1979. 35.

[18] Zajcava, S. M., Orlova, N. V.: US P 124436 (1959).

[19] Ettler, P.: Conference on Actinomycetes. Liblice, CSSR, 1982.

[20] Dmitrieva, S. V., Ivankova, T. A., Zaslavskaja, P. L., Kovalev, V. N., Listvonova, S. N.: Antibiotiki **27** (1982), 502.

[21] Kovalev, V. N., Ivankova, T. A., Bylinkina, E. S.: Antibiotiki **4** (1982), 263.

[22] Procházka, P., Nohýnek, M., Vaněk, Z., Rokos, J.: Folia Microbiol. 28 (1983) 5, 406.

Acta Biotechnol. 7 (1987) 1, 9–16

Studies on Biosynthesis of Hydrolases by Trichoderma sp. M_7 on Submerged and Solid-State Cultivation Conditions

ATEV, A. P., PANAYOTOV, Ch. A., BOBAREVA, L. G., DAMYANOVA, L. D., NICOLOVA, U. D.

Sofia University „Kliment Ohridski"
Department of Engineering Biology
8, Dragan Tzankov Bld, Sofia, Bulgaria

Summary

The biosynthesis of cellulases and xylanase by the mould strain *Trichoderma* sp. M_7 on submerged and solid-state cultivation conditions has been studied. The effect of different inducers on the enzyme biosynthesis on the conditions used was determined. The relation between the enzyme biosynthesis and the morphological state of the producing strain was studied. The advantages of the submerged cultivation conditions with regards to the efficiency of the enzyme inducer are shown.

Introduction

The submerged and the solid-state methods of cultivation of cellulases and hemicellulases producing strains are a topical trend of the present day biotechnology. Both methods have their advantages and disadvantages and therefore, they are used concurrently for production of hydrolases by filamentous growing strains on insoluble in water substrates [1–6].

While the submerged culture — batch and continuous — has its theory, the solid-state cultivation theory is not yet developed [1] nevertheless that this method has been known since ancient times [1, 4, 7]. There are many problems to be solved with the solid-state cultivation. Of special importance are considered those connected with physico-mechanical properties of the substrates and the morphological special features of the producing strains [1, 6, 8, 9].

According to some authors [8, 10–12] it is more expedient to apply solid-state cultivation technique when cellulase enzyme preparations are to be used in animal feeding. DESCHANPS and HUET [10] developed a solid-state fermentation method to produce a combine product of hydrolases and substrate after drying at 40 °C. This product can be used in dry form or as a water suspension. The solid-state fermentation products are applicable in silage of row fodder and for protein enrichment of lignin-cellulose substrates [6, 10, 12].

The difficulties of carrying out a solid-state fermentation are usually connected with the transfer of the heat evolved during the process of cultivation, pH monitoring and maintaining constant humidity of the system [1–3, 5].

The aim of this study is to investigate and to compare the biosynthesis of some cellulases and hemicellulases by *Trichoderma* sp. M_7 in submerged and solid-state cultivation.

Materials and Methods

The organism used was *Trichoderma* sp. M_7, a mould strain taken from the Microbiological Collection of Engineering Biology Department, Sofia University. The strain was maintained on potato-dextrose agar slants.
The biosynthesis of the examined enzymes was studied using inducers — desintegrated lignin-cellulose substrates: wheat strow, maize, stems, wheat bran and micricel*; single and in combination.
In this work, a 20 h vegetative culture of the strain was used as inoculum at 20% (v/v). For producing the inoculum a 14 days spore culture of the strain with cell number $(1-3) \times 10^6$ spores/ml was grown in medium containing glucose and maize extract [13].
The submerged cultivation was carried out in Erlenmeyer flasks of 750 cm^3, containing 50 cm^3 fermentation medium with a composition (in g/l): cellulose substrate —30; NH_4Cl — 1.4; Urea — 0.3; KH_2PO_4 — 2; $(NH_4)_2SO_4$ — 1.4; $MgSO_4 \cdot 7H_2O$ — 0.3; $CaCl_2 \cdot 2H_2O$ — 0.4; in tap water. The cultivation was performed on a flask shaker with 220 rpm at 28 °C.
The solid-state cultivation was carried out in metal cuvettes containing 150 g of the respective cellulose containing substrate, damped with the mineral part of the above described medium in ratio 2 : 1.5 (w/v). The cultivation was performed in conditioning chamber at 28 °C and relative air humidity of 90%. Every 6 h the cuvettes were shaked for 1 min.
The enzyme activity determinations were carried out on culture-broth filtrate for the submerged culture and on water extract for the solid-state culture. The water extract was prepared by pouring out of 15 g of the solid-state cultivation mixture with 50 cm^3 destilled water, kept in contact for 30 min with periodical stirring and after that the extract separated from the solid phase by a hand press. The extract obtained was centrifuged for 10 min at 7000 rpm.

Enzyme Activity Assay

The activity of endoglucanase (cellulase, Cx-enzyme EC. 3.2.1.21) was assayed using as substrate a 1% solution of sodium carboxymethylcellulose in 0.1 M citrate-phosphate buffer of pH 5.0. The reaction medium was incubated for 30 min at 50 °C and the yield of liberated reducing substances was measured according to Shomogyi-Nelson [14].
The xylanase activity was measured using a 1% solution of xylane in acetate-phosphate buffer of 4.0. The reaction medium was incubated for 60 min at 40 °C. The liberated reducing substances were measured according to Shomogyi-Nelson [14].
As a unit of enzyme activity we accepted this enzyme quantity which liberates 1 mg reducing sugars, expressed as glucose equivalent.
Aryl-β-glucosidase activity (EC.3.2.1) was measured using 0.01 M p-nitrophenyl, β-D-glucopyranozide as substrate in 0.1 M acetate-phosphate buffer of pH 4.5. The reaction medium was incubated for 30 min at 40 °C. As a unit of enzyme activity it was accepted this quantity which on the condition of the determination liberates 1 mg p-nitrophenol [15].
In the case of submerged culture the enzyme activity is expressed in units per ml, and in the case of solid-state cultivation — in units per g dry weight substrate**.

The exocell protein was determined according to Lowry [16].

Results and Discussion

Most studies on solid-state cultivation are usually measurements of microbial growth when a free liquid phase is lacking but without mentioning the exact range of water content of the system [1, 3, 9]. When contact between hyphes of the producing strain

* Micricel is a licensed product of AVICEL (Merk) — microcristallin e cellulose, produced by TPO "PHARMACHIM", Bulgaria.

** The term "substrate" in solid-state cultivation is an adequat term for inducer in submerged cultivation.

and the solid-substrate particles is established, a biosynthesis of the subsequent enzyme is induced. The biosynthesized enzymes attack the substrate particles and distruct them into digestible monomers. The rate of enzymic destruction of the substrate is governed by the enzyme activity, but this activity is a function of the inducing properties of the substrate itself.

The biosynthetic induction of the examined enzymes of *Trichoderma* sp. M_7 in different conditions and media with the substrates used is presented in Table 1. The results show that more effective induction happens when combined substrates are used, compared with a single substrate application. In solid-state cultivation the most effective inducer of enzyme biosynthesis is a mixture of equal parts of wheat bran + wheat straw. In this case the strain biosynthesizes C_x-enzyme having an activity of 940 U/g substrate and xylanase activity of 430 U/g substrate.

Tab. 1. Biosynthesis of enzymes by *Trichoderma* sp. M_7 on submerged and solid-state cultivation conditions on different substrates for 120 h cultivation time

Substrates (inducers) of the enzyme biosynthesis	Submerged cultivation			Solid-state cultivation		
	Enzyme activity [U/ml]			Enzyme activity [U/g]		
	C_x	β-glucosidase	Xylanase	C_x	β-glucosidase	Xylanase
Wheat straw	70	0.8	96	168	12	210
Wheat bran	62	1.0	108	104	6.5	125
Maize stems	43	0.6	83	96	8.0	156
Micricel	150	0.7	41	14	1.0	27
Wheat straw + Wheat bran	228	0.8	110	940	13.0	420
Maize stems + Wheat bran 1 : 1	189	1.2	108	660	14	314
Wheat straw + Maize stems + Wheat bran 1 : 1 : 1	167	1.5	113	548	9	338
Wheat bran + Micricel 1 : 1	345	2.1	98	143	4.5	58

The oxygen supply is one of the most substantial parameter of the microorganism's growth and multiplication especially in the case of solid-substrate cultivation. On the conditions used here the oxygen supply is accomplished through the cavities formed among the solid-substrate particles [1, 2]. This explains the very low enzyme activity of the strain when Micricel was used — single or in combination with other substrates.

In submerged cultivation of *Trichoderma* sp. M_7 the most effective inducer of enzyme biosynthesis is the mixture of Micricel and wheat bran in ratio 2 : 1. In this case the strain produces C_x-enzyme having an activity of 345 U/ml, β-glucosidase — 2.1 U/ml and xylanase — 98 U/ml of culture broth.

The biosynthesized hydrolytic enzymes by *Trichoderma* sp. M_7, especially in solid-state cultivation conditions, are of high activities. On conditions of cultivation analogous to ours, LONGINOVA and TOSHPULATOV [8] found a xylanase activity of 250 U/g and Cx-enzyme activity of 150 U/g substrate with a culture of *Aspergillus terreus* 17 P. Lower are the activities of the synthesized enzymes in solid-state cultivation of *Trichoderma lignorum* 6 C [11].

The time courses of biosynthesis of the studied enzymes by *Trichoderma* sp. M_7 in submerged culture with Micricel + wheat bran as inducer (Fig. 1), and in solid-state cultivation with a substrate of wheat bran + wheat straw — in ratio 1 : 1 (Fig. 2) are shown.

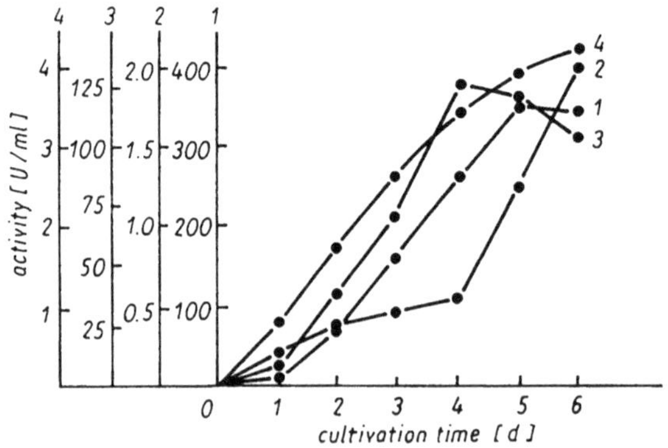

Fig. 1. Influence of cultivation time on biosynthesis of C_x-enzyme (1), β-glucosidase (2), xylanase (3) and exocell protein synthesized (4) in submerged cultivation of *Trichoderma* sp. M_7.

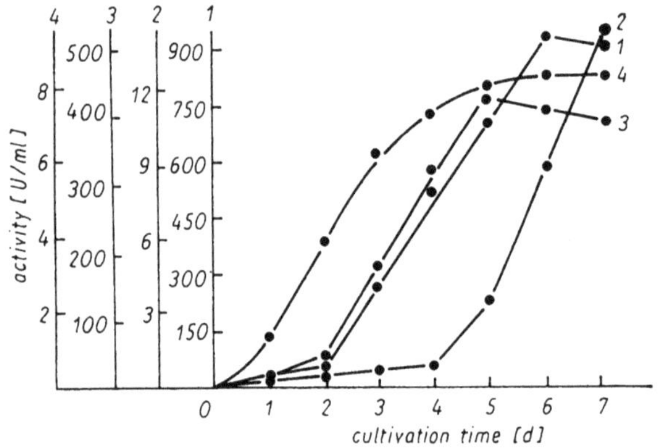

Fig. 2. Biosynthesis of C_x-enzyme (1), β-glucosidase (2), xylanase (3), exocell protein synthesized (4), in solid-state cultivation of *Trichoderma* sp. M_7.
Please read [U/g] instead of [U/ml]

The results point out that the dynamics of enzyme biosynthesis of the strain used is in close relation to the kind of cultivation. In submerged cultivation the biosynthesis is more intensive, substantiated by the obviously better conditions for growth and multiplication of the culture. In this case the maximum enzyme activities are obtained in an earlier stage of cultivation of the strain — the 4th day (for xylanase) and the 5th day (for C_x-enzyme) compared with the solid-state cultivation, but the trend of the curves expressing the biosynthesis of the enzymes studied are identical for both kinds of cultivation.

The analysis of the results confirms the fact that cellulase producing strains of *Trichoderma* sp. M_7 have a low β-glucosidase activity. During the first days of cultivation the activity of this enzyme is very low for both kinds of cultivation conditions. Nearly by the end of the cultivation time the biosynthesis of β-glucosidase activates, when the activities of xylanase and C_x-enzyme start to decrease. Substrate for the action of the β-glucosidase is the cellobiose, being an inducer, too [17]. This explains the retarded biosynthesis of the enzyme compared with the others studied (Fig. 1 and Fig. 2).

According to a great number of researchers, the solid-state cultivation provides bigger quantity of enzyme preparation per unit of fermentation volume. RAO et al. [9] pointed out that cultivation of *Pastalipsis versicolor* in solid-state conditions produces higher enzyme activities compared with a submerged cultivation of this cellulase producing strain.

The cultivation of *Trichoderma* sp. M_7 on submerged and solid-state conditions leads to biosynthesis of hydrolases depending on the kind of cultivation and the kind of inducer used as Table 1. If one calculates the enzyme activity produced to the total volume of the fermentation medium for the case of submerged culture (50 ml) and to the whole amount of the substrate for the solid-state case (15 g) it will be seen that the total enzyme activity is higher for the experiments with the submerged culture but the total quantity of the exocell protein synthesized is considerably higher in the solid-state cultivation (Table 2). The higher amount of exocell protein and the lower enzyme activity with the

Table 2. Biosynthesis of the studied enzymes in submerged and solid-state cultivation of *Trichoderma* sp. M_7

Inducers (Substrates)	Enzymes synthesized	Submerged cultivation			Solid-state cultivation		
		exocell protein synthesized [mg/50 ml]	enzyme activity		exocell protein synthesized mg/50 ml extract	enzyme activity	
			per g inducer*	per 50 ml flask		per g substrate*	per substrate far extract
Wheat straw +	C_x-enzyme		7600	11400		940	9400
Wheat bran	β-glucosidase	160	30	45	325	13	130
	xylanase		3600	5500		460	4600
Wheat bran +	C_x-enzyme		11500	17250		143	1430
Micricel (1: 2)	β-glucosidase	190	70	105	108	4.5	45
	xylanase		3266	4900		58	580

* "inducer" is adequate to "substrate" in solid-state cultivation

solid-state cultivation of the already synthesized enzyme and a possible synthesis of other enzymes, due to the complex chemical composition of the solid-substrate used. It is possible the strain to synthesize amylase also, because of the starch content in the wheat bran.

A matter of fact the advantages of the submerged fermentation in relation to the activity of the enzymes are synthesized. This is best realised when the enzyme activities are expressed per g inducer (substrate) Table 2. From these data it is clearly seen that the cellulase containing substrates provoke much better the enzyme biosynthesis in the case of submerged fermentation.

The advantages of the solid-state cultivation of enzyme producers, and especially of those strains producing cellulase, are the simpler technology, the open character of the fermentation, the possible use of the dry mixture of substrate, miscellum and enzyme [1, 3, 6, 11, 18], as an "inoculum" and a technical (unpurified) enzyme preparation for mass application in agriculture and animal feeding. This kind of fermentation avoides the substantial energy expenditures for evaporation of the liquid phase and the technological operations connected with the production of pure enzyme preparations.

According to some authors [1, 5, 11] it is not recommendable to use sporolating strains in the industrial production of microbial protein. The spore-forming strains are of potential danger for the human health and an environmental polluting factor, too. The spores formed during the cultivation of these strains make the technology of enzyme production more difficult.

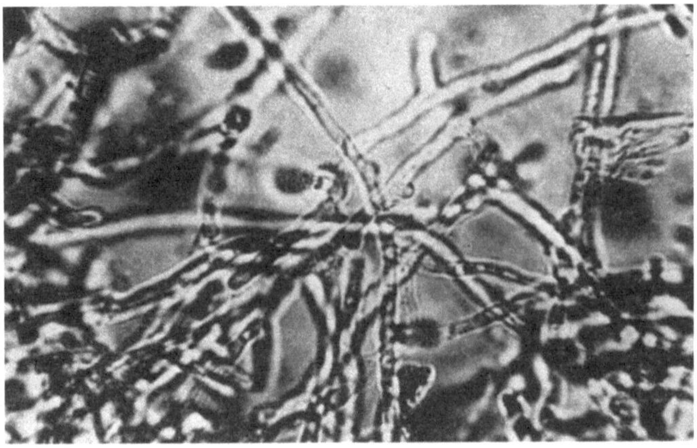

Fig. 3. Miscellium of the enzyme producing strain *Trichoderma* sp. M_7.

The present studies show that in submerged cultivation of *Trichoderma* sp. M_7 the young hyphes with homogeneous protoplasm (Fig. 3) thin out and tear in pieces. The destruction of the hyphes is accompanied by an increase of the enzyme activity. By the end of the fermentation a hyphes' lysis takes place and a formation of s.c. endospores — chlamydospores (Fig. 4) is observed. In submerged cultivation of the strain conidia spores are not formed.
In the solid-state cultivation of the strain a sporolation of the culture is observed after the 60th h from the start. It is due to the unoptimized technology, lacking best fermentation equipment design, constant humidity, aeration, etc. Therefore it was substantial to investigate the enzyme activities till the time of sporolation of the culture. The results

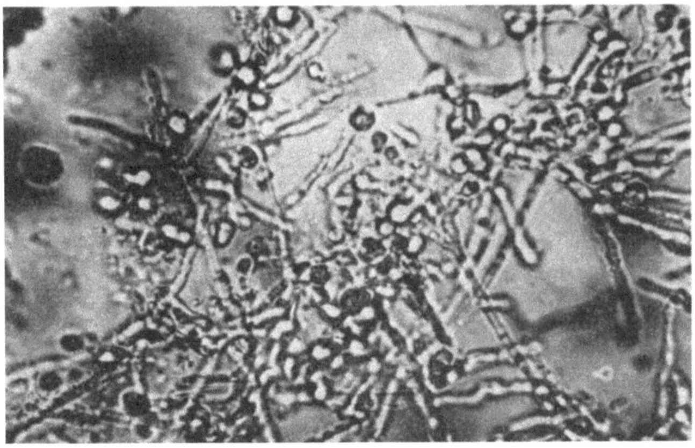

Fig. 4. Chlamydospores of *Trichoderma* sp. M_7 in submerged cultivation of the strain.

of these experiments are shown in Table 3. It can be seen that before sporolation the strain synthesized C_x-enzyme of activity 136 U/g and xylanase of activity 106 U/g substrate.

For protein enrichment treatment of raw fodder and preservation of not readily processing silages it is appropriate to use a "start inoculum" containing cellulases and hemicellulases, substrate and cell biomass up to the 54th h of the presented technique of solid-state cultivation. According to this technique the activity of enzymes synthesized is higher, compared with those described for some production micromycetic strains cultivated in solid-state conditions [3, 8, 11].

Table 3. Enzyme biosynthesis of *Trichoderma* sp. M_7 in solid-state cultivation until the time of sporolation

Enzyme synthesized	Enzyme activity				Units per g substrate		
	24 h	30 h	36 h	42 h	48 h	54 h	60 h
C_x-enzyme	25	45	60	74	90	107	136
β-glucosidase	0	1	1	1.2	1.5	1.5	2.0
Xylanase	15	20	25	36	48	60	106

Conclusions

The industrial production of enzyme preparations of the present day is based exclusively on the submerged cultivation of the producing strains. Nevertheless, the solid-state method of cultivation is older and not yet well-developed, it attracts a growing attention due to some economical, technological and biological reasons.

The examined exocell producing strain of hydrolases, *Trichoderma* sp. M_7, biosynthesizes cellulases and hemicellulases of high activity as on submerged as well as on solid-state cultivation conditions. In spite of the not yet optimized process of cultivation the strain synthesizes C_x-enzyme with activity of 345 U/ml and 940 U/g substrate and xylanase with activity of 98 U/ml and 420 U/g substrate. This makes the strain a prospective enzyme producer.

Received October 15, 1985

References

[1] BEKER, M. E.: Transformatsiya produktov fotosinteza. Riga: Zinatne, 1984, 95 (in Russian)
[2] GRACHOVA, I., POPOVA, Ya.: Enzimni preparati. Sofia: Technika, 1982, 34.
[3] KALUNYANTS, K. A., GOLGER, L. Z.: Mikrobnyye fermentnyye preparaty. Moskow: Kolos, 1960 (in Russian).
[4] HESSELTINE, C. W. — In: Development in Industrial Microbiol. **22** (1981), 1.
[5] SENEZ, J. C.: Acta Biotechnol. **4** (1984), 83.
[6] TOYOMA, N., OGAWA, K.: Proc. IV IFS: Ferment. Technol. Today, 1972, 743.
[7] STEINKRAUS, K. H.: Acta Biotechnol. **4** (1984), 83.
[8] LOGINOVA, L. G., TASHPULATOV, Zh.: Prikl. Biokhim. Mikrobiol. **14** (1978), 361 (in Russian).
[9] RAO, M. N., MITHAL, B. M., THAKKUR, R. N., SASTRY, K. S.: Biotechnol. Bioeng. **25** (1983), 859.
[10] KALUNYANTS, K. A., EZAKOV, N. V., PIVNEN, I. B.: Primeneniye produktov mikrobiologitscheskogo sinteza w shiwotnowodstwe. Moscow: Kolos, 1960 (in Russian).

[11] LOSYAKOVA, L. S., KOSHEMYAKINA, O. P., SHILOVA, A. A.: Omnibus volume. Zellulazy mikroorganismow. Moscow: Nauka, 1981, 125.
[12] TASHPULATOV, Zh.: Mikrobiol. **69** (1980), 342 (in Russian).
[13] ATEV, A. P., SPASSOV, S. D., OBRESHKOVA, BOUBAREVA, L. G.: Comp. rend. de l'Acad. bulgare des Sciences **36** (1983), 941.
[14] SOMOGYI, M.: J. Biol. Chem. **195** (1954).
[15] ATEV, A. P., SPASSOV, S. D., PANAYOTOV, H. A., BAKALOVA, N. G.: Comp. rend. de l'Acad. bulgare des Sciences **36** (1983), 533.
[16] STEINKRAUS, K. H.: Acta Biotechnol. **4** (1984), 83.
[17] ENARY, Tor-Magnus-In: Microbial Enzymes and Biotechnology. London and New York: Appl. Sci. Publ. 1983, 183.
[18] DESCHANPS, F., HUET, M. C.: Biotechnol. Lett. **6** (1984), 55.

Acta Biotechnol. 7 (1987) 1, 17–21

Zur mixotrophen Kultivation von Chlorella vulgaris in homokontinuierlicher Chemostatkultur

ROTH, P., BÜRGER, S.

Akademie der Wissenschaften der DDR
Institut für Biotechnologie, Leipzig
Permoserstraße 15, Leipzig, 7050, DDR

Summary

Cells of *Chlorella vulgaris*, BEIJ. Greifswald 9, were grown on autotrophic and mixotrophic conditions using glucose and acetate as organic substrates. It was shown that these C-sources applicated in a suitable range of concentrations increase the growth rate and the productivity of the algal cultures. The cells grown on mixotrophic conditions have a higher total pigment content and exhibit variations in the ratio chlorophyll a/chlorophyll b.
In addition the contents of proteins, lipids, carbohydrates, and nucleic acids of the biomass were shown to be dependent on the kind of the organic substrate used.

Einleitung

Bestrebungen, phototrophe Mikroorganismen zur Konversion von Lichtenergie in die chemische Energie von Biomasse im Rahmen industrieller Prozesse zu nutzen, befinden sich noch in der Anfangsphase. Einerseits lassen Prädispositionen wie etwa hohe Gehalte an ungesättigten Fettsäuren und Rohprotein bei einigen Chlorophyta und Cyanophyta oder attraktive Polysaccharidkompositionen bei Rhodophyta die Kultivation von Algen in kontinuierlichen oder diskontinuierlichen Verfahren lukrativ erscheinen, andererseits stellt das angespannte Energieaufwand-Nutzen-Verhältnis ihre Übertragung in industrielle Maßstäbe vielfach in Frage. Die durch die diurnale Rhythmik nicht permanent zur Verfügung stehende Sonnenenergie (in unseren Breiten ca. 110 $W \cdot m^{-2}$ im Jahresdurchschnitt) [1] muß, falls Licht bei rein autotropher Prozeßführung nicht zum limitierenden Faktor des Wachstums werden soll, durch Kunstlicht ergänzt werden, was wiederum die ungünstige Kostenentwicklung der Produkte begünstigt. Selbst wenn es in der Suspensionskultur gelingt, Verluste durch Reflexion, Transmission, inaktive Absorption und respiratorische Vorgänge gegenüber terrestrischen Kulturen zu minimieren, bleibt die praktisch erzielbare Effizienz von je nach Anschauung 5 – 12% [2] weit unter dem thermodynamisch möglichen Wirkungsgrad der Energiewandlung von 70–95% [3, 4, 5].
Nach Angaben von GOLDMANN [6] zur Massenkultur von Algen, deren Ertrag bei 40 g Trockenmasse $m^{-2} \cdot d^{-1}$ lag, und der Annahme von 23 $kJ \cdot g^{-1}$ [7] sowie einer durchschnittlichen Einstrahlung von 250 Wm^{-2} ergibt sich eine Effizienz von 4,3% gegenüber Spitzenwerten landwirtschaftlicher Hochleistungskulturen von 3,2% und einem globalen Durchschnitt von 0,1%.

In diesem Zusammenhang war für uns von Interesse, inwieweit die zusätzliche Bereitstellung organischer Substrate die Produktivität der Kulturen schlechthin beeinflußte und zu Modifikationen im Verhältnis der zellulären Produkte zueinander führte.

Material und Methoden

Stellvertretend für andere, noch zu testende Arten und Stämme wurde von uns die gut handhabbare einzellige Grünalge *Chlorella vulgaris*, BEIJ. Greifswald 9 [8] eingesetzt. Unter Laborbedingungen ist bei entsprechender Wahl des Bedingungsgefüges autotrophe, mixotrophe und heterotrophe Lebensweise [9] realisierbar. Die Reproduktion des asynchronen Organismenmaterials erfolgte in thermostatisierbaren 170 ml Flüssigkeit fassenden Glaszylindern unter Dauerlichtbedingungen. Die Begasungsluft (35 l · h^{-1}) wurde auf einen Gehalt von 2 Vol.% mit Kohlendioxid angereichert. Die Temperatur betrug 37 °C. Das Nährmedium entsprach dem von Böhm, Doucha [10] vorgeschlagenen, ergänzt durch Ammoniumacetat bzw. Glucose im Falle der mixotrophen Kulturvariante. Die Lichtenergie lieferten sechs waagerecht angeordnete Leuchtstoffröhren LS 18 coolwhite 20, VEB Narva Brand-Erbisdorf. Die Bestimmung der Trockenmasse sowie die des relativen Chlorophyllgehaltes mit diffusem Licht wurden bereits an früherer Stelle [11] beschrieben. Bei der Ermittlung des Gesamtchlorophyllgehaltes sowie des Verhältnisses Chl. a : Chl. b bezogen wir uns auf die Angaben von Boeger [12], nach Extraktion der Farbstoffe bei 55 °C und einer Dauer von 45 min. Zur Bestimmung der Absolutzellzahlen kam eine Zählkammer nach Thoma, Tiefe 0,1 mm, zum Einsatz. Die angegebenen Parameter beziehen sich jeweils auf eingestellte steady-state-Verhältnisse in den Chemostatkulturen, deren Stabilität mit Hilfe submers installierter optoelektronischer Sensoren des Typs SP 201 (Werk für Fernsehelektronik Berlin) kontrolliert wurde.

Ergebnisse

Die getesteten Substrate stimulieren in den eingesetzten Konzentrationen die Biomasseentwicklung des planzlichen Einzellers bei allen Verweilzeiten. Diese Erscheinung konnte bereits nach einigen Stunden visuell anhand intensiver Grünfärbung der Kulturen wahrgenommen werden. Bei Verwendung von Acetat nimmt die stimulatorische Wirkung bei Verweilzeiten von 4,5 h wieder ab (Abb. 1). Unter dem Einfluß der Glucose gilt die Einschränkung nicht, ihre fördernde Wirkung wächst mit steigender Durchflußrate, zumindest bis $D = 0,3 \cdot h^{-1}$.

Wie Pipes, Koutsovannis [13] sowie Tamiya [14] nachwiesen, zeigen autotroph wachsende Grünalgen konstante Produktivitäten im Bereich der Lichtlimitation. Diese Aussage ließ sich unter unseren Bedingungen auf die durch 0,5 g Acetat · l^{-1} stimulierte Kultur, bei insgesamt höheren Produktivitäten als bei der autotrophen Kultur, erweitern. Der Plateaubereich verkleinert sich bei Einsatz von 1 g Acetat · l^{-1}, wodurch die Kurve Ähnlichkeit mit den Verhältnissen in substratlimitierten Bakterienkulturen [15, 16] annimmt.

Auch wenn auf eine experimentelle Bestimmung der maximalen spezifischen Wachstumsrate nach der "wash-out"-Methode verzichtet wurde, geht aus Abb. 1 hervor, daß zumindest der Einsatz von Glucose als Substrat eine höhere Wachstumsrate erlaubt, als die rein autotrophe Kultivierung, was allerdings unter der Zielstellung einer möglichst hohen Trockenmasseproduktion je Zeit- und Volumeneinheit nur von untergeordnetem Interesse ist.

Für die vergleichende Untersuchung des Substrateinflusses auf verschiedene Parameter der Suspensionskulturen entschieden wir uns für eine mittlere Verweilzeit von 5 h. Wichtig für die Deutung der Effekte erschien uns der Steigerungsfaktor 5,4 für die Biomassekonzentration beim kombinierten Einsatz von 0,5 g Acetat und 0,5 g Glucose · l^{-1} verglichen mit 3,5 bzw. 3,9 bei alleinigem Einsatz von 1 g Glucose bzw. Acetat · l^{-1}

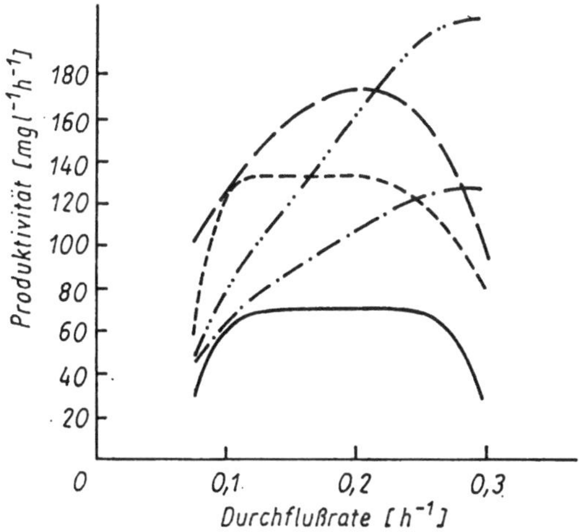

Abb. 1. Zusammenhang zwischen Produktivität und Durchflußrate in homokontinuierlicher Chemostatkultur bei unterschiedlichem Substratangebot.

——— autotr. Kultur, — · — 0,5 g · l^{-1} Glucose
- - - - 0,5 g · l^{-1} Acetat, — · · — 1,0 g · l^{-1} Glucose
— — 1,0 g · l^{-1} Acetat,

(Abb. 2). Trotz Zunahme der Biomassekonzentration mußte bei Einsatz von 0,5 g Glucose · l^{-1} eine leichte Depression der Zellzahl verzeichnet werden. Dieser Umstand ist mit einer erheblichen Vergrößerung der Einzelzellen verbunden, die eine Folge der Ausweitung der Generationszeit bzw. der Reduktion der Zahl der je Mutterzelle gebildeten Autosporen ist.

Die Bestimmung des Gesamtchlorophyllgehaltes erbrachte übereinstimmend sowohl nach absoluter als auch nach relativer Methode in Gegenwart der Substrate durchweg erhöhte Farbstoffgehalte sowohl je Volumeneinheit der Kulturen als auch je Einzelzelle. Neben der Zunahme der Pigmentkonzentration je Einzelzelle führt der Einsatz der organischen Substrate auch zu Variationen im Verhältnis der Chlorophyllformen zueinander. So hatte die Kultivierung in Gegenwart von 1 g Acetat bzw. 1 g Glucose · l^{-1} Nährlösung annähernd den gleichen Effekt wie die Verdoppelung der Verweilzeit in der

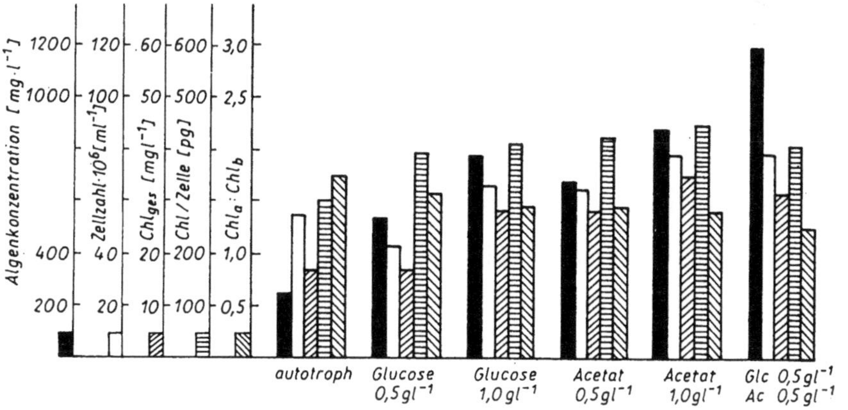

Abb. 2. Einige Parameter kontinuierlicher Chlorellakulturen (Verweilzeit 5 h) in Abhängigkeit von Substratqualitäten und -quantitäten.

autotrophen Kultur von 5 auf 10 h (Abb. 3). Ebenso wie bei der Verringerung der Durchflußrate in der autotrophen Kultur sinkt bei Substratzugabe das Verhältnis Chl. a : Chl. b von 1,7 auf 1,5, was in beiden Fällen für die relative Zunahme der Chlorophyll b-Form spricht (Abb. 2 und 3).

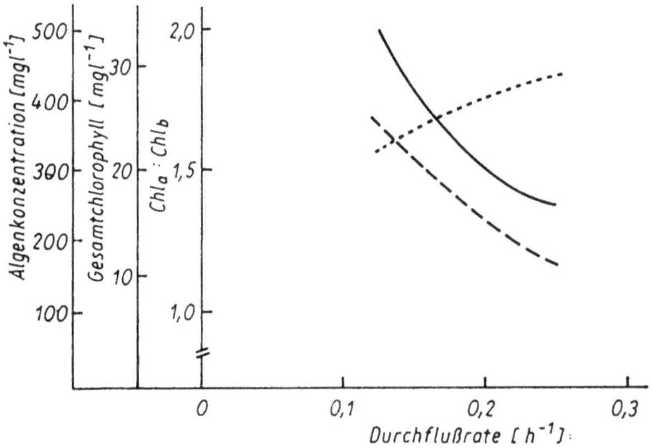

Abb. 3. Abhängigkeit des Trockenmasse- und Chlorophyllgehalts autotropher Kulturen von der Durchflußrate

——— Algenkonzentration, --- Gesamtchlorophyll

······ $Chl_a : Chl_b$

Die Erhöhung der Substratkonzentration über die angegebenen Werte hinaus erweist sich als uneffektiv, da z. B. 2 g Glucose $\cdot$ l^{-1} keine signifikante Steigerung der Biomassekonzentration bei sinkendem Ertragskoeffizienten bewirkt oder gar inhibitorische Effekte, wie sie beim Einsatz von 2 g Acetat $\cdot$ l^{-1} registriert wurden.
Stellvertretend für weitere Veränderungen unter dem Einfluß der organischen Substrate seien hier die im Vergleich zur autotrophen Kultur um 50% höher liegenden Gesamtkohlehydratanteile der glucosestimulierten Kulturen bei etwa 20% verminderten Proteingehalten oder die höheren Nucleinsäure- und Lipidgehalte der Organismen unter Acetateinfluß genannt.

Diskussion

Von beiden Substraten ist bekannt, daß sie von einigen Organismenarten nach entsprechender ATP-abhängiger Umwandlung in den Zellstoffwechsel integriert werden.
Unter mixotrophen Bedingungen stammt die nötige Energie in erster Linie aus der zyklischen Photophosphorylierung der Lichtreaktion.

$$C_6H_{12}O_6 + ATP \xrightarrow{\text{Hexokinase}} C_6H_{11}O_5{-}O{-}PO_3H_2 + ADP \qquad [17, 18]$$

$$CH_3{-}COOH + ATP + CoA{-}SH \xrightarrow{\text{Acetyl-CoA-Synthetase}} CH_3CO{-}S{-}CoA + AMP + P{-}P \qquad [19-21]$$

Für die zentrale Stellung des Glucose-6-phosphates bei der Synthese pflanzlicher Di- und Polysaccharide sprechen die um 50% höheren Gesamtkohlehydratanteile bei Kultivierung in Gegenwart von Glucose. Wie wir weiter feststellen konnten, ist der von uns getestete Stamm der Species *Chlorella vulgaris* in der Lage, das gebotene Acetat über den Glyoxylsäurezyklus zu assimilieren. Die Hauptbedeutung dieser mit der Initial-

reaktion zwischen Acetat und Oxalacetat beginnenden und Bernsteinsäure liefernden Reaktionskette besteht in der Möglichkeit der Synthese von C_4-Körpern aus C_2-Verbindungen.
Neben dem stimulatorischen Effekt der Belichtung auf den Einbau des Acetatkohlenstoffs in Polysaccharide und Proteine [22] ist besonders die gesteigerte Lipidkonzentration im Ergebnis der rapiden Synthese langkettiger Fettsäuren [23] hervorzuheben. Die Doppelfunktion der untersuchten Thylakoidpigmente besteht in der Absorption des eingestrahlten Lichtes und seiner Umwandlung in chemische Energie. Zur photochemischen Primärreaktion sind allein die Chlorophyll-a-Modifikationen P_{700} (PS I) und Chl a_2 (PS II) befähigt.
Den ebenfalls zu den photosynthetischen Einheiten gehörenden, auch strukturell vom Chlorophyll-a unterscheidbaren akzessorischen Chlorophyll-b-Formen kommt hingegen nur die Funktion der Absorption des angebotenen Lichtes zu. Erkenntnisse zur Wirkung der Beleuchtungsstärke auf den Gehalt an Thylakoidfarbstoffen gehen bereits auf WARBURG und NEGELEIN [24] zurück, sind jedoch auch Gegenstand moderner Arbeiten [25, 26]. Die bei Abnahme des Lichtangebotes — in unserem Falle durch Erhöhung der Algenkonzentration hervorgerufen — zu beobachtende Chlorophyllzunahme je Einzelzelle steigert deren Energieaufnahme und engt über die verhältnismäßig größere Zunahme des Chlorphyll-b-Anteils den Bereich der Grünlücke ein, wodurch Licht eines Spektralbereiches mit in die Nutzung einbezogen wird, auf das bei höherer Beleuchtungsstärke verzichtet werden kann.

Eingegangen: 31. 10. 1985

Literatur

[1] STOY, B.: Wunschenergie Sonne. Heidelberg: Energie-Verlag, 1980.
[2] BOARDMANN, K.: Phil. Trans. Soc. Lond. A **295** (1980), 477.
[3] DUYSENS, L. N. M., AMESZ, J.: Plant Phys. **34** (1959), 210.
[4] MORTIMER, R. G., MEZO, R. M.: J. Chem. Phys. **35** (1961), 1013.
[5] SHOCKLEY, W., QUEISSER, H. J.: J. Appl. Phys. **32** (1961).
[6] GOLDMANN, J. C.: Water Res. **13** (1979), 119.
[7] GOEDHEER, J. C., HAMMANS, J. W. K.: Nature **256** (1975), 333.
[8] Algenkatalog Ernst-Moritz-Arndt-Universität. Greifswald, 1981.
[9] FOTT, B.: Algenkunde. Jena: VEB Gustav-Fischer Verlag, 1971.
[10] BÖHM, H., DOUCHA, J.: Wiss. Hefte d. Pädagog. Hochschule Köthen **6** (1979), 155.
[11] BÜRGER, S.: Dissertation (A), Pädagog. Hochschule Köthen 1984.
[12] BOEGER, P.: Flora **154** (1964), 174.
[13] PIPES, W. O., KOUTSOVANNIS, S. P.: Appl. Microbiol. (Baltimore) **10** (1962), 1.
[14] TAMIYA, H.: Ann. Rev. Plant Physiol. **8** (1957), 309.
[15] HARDER, W., KUENEN, J. G., MATIN, A.: J. Appl. Bact. **43** (1977), 1.
[16] PFENNIG, N., JANNASCH, H. W.: Ergebn. Biol. **25** (1962), 92.
[17] KANDLER, O.: Z. Naturforschung **96** (1954), 625.
[18] TANNER, W., LOOS, E., KANDLER, O. — In: Currents in Photosynthesis. Hrsg.: THOMAS, J. B., GOEDHEER, J. C. Rotterdam: A. D. Donker Publ. 1966, 243.
[19] PRINGSHEIM, E. G., WIESSNER, W.: Nature (London) **188** (1960), 919.
[20] WIESSNER, W., GAFFRON, H.: Nature (London) 201 (1964), 725.
[21] WIESSNER, W.: Nature (London) **205** (1965), 56.
[22] SYRETT, P. J., SHEILA, M. B., MERRETT, M. J.: Exp. Bot. **15** (1964), 35.
[23] STUMPF, P. K., JAMES, A. T.: Biochem. Biophys. Acta **57** (1962), 400.
[24] WARBURG, O., NEGELEIN, E.: Z. physik. Chem. **102** (1922), 235.
[25] FRENCH, C. S., WIESSNER, W., LAWRENCE, M. C.: Carnegie Inst. of Washington Year Book **69** (1971), 662.
[26] BROWN, J. S.: Ann. Rev. Plant Physiol. **23** (1972), 73.
[27] WIESSNER, W.: Bioenergetik bei Pflanzen. Jena: VEB Gustav-Fischer Verlag, 1975.

Acta Biotechnol. 7 (1987) 1, 22

Book Review

W. CRUEGER, K. ESSER, P. PRÄVE, M. SCHLINGMANN, R. K. THAUER, F. WAGNER

Jahrbuch Biotechnology 1986/87

München: Carl Hanser Verlag, 1986. 532 S.; 138 Abb.; 64 Tab.; 86 DM

Der wissenschaftliche Teil des Jahrbuches der Biotechnologie stellt wichtige neuere Teilgebiete der Biotechnologie vor. Gleichzeitig soll damit der interessierte Leser angeregt werden, die einzelnen Teile des Buches als Arbeitsfeld zu erkunden. In einem zweiten Abschnitt wird die Praxis der Biotechnologie in ausgewählten Einzelthemen dargestellt. Da Biotechnologie interdisziplinär und in vielen Bereichen eine sehr junge Wissenschaft ist, muß der dort Tätige oft mühsam methodische Grundlagen zusammentragen. Hier soll dieser praktische Teil methodische Hilfestellung geben, wobei vor allem die heute wichtigen, neuen Methoden berücksichtigt wurden.
Ein dritter Teil beinhaltet Informationen, die immer wieder benötigt werden, sowohl von in den Hochschulen, als auch in der Industrie tätigen Biotechnologen und anderen Fachleuten, die an diesem Arbeitsgebiet interessiert sind.
Das Jahrbuch Biotechnologie richtet sich an den Forschenden wie an den in der Praxis tätigen Biotechnologen gleichermaßen. Es will aber auch Wissenschaftler und Studenten am Rande liegender Fachbereiche ansprechen, die sich für das Arbeitsgebiet Biotechnologie interessieren.
Das neue Jahrbuch der Biotechnologie wird damit ein wertvoller Ratgeber für die tägliche Arbeit.

Acta Biotechnol. 7 (1987) 1, 23–29

Saccharification of Cassava Peels Waste for Microbial Protein Enrichment

ODUNFA, S. A., SHASORE, S. B.

Department of Botany and Microbiology
University of Ibadan
Ibadan, Nigeria

Summary

Cassava waste peels may constitute up to 55% of the original tuber. These waste peels were found to contain 41.8% carbohydrate, 1.1% protein, 12.5% ether extract and 4.9%, 4.9% total ash, and 20.8% crude fibre. Studies were conducted to formulate a fermentation medium to convert the waste peels to reducing sugars and to enrich the peels with microbial protein. Amylase producing microorganisms were isolated from rotten cassava tuber discs buried in the soil at different locations. The microorganisms isolated were *Aspergillus fumigatus*, *A. flavus*, *A. niger*, and a *Pseudomonas* sp. and *A. niger*; the level of reducing sugar was 20.5 mg/ml. The lowest was by *B. subtilis* an isolate from fermenting locust bean. Generally the levels of saccharification were higher when the waste media were supplemented with different nitrogen sourses. The crude protein yield in the cassava peel waste media by different microorganisms varied from 5.6% to 17.5%. The highest protein yield was in the waste medium fermented by *A. fumigatus* followed by *A. niger*, *B. subtilis*, *Pseudomonas* sp. in decreasing order.

Introduction

Cassava (*Manihot esculenta* Cranz) is the most important root crop in Nigeria and many tropical countries. Its production [1.28×10^8 tons, FAO, 1982] is much greater than all other root crops put together. It is processed to produce various traditional foods such as *gari*, *fufu*, or *lafun* flour, most of which are fermented products [1].
A preliminary step in the processing is the peeling. With hand peeling the peels constitute 20–35% of the total weight of the tuber [2]. However, with the recently developed peeling machine, the wastes constitute 38% for sized roots and 55% for unsized roots [3]. Although the former method is less wasteful, it is a very laborious operation and is unsuitable for large-scale industrial processing.
With the development of gari-processing machines [4, 5] and the popularity of their use, the peeling machine is a more attractive option. The wastes generated at present pose a disposal problem and would even be more problematic in the future with increased industrial production of cassava products. Part of the problem arises from the limitations of cassava peels for use directly as animal feed [6] because they contain poisonous cyanogenic glucoside, linamarin. Within any root tuber the linamarin content is highest in the peels. However, since the peels contain a high portion of starch they are appropriate for use as substrates for α-amylase production and or they may be saccharified for microbial protein enrichment. Amylase is an important enzyme with

numerous industrial applications [7] especially in the brewing industry and in the manufacture of glucose syrup. No information is available about the suitability of cassava or its peels for these purposes.
Another possibility is the enrichment of the cassava peels with microbial protein for use as animal feed; during the microbial growth on the starchy peels, the linamarin is detoxified [8]. Previous studies along this line have been on the use of cassava flour itself [9—11]. Unfortunately it seems most unlikely that there will be enough cassava now or in the near future in most producing countries (except Thailand where its use as food is limited) to spare for use as animal feed. Hence the use of peels is a more economically attractive option.
The objective of this investigation is therefore two fold; to formulate a fermentation medium for the production of α-amylases to effect saccharification of the starch in the peels and secondly to utilise the peel as a substrate for microbial protein enrichment.

Materials and Methods

Materials

The cassava peels wastes were obtained from the Federal Institute of Industrial Research, Oshodi, Lagos, Nigeria. The peels were dried in an oven at 105 °C for 2 days and ground. The waste was analysed and found to contain 1.1% crude protein, 41.8% total carbohydrate (mostly starch), 12.5% ether extract, 4.9% ash, and 20.8% crude fibre.
Isolation of microorganisms. Pieces of cassava tubers were buried in the soil. The decayed cassava tuber were weighed and ground and suspended in sterile water. Decimal dilutions of the resulting suspensions were prepared and 0.1 ml portions were spread on the surface of Plate Count Agar Oxoid, U.K.) and Potato Dextrose Agar plates (with 30 mg/ml streptomycin sulphate). Plates were incubated at 30° and 40 °C and microbial colonies arising were purified by steaking.
Screening for amylolytic organisms. Pure isolates obtained above were grown on nutrient starch agar plates (with soluble starch, 2%) which were subsequently flooded with Lugol's iodine to detect zones of hydrolysis. Characterization and identification of amylolytic isolates were done with reference to BERGEY's Manual [12] for bacteria and to BARNETT [13] for fungi.

Determination of Optimal Temperature for Amylase Activity

The amylolytic isolates were inoculated into appropriate nutrient broth containing 0.5% soluble starch (Nutrient agar for bacteria and Czapek — Dox broth, without sucrose for fungi). After incubating the cultures for 3 days at 30 °C, the fungal cultures were filtered through Whatman No. 1 filter paper while the bacterial culture was centrifuged with an MSE high-speed refrigerated centrifuge at 5000 r.p.m. for 30 min.
The α-amylase activity in the filtrate was determined following the method of BERNFIELD [14] with the following modification. The reaction mixtures containing the filtrate and soluble starch solutions were incubated at 30°, 35°, and 40 °C.

Microorganisms Used

The following microorganisms were isolated from decomposing cassava tuber buried in the soil: *Aspergillus fumigatus, A. niger*, and a *Pseudomonas* sp.
In addition to these, *Aspergillus niger* (NRRL 3122) was obtained from Northern Regional Research Center, Peoria, Illinois. *Bacillus subtilis* was obtained from the microbial culture collection of the Department of Botany and Microbiology, University of Ibadan.

Fermentation Media

The medium used was that described by READE and GREGORY [15]. The composition is as follows: [grams per litre] KH_2PO_4 1 g; $MgSO_4 \cdot 7H_2O$, 0.05 g; $CaCl_2 \cdot H_2O$, 0.01 g; $FeSO_4 \cdot 7H_2O$, 0.01 g; $ZnSO_4 \cdot 7H_2O$, 0.01 g; nitrogen source (in urea or a salt) 0.8 g; ground cassava peels 40 g. The medium was autoclaved at 1.05 kg/cm^2 for 15 minutes.

Assay of Saccharification

The media were inoculated with a 5 mm disc from the growing edge of the fungal culture or 1 ml aliquot of a 24-hour broth bacterial culture. In co-inoculation studies, 0.5 ml of each culture was used.
Fermentations were done in shake flask cultures using 30 ml of medium in 150 ml conical flasks on a rotary shaker incubator (New Brunswick).
At intervals the waste fermentation medium was centrifuged at 4000 rpm in an MSE refrigerated high speed centrifuge for 30 min. The sediment waste residue was dried in an oven at 105 °C for 24 hours and ground in a mortar. The crude protein and total carbohydrate and reducing sugars were determined on duplicate samples.
Reducing Sugars. The level of reducing sugars in duplicate samples were determined using 3,5 — dinitrosalicylic acid reagent (DNSA) [14]. One millilitre of the suspension was appropriately diluted, and to this was added 1 ml of DNSA. The mixture was heated in a boiling water bath for 5 min and then allowed to cool. Water (10 ml) was added. The absorbance was measured at 540 nm in a Pye Unicam SP 6 250. The reducing sugar level was determined by comparison with a standard curve prepared with maltose.
Total carbohydrate. A modification of the anthrone method described by Clegg [16] was used.
Crude protein. Nitrogen was determined by Kjeldahl procedure [17] and the factor 6.25 was used to calculate crude protein.

Results

Table 1 shows the optimum temperatures for saccharification. The optima for *Aspergillus flavus*, *Pseudomonas* sp. and *Rhizopus* sp. is 35 °C, for *A. fumigatus* it is 40 °C while for *A. niger* it is 30 °C. The industrial strain of *A. niger* (which is designated *A. niger* 1) was found to produce the higher amount of reducing sugars at 35 °C.

Table 1. Amylase production by the different microbial isolates at different temperatures

Microorganisms	Concentration of reducing sugars [mg/ml]		
	30 °C	35 °C	40 °C
Aspergillus fumigatus	1.7	2.1	2.5
Rhizopus sp.	0.9	1.9	1.7
Aspergillus niger	1.5	0.8	0.9
Aspergillus flavus	0.5	0.8	0.5
A. niger NRRL 3122	1.4	2.2	0.7
Pseudomonas sp.	1.2	2.7	1.5

Figures 1—4 show the levels of saccharification by each of the microorganisms tested. The highest level of saccharification by each of the microorganisms tested. The highest level of saccharification was by *Pseudomonas* sp.; the level of reducing sugars produced was 20.5 mg/ml. The lowest saccharification was by *B. subtilis*.
The highest level of saccharification was produced when A. *niger* and *B. subtilis* were co-inoculated into the cassava peel waste medium (Fig. 5); the level of reducing sugars produced was 21.25 mg/ml in 48 hours. The level achieved is the same as achieved by *A. niger* in 24 h.
Figures 6—9 show the reducing sugars produced by the different microbial isolates when the cassava peel waste media were supplemented with different nitrogen sources. Generally the levels of saccharification were significantly higher when the waste media were

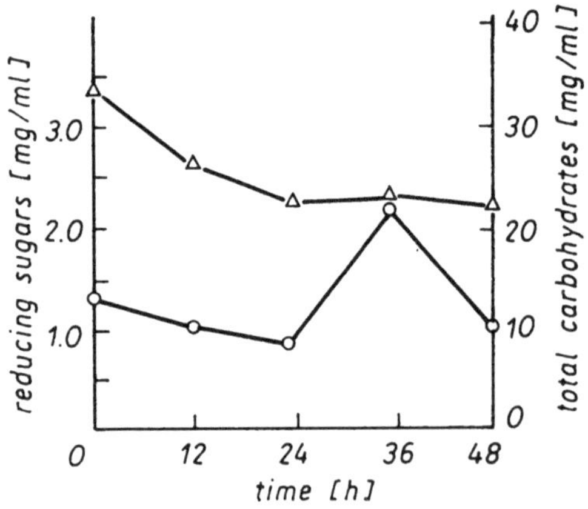

Fig. 1. Changes in the total carbohydrates and reducing sugars in cassava peel waste media fermented by *Aspergillus fumigatus*.

△ — Total carbohydrates, ○ — Reducing sugars.

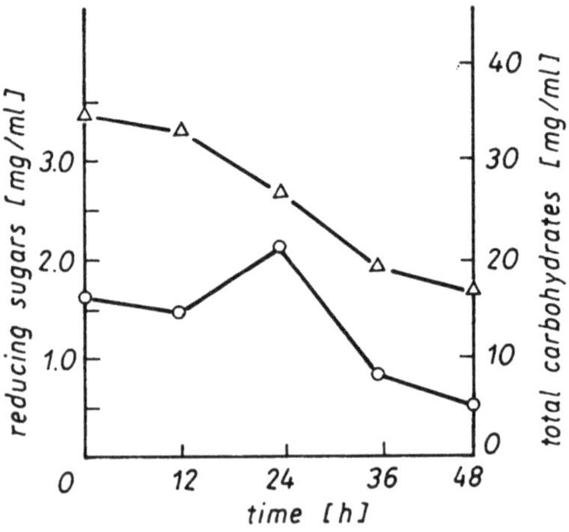

Fig. 2. Changes in the total carbohydrates and reducing sugars in cassava peel media fermented by *Aspergillus niger*.

△ — Total carbohydrates, ○ — reducing sugars

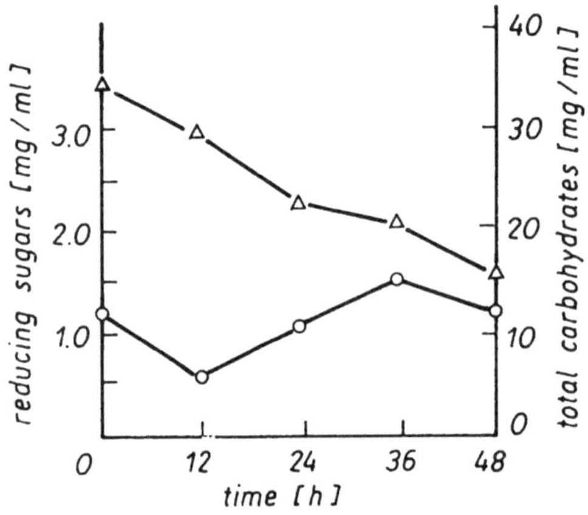

Fig. 3. Changes in the total carbohydrates and reducing sugars in cassava peel media fermented by *Bacillus subtilis*.

△ — Total carbohydrates, ○ — reducing sugars.

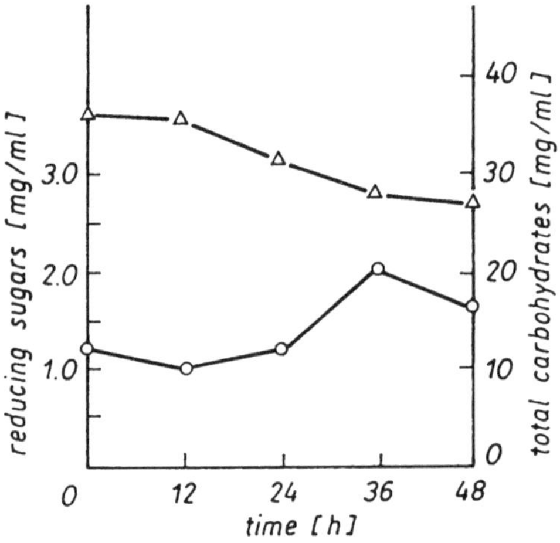

Fig. 4. Changes in the total carbohydrates and reducing sugars in cassava peel media fermented by *Pseudomonas* sp.

△ — Total carbohydrates, ○ — reducing sugars

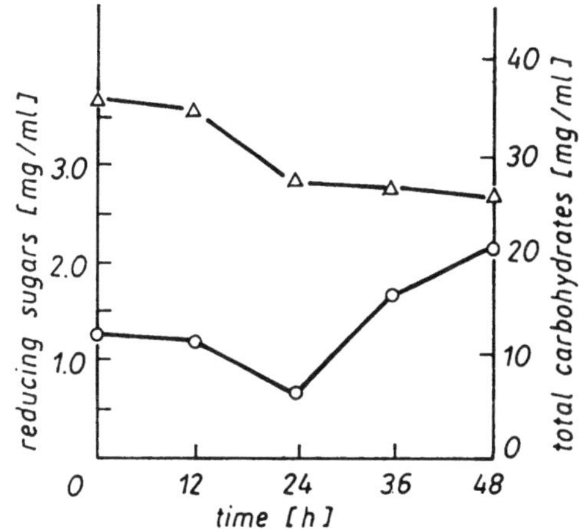

Fig. 5. Changes in the total carbohydrates and reducing sugars in cassava peel media fermented by *B. subtilis* and *A. niger*.

△ — Total carbohydrates, ○ — reducing sugars.

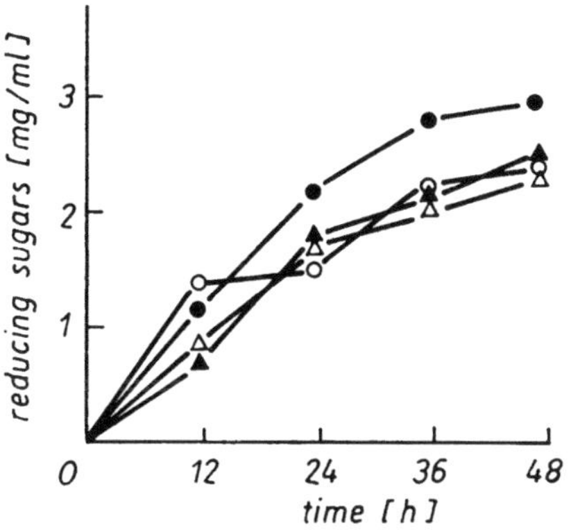

Fig. 6. Effect of different nitrogen sources on the saccharification of cassava peels by *Aspergillus niger*.

● — $(NH_4)_2HPO_4$, ▲ — $NaNO_3$, ○ — Urea, △ — NH_4Cl.

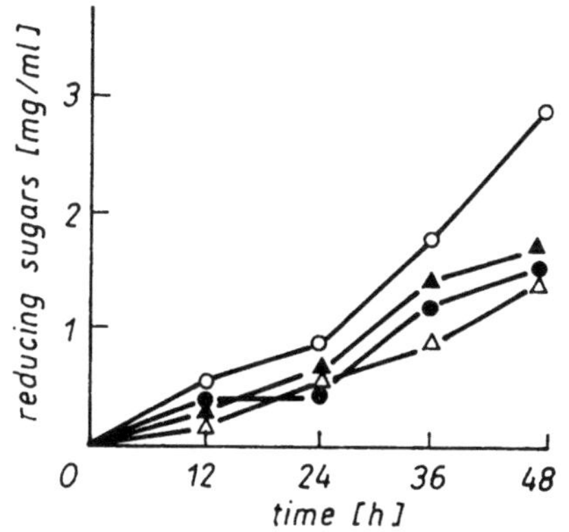

Fig. 7. Effect of different nitrogen sources on the saccharification of cassava peels by *Aspergillus fumigatus*.

● — $(NH_4)_2HPO_4$, ▲ — $NaNO_3$, ○ — Urea, △ — NH_4Cl.

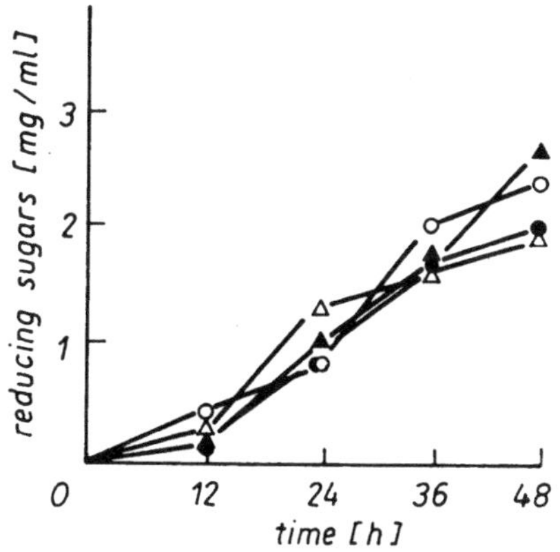

Fig. 8. Effect of different sources on the saccharification of cassava peels by *Bacillus subtilis*.

● — $(NH_4)_2$ HPO_4, ▲ — $NaNO_3$, ○ — Urea, △ — NH_4Cl.

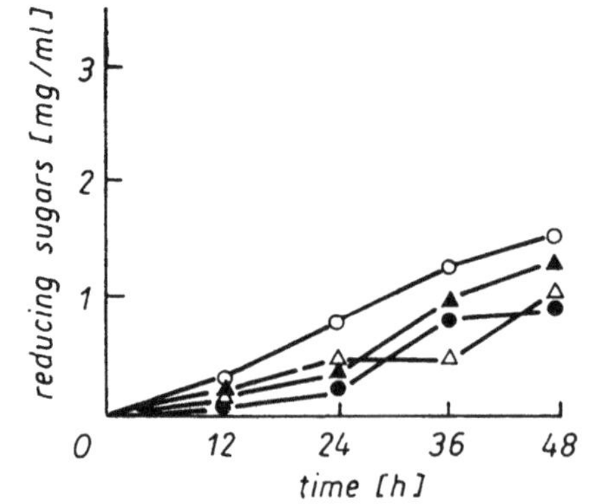

Fig. 9. Effect of different nitrogen sources on the saccharification of cassava peels by *Pseudomonas* sp.
●— $(NH_4)_2HPO_4$, ▲—$NaNO_3$, ○— Urea, △—NH_4Cl.

supplemented with the nitrogen sources than when they was not. With *A. niger* the most suitable nitrogen source was diammonium hydrogen phosphate (Fig. 6), for *A. fumigatus* it was urea.
There were no significant differences in the saccharification levels in *B. subtilis* and *Pseudomonas* sp. with the different nitrogen sources although urea was generally the better nitrogen source.
The crude protein yield in the cassava peel waste media by the different microorganisms are shown in Table 2. The highest protein yield was in the waste medium fermented by *A. fumigatus* followed by *A. niger*, *B. subtilis*, *Pseudomonas* sp. in decreasing order.

Table 2. Protein and carbohydrate yield of cassava waste peels inoculated with different microbial isolates

Organism[s]	Percentage total Carbohydrate	Percentage crude protein [N × 6.25]
Aspergillus fumigatus	31.25 ± 1.76	19.0 ± 0.28
A. niger [NRRL 3122)	34.5 ± 0.71	11.9 ± 0.77
Pseudomonas sp.	37 ± 0.8	7.2 ± 0.42
Bacillus licheniformis	38.4 ± 0.84	8.5 ± 0.49
Candida utilis and *B. licheniformis*	37 ± 1.2	8.37 ± 0.38
Control	41.5 ± 2.12	1.55 ± 0.07

Discussion

There was a noticeable decrease in the level of reducing sugars within the first 24 hours in all the saccharification experiments. This initial decrease is presumably due to the fact that some of the reducing sugars were utilized by the microorganisms for growth; subsequently enough amylases were produced which subsequently hydrolysed the starch substrate.
Generally the fungi were more effective in saccharifying the starch than the bacteria. They also yielded more protein in the waste media. This observation is in agreement with other studies. The fungi are generally good sources of amylases [7]. Also they generally yield more biomass which increases the protein yield.
Although *A. fumigatus* is effective in saccharification and producing high microbial protein, its use for producing protein enriched animal feed will be limited by its toxicity. It has been implicated in animal diseases [18] and mycotoxin production [19]. *Aspergillus fumigatus* has also been previously found to produce high microbial protein;

READE and GREGORY [15] used a thermophilic strain of *A. fumigatus* to produce cassava meal enriched with 36.9% protein.
The results of microbial protein yield in this study is comparable to that in other studies. HUTAGALUNG [20] obtained 9.8% and 10.8% protein from cassava chips fermented by *Rhizopus* and *Aspergillus* mycelia, respectively. An overall protein yields of 7.2—19% were obtained in this study.
As shown in this study the cassava peels waste with salts can not be used alone to support microbial growth and fermentation because of its low nitrogen content. Supplementation with a nitrogen source is needed to formulate appropriate fermentation media. Generally cassava is known to be very low in nitrogen.
Although cassava peel contains some microbial inhibitors which enable it to withstand deterioration in the soil [2], these inhibitors do not seem to affect microbial fermentation adversely.
From this study it appears that a protein enriched cassava starch can be obtained by using *Aspergillus niger* supplemented with diammonium hydrogen phosphate and cassava peels can be used as a source of producing reducing sugars such as glucose and fructose syrups.

Received June 24, 1985

References

[1] ODUNFA, S. A. — In: Microbiology of fermented foods. Ed.: WOOD, B. J. B. Amsterdam: Elsevier Science Publishers, 1985, pp. 155—191.
[2] EKUNDAYO, J. A. — In: Fungal Biotechnology. Eds.: SMITH, J. E., BERRY, D. R., Kristiansen, B. London: Academic Press, 1980, pp. 244—270.
[3] NWOKEDI, P. M. — In: Tropical root crops: production and uses in Africa. Eds.: TERRY, E. R., KOKU, E. V., ARENE, O. B., MAHUNGU, N. M. Ottawa, Canada: IDRC, 1984, pp. 108—110.
[4] ANON: Mechanized processing of cassava into gari. Technical Information Bulletin for Industry 3 No. 2. Federal Institute of Industrial Research, Oshodi, Lagos, Nigeria, 1974.
[5] MEUSER, F., SMOLNIK, D.: Stärke **32** (1980), 116.
[6] ADEYANJU, S. A., PIDO, P. P.: Nutr. Rep. Int. **18** (1978), 79.
[7] UNDERKOFLER, L. A. — In: Industrial Microbiology. Eds.: MILLER, B. M., LITSKY, W. New York: McGraw-Hill Book Coy, 1976, pp. 128—164.
[8] EJIOFOR, M. A. N., OKAFOR, N. — In: Tropical root crops: production and uses. Eds.: TERRY, E. R., DOKU, E. V., ARENE, O. B., MAHUNGU, N. M. Ottawa, Canada, 1984, pp. 105—107.
[9] AZOULAY, E., JONANNEAU, F., BERTRAND, J. C., RAPHAEL, A., JONSSENS, J., LEBEAULT, J. N.: Appl. Environ. Microbiol. **39** (1980), 41.
[10] MUINDI, P. J., HANSSEN: J. Sci. Food Agric. **32** (1981), 655.
[11] GREGORY, K. F.: Conservation and Recycling **5** (1982), 33.
[12] BUCHANAN, R. E., GIBBONS, N. E.: Bergey's Manual Determinative Bacteriology. 8th Ed. Baltimore: Wiliam and Wilkins Coy, 1974.
[13] BARNETT, H. L.: Illustrated genera of imperfect fungi. Minnesota: Burgess Publishing Co., 1960.
[14] BERNFIELD, P. — In: Methods in Enzymology. Vol. 1. Eds.: COLOWICS, S. P., KAPLAN, N. O. New York: Academic Press, 1955, pp. 149—158.
[15] READE, A. E., GREGORY, K. F.: Appl. Microbiol. **30** (1975), 897.
[16] CLEGG, K. M.: J. Sci. Food Agric. **7** (1956), 40.
[17] PEARSON, D.: The Chemical Analysis of Food. 7th Ed. Churchill and Livingstone, U. K.
[18] RICHARD, J. L., THURSTON, J. R., PEDEN, W. M., PINELLO, C.: Mythopathologia **87** (1984), 3.
[19] MOSSEL, D. A. A.: Microbiology of Foods. The ecological essentials of assurance and assessment of safety and quality. Netherlands: The University of Utrecht, 1982.
[20] HUTAGALUNG, R. I. — In: The Use of Organic Residues in Rural Communities. Ed.: SHACKLADY, C. A. Tokyo: United Nations University, 1979.
[21] OGUNDANA, S. K., COXON, D. T., DENNIS, C. — In: Tropical root crops: production and uses. Eds.: TERRY, E. R., DOKU, E. V., ARENE, O. B., MAHUNGU, N. M.: Ottawa, Cannada: IDRC, 1984, 133—135.

Acta Biotechnol. 7 (1987) 1, 30

Book Review

Robert E. KESTING

Synthetic Polymeric Membranes — A Structural Perspective

New York, Chichester, Brisbane, Toronto, Singapore: John Wiley & Sons, 1985 (Second Edition) 348 S., 55.75 Ł

Die vorliegende stark überarbeitete zweite Ausgabe des Buches befaßt sich, wie schon die 1. Auflage (1971, McGrow-Hill, New York), im Gegensatz zur umfangreichen anwendungstechnischen Literatur auf dem Membrangebiet, hauptsächlich mit der synthetischen Polymermembran selbst, welche in Form eines festen oder flüssigen Films als semipermeable Membran für gasförmige, flüssige oder feste Stoffe wirkt. Das ist nicht nur günstig für die Entwickler von Membranen, sondern dient ganz allgemein dem besseren Verständnis der Wirkungsweise der Membranen auf Grund einer vertieften Betrachtungsweise der Struktur-Funktions-Beziehungen.
In 10 Kapiteln wird 1. ein historischer Überblick gegeben; 2. die treibenden Kräfte bei Membranprozessen behandelt; 3. eine Reihe neuer Anwendungen vorgestellt; 4 .Membranpolymere und 5. Polymerlösungen charakterisiert; 6. dichte Filme; 7. Phaseninversionsmembranen und 8. andere poröse Membranen beschrieben; 9. flüssige und dynamisch gebildete Membranen sowie 10. biologische Membranen besprochen, wobei jedes Kapitel mit ausführlichen Literaturangaben versehen ist. Es ist ein Buch mit hohem Informationsgehalt für alle, die sich mit Membranherstellung und Charakterisierung befassen. Man könnte sich allerdings eine ausführlichere Darstellung der beiden letzten Kapitel auf Grund der schnellen Entwicklung auf diesen Gebieten vorstellen.

W. Zirkler

Acta Biotechnol. 7 (1987) 1, 31—38

Assimilation of Methanol by Yeasts

JWANNY, E. W., RASHAD, M. M.

Biochemistry Department
National Research Centre
Dokki, Cairo, Egypt

Summary

Candida lipolytica was cultured on a methanol containing medium as the only C source. The influence of different concentrations of methanol and ammonium sulphate and the effect of addition of some biological active materials to the methanol containing medium were studied.
Dry cell yield of *C. lipolytica* and protein content were determined to be 23.1 g of cells/100 g of methanol added and 40.2%, respectively. Total lipids, phospholipids, keto acid and amino acid composition were estimated. The amino acid composition of the protein was comparable to the composition of the other reported single cell proteins (SCP). These results indicate that *C. lipolytica* is potentially important for fodder use.

Introduction

Methanol has attracted much attention as a convenient raw material for industrial fermentation. Consequently investigations of microorganisms which grow on reduced C_1 compounds, e.g., methane and methanol, as the sole source of carbon and energy have increased. Investigations have been limited to studying the unique metabolic pathway of these compounds. Results of this research have made methanol of practical use in the fermentative production of cells and metabolites. This is a typical pattern in the development of applied microbiology, in which applied and fundamental studies are interrelated.
The utilization of reduced C_1 compounds by yeasts began with the investigations of *Kloecker* sp. 2201 by OGATA et al. [1]. The history of research on methanol-utilizing yeasts is short in comparison with that on bacterial methylotrophs. However, after the brief review by COONEY and LEVINE [2], research on methanol utilization by yeasts has made rapid progress (SAHM [3]).
In the present study, the effect of various culturing conditions for cell production by the yeast *Candida lipolytica* on methanol medium were established.

Material and Methods

Candida lipolytica was obtained from the Biochemistry Laboratory of the French Petroleum Institute.
The medium composition for the growth was as follows (g per l): KH_2PO_4, 2.0; $(NH_4)_2SO_4$, 3.0; $MgSO_4 \cdot 7H_2O$, 0.5; yeast extract, 0.3; $FeSO_4 \cdot 7H_2O$, 10 mg; $MnSO_4 \cdot 4H_2O$, 10 mg; biotin, 1 mg; thiamin-HCl, 10 mg; and methanol (v/v), 20 ml, in distilled water (KONO et al. [4]).

Cultivation was carried out at 30 °C and pH 4.6 in 250 ml Erlenmeyer flasks containing 50 ml medium. Flasks were kept on a rotary shaker at 190 r.p.m. The flasks were inoculated with 5 ml, 48 hours cultures having about 0.15 O. D. Methanol was added without sterilization in the form of vapour supply (MINAMI et al. [5]).
Growth and cell concentration was monitored by measuring the optical density (O.D.) of the culture fluid with Spekol Carl Zeiss at 610 nm. Cells were harvested by centrifugation at 4000 r.p.m. and washed three times by water, then lyophilized.
Cellular protein content was estimated by both LOWRY method [6] (after treatment of a sample of packed cells with 1N NaOH at 80 °C for 10 min), and by KJELDAHL method.
The amino acid profile was determined for the acid hydrolysate of the dry cells by using an amino acid analyzer BECKMAN model 118 CL. Tryptophan was determined in the alkaline hydrolyzate colourimetrically by BLAUTH et al. [7].
Lipid extracts of yeast cells were obtained by the method of PEDERSEN [8], purified by FOLCH et al. [9] and evaporated under reduced pressure till constant weight. The percentage of lipids was then calculated and raised to a known volume with chloroform.
The fractionation of lipid extracts into classes was conducted on glass plates coated with 250 to 270 μm layer of Stahl silica gel G according to CHENOUDA and JWANNY [10], activated at 105 to 110 °C for 2 hours. The plates were developed with a solvent system of petroleum ether (40–60)-diethyl ether-glacial acetic acid (60 : 40 : 1). The individual components visualized on exposure to iodine vapour were recovered by scrapping off and extraction with a mixture of chloroform and methanol (2 : 1) in a SOXHLET apparatus for 3 hours. After filtration and evaporation, the amount of extracted lipid fraction was determined gravimetrically.
Acid hydrolysates were prepared by MICHAELIS and HOCKETT [11] method for carbohydrate estimation DUBOIS method [12] was used.
Cellular keto acids: The hydrazones of certain amount of freshly harvested yeast cells are prepared according to FRIEDEMANN and HAUGEN [13], then chromatographically separated and estimated according to the method of EL-HAWARY and THOMPSON [14] and ISHERWOOD [15].

Results and Discussion

The first aim of the microbial utilization of methanol has been to produce single cell protein (SCP) and other metabolites. Several efforts were made by using methanol medium with different concentrations of methanol and ammonium sulphate, or supplementing the medium by different amino acids, vitamins or some biological active materials. The results in Fig. 1 show the effect of initial methanol concentrations (0.5–6.0% v/v)

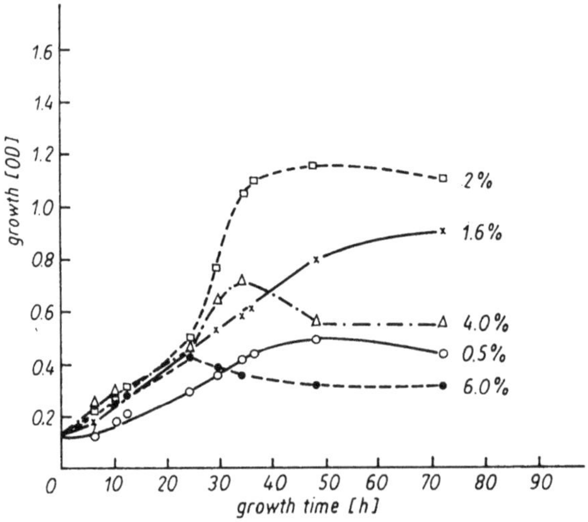

Fig. 1. Effect of initial methanol concentration on growth of *C. lipolytica*.

which suggest that when the initial concentration is adjusted to 2.0% (v/v) rapid growth without long adaptation occur. Higher concentrations lowered the biomass content and the growth was inhibited.
These results are supported by the findings of SAHM and WAGNER [16]; MIMURA et al. [17] and ALLAIS and BARATTI [18] who indicated that the maintenance of methanol as low as possible (0.4—4.0%) gave good biomass yield while higher methanol concentrations (5—6%) were found to be toxic and prevents growth.
Ammonium sulphate at concentration 636 mg N/l was the best nitrogen source with regard to biomass and total protein content (Table 1). This finding coincide with those of SUZUKI et al. [19] and MIMURA et al. [17].

Table 1. Effect of different concentrations of ammonium sulphate on growth and protein formation by *C. lipolytica* grown on methanol medium.

Concentration of $(NH_4)_2SO_4$	O.D.	Cell yield [mg dry wt/l]	mg protein/l
848 mg N/l	0.65	950	90.6
636 mg N/l	1.15	1150	254.2
424 mg N/l	0.315	240	78.5
63.6 mg N/l	0.71	590	25.4

The results illustrated in Table 2 showing the effect of supplementing the methanol medium with individual vitamins or with mixtures of 2 vitamins or of all vitamins indicated that the growth of *C. lipolytica* requires as a growth factor thiamin-HCl in a con-

Table 2. Effect of some vitamins on growth and protein formation by *C. lipolytica* (after 2 days of growth).

Vitamins tested	Concentration [mg/l]	Cell yield [mg dry wt/l]	mg protein/l
Mixture of all vitamins	*	490.7	92.7
Folic acid	0.01	213.0	50.7
Pyridoxine-HCl	1.00	281.0	51.1
P-aminobenzoic acid	0.20	311.7	66.8
Riboflavin	1.00	331.6	68.0
Nicotinic acid	1.00	332.9	63.9
M-Inositol	2.00	272.4	109.1
Ca-pantothenate	1.00	479.5	94.4
Ascorbic acid	1.00	591.6	128.4
D-biotin	1.00	467.0	98.5
D-biotin	0.01	585.7	104.3
Thiamin-HCl	10.00	1520.0	325.3
Thiamin-HCl	1.00	1015.5	212.2
Thiamin-HCl + D-biotin	10.00 + 1.00	1326.0	371.3
Thiamin-HCl + D-biotin	1.00 + 0.01	729.0	134.9
Thiamin-HCl + D-biotin	10.00 + 0.01	732.5	156.0

The media contain/l: KH_2PO_4, 2 g; $(NH_4)_2SO_4$, 3 g; $MgSO_4 \cdot 7H_2O$, 0.5 g; yeast extract, 0.3 g; trace amounts of $FeSO_4 \cdot 7H_2O$; $MnSO_4 \cdot 4H_2O$ + 20 ml methanol.

* Mixture of all vitamins: Thiamin-HCl, 10; D-biotin, 1.0; folic acid, 0.01; pyridoxine-HCl, 1.0; p-aminobenzoic acid, 0.2; riboflavin, 1.0; nicotinic acid, 1.0; m-inositol, 2.0; Ca-pantothenate, 1.0; and ascorbic acid, 1.0 mg/l.

centration of 10 mg/l. It can be noticed also that a mixture of biotin and thiamin-HCl in a ratio of 1 : 10 gave the best protein content.

In agreement with these results are the findings of FUJII et al. [20], MINAMI et al. [5] and ALLAIS and BARATTI [18], who published that biotin and thiamin-HCl were essential for growth, while MIMURA et al. [17] found that *C. methanophilum* Y-97 require biotin only for maximum growth.

The results obtained in Table 3 show good protein productivity by *C. lipolytica* when supplementing the medium with each of the 18 amino acids. This may be explained by the suggestions of THORNE [21] and BARTON-WRIGHT and THORNE [22] that the amino acids utilization takes place, at least to some extent, by the assimilation mechanism used by yeast for building proteins direct from the amino acids taken up. Also the results of JONES et al. [23] indicated that the amino acids are not incorporated into yeast protein intact, but participate in reactions which include transamination prior to their involvement in protein synthesis.

Table 3. Effect of supplementing different amino acids (concentration 0.5 g/l) to the culture medium of *C. lipolytica* (after 2 days of growth).

Amino acids added	Cell yield [mg dry wt/l]	mg protein/l	mg protein/ 100 mg dry wt
Control	1141.1	321.8	28.2
L-aspartic acid	267.0	86.5	32.4
Cystine	722.1	166.8	23.1
Cysteine	572.7	149.5	26.1
Phenyl alanine	308.3	127.0	41.2
DL-lysine	360.9	114.4	31.7
DL-valine	940.7	344.3	36.6
Histidine	608.6	245.3	40.3
Arginine	409.1	178.4	43.6
Threonine	511.3	194.3	38.0
Leucine	721.9	349.4	48.4
Isoleucine	644.3	180.4	28.0
Tryptophan	1008.7	319.8	31.7
Tyrosine	448.6	152.8	34.1
B-aminobutyric acid	390.0	182.1	46.7
γ-aminobutyric acid	1010.0	243.4	24.1
DL-glutamic acid	421.2	151.2	35.9
Methionine	964.9	351.2	43.4
Glycine	476.3	192.4	40.4
Serine	663.4	306.6	46.2
Proline	675.6	277.0	41.0
DL-alanine	918.5	347.2	37.8

Control medium contains/l: KH_2PO_4, 2 g; $MgSO_4 \cdot 7H_2O$, 0.5 g; $(NH_4)_2SO_4$, 3 g; yeast extract, 0.3 g; traces of $FeSO_4 \cdot 7H_2O$; $MnSO_4 \cdot 4H_2O$. Thiamin-HCl + biotin 10 : 1 mg/l, and 20 ml methanol — No amino acids.

Effect of some biological active base, as additives, on growth and protein formation by *C. lipolytica*:

The results of supplementing the methanol medium with some biological active bases at 0 or 24 hours (Table 4) indicated the stimulation of growth and increase in protein production by *C. lipolytica* when adding adenine, guanine-HCl, 2,6-dichlorophenol indophenol and indole 3-acetic acid. This may be due to the role of the nucleotides that parti-

Table 4. Effect of some biological active base, as additives (concentration 10 mg/l) on growth and protein formation by *C. lipolytica* (after 48 h).

Biological active base tested	Cell yield [mg dry wt/l]	Protein content [mg/l]	Protein [%]
Control	1156.9	335.5	28.99
Adenine	1593.9	503.7	31.60
Xanthine	1107.5	351.1	31.70
Guanine-HCl (0 h)	2042.7	805.1	39.40
Guanine-HCl (after 24 h old)	2636.0	1035.9	39.3
Indole 3-propionic acid	966.1	326.5	33.80
Indole 3-acetic acid	1232.5	461.0	37.40
Cyanocobalamine	809.9	285.9	35.30
Hydrogen peroxide	808.6	270.1	33.40
2,6-Dichlorophenol indophenol (0 h)	1706.1	751.2	44.03
2,6-Dichlorophenol indophenol (after 24 h)	2898.4	1084.0	37.4
Guanine-HCl + 2,6 di-chlorophenol indophenol (after 24 h)	3450.0	1469.7	42.6

Control medium contains/l: KH_2PO_4, 2 g; $MgSO_4 \cdot 7H_2O$, 0.5 g; $(NH_4)_2SO_4$, 3 g; yeast extract, 0.3 g; traces of $FeSO_4 \cdot 7H_2O$; $MnSO_4 \cdot 4H_2O$ + 20 ml methanol and thiamin-HCl + biotin 10.0 : 1.0 mg/l.

cipate in a wide variety of biochemical processes and in biological systems. This also may be due to reactions involving oxidation and reduction, the free energy exchange that is proportionate to the tendency of reactants to donate or accept electrons. It is clearly evident that the latter 2 additives seem to act as electron mediators together with the dehydrogenase systems of the yeast resulting in profound increase in the contents of protein and biomass of yeast cells.

Table 5. Cellular yield, protein, lipid and other metabolites of *C. lipolytica* cells after 3 days of growth on the indicated medium below*.

Metabolites		Keto acids in yeast cells	mg in cells of 1 liter medium
Biomass content [g/l]	3.65	α-Ketoglutaric acid	51.10
Cellular yield [g cell/g methanol]	0.23	Oxal acetic acid	3062.35
Protein [% of dry cells]	40.2	Pyruvic acid	427.05
Total lipids [% of dry cells]	12.3	Dihydroxyacetone	912.50
Carbohydrates [% of cells]	26.0		

* As that indicated in the control medium in Table 4 and addition of guanine-HCl and 2,6-dichloro-phenol after 24 h culturing.

Analysis results of yeast cells obtained from flask cultures in Table 5, could be considered good enough if compared to other studies made in fermenters or using continuous cultures.

COONEY and MAKIGUCHI [24] and MINAMI et al. [5] and other authors indicated that cellular yields ranged from 0.29—0.45 g cell/g methanol which is near our result 0.23. Also RATTRAY and HAMBLETON [25] and URAKAMI et al. [26] and URAKAMI and TAKANO [27] indicated that the protein content ranged from 16.8—56% which supports our result 40.2%.

Oxaloacetic acid and dihydroxyacetone existed in large amounts, the first may be due to the conversion of the large quantities of estimated aspartic acid to oxaloacetic acid by successive actions of asparaginase and transaminase. The presence of dihydroxyacetone in these results coincide with that found by van DIJKEN et al. [28] and KATO et al. [29] who suggested that dihydroxyacetone is a key intermediate and its synthese is responsible for catalyzing the transfere of glycol aldehyde group from xylulose-5-phosphate to formaldehyde to form glyceraldehyde 3-phosphate and dihydroxyacetone.

Table 6. Profiles of amino acids in the hydrolyzates of *C. lipolytica* dry cells.

Amino acids	g/100 g of biomass	g/100 g of protein
Lysine	2.66	6.62
Histidine	0.67	1.67
Arginine	1.12	2.79
Aspartic acid	9.14	22.74
Threonine	2.16	5.38
Serine	0.18	0.45
Glutamic acid	4.00	9.94
Glycine	2.03	5.05
Alanine	2.70	6.71
Cystine + methionine	1.26	3.13
Valine	2.31	5.74
Leucine	2.87	7.13
Isoleucine	1.60	3.97
Tyrosine	0.27	0.67
Phenylalanine	1.61	4.02
Tryptophan	0.84	2.20
Phosphoserine	0.08	0.20
γ-Aminobutyric acid	0.56	1.40
Ornithin	0.03	0.08

The profile of amino acids obtained in Table 6 was more or less the same as that obtained by LEVINE and COONEY [30], MIMURA et al. [17], LIOYED [31] and URAKAMI et al. [26].

It can be observed that the essential amino acids are present in good percentages resembling those of the FAO standard and other amino acids of cells of other yeasts grown on methanol as the only C source (Table 7). However, cereal grains and forage feed crops are relatively low in protein content and amino acids. Since synthesized amino acids cost more than those obtained from natural sources, the latter can be supplied in the form of dried microbial cells.

The lipid content and its composition (Table 8) does not appear to contain any components that might be considered undesirable if the yeasts were used as a dietary source of SCP.

From the results of this work, it can be concluded that the biomass of *C. lipolytica* grown on methanol could be used as another feeding source for domestic animals and poultry.

Table 7. Comparison of amino acid composition of cells and the FAO standard.

Amino acids	FAO* standard	*C. lipolytica* [10]		*P. methanolica** Y-1023		*T. sonorensis** MT-8		*H. polymorpha** 1—3	
	[g/100 g of protein]	[g/100 g of biomass]	[g/100 g of protein]	[g/100 g of biomass]	[g/100 g of protein]	[g/100 g of biomass]	[g/100 g of protein]	[g/100 g of biomass]	[g/100 g of protein]
(Essential amino acids)									
Lysine	4.2	2.7	6.6	4.0	7.7	3.6	7.2	3.3	6.6
Threonine	2.8	2.2	5.4	2.6	5.0	2.7	5.4	2.7	5.4
Valine	4.2	2.3	5.7	3.1	6.0	2.9	5.8	3.1	6.2
Methionine	2.2	1.3	3.1	1.0	1.9	0.9	1.8	0.9	1.8
	+ cystine	+ cystine							
Isoleucine	4.2	1.6	4.0	2.7	5.2	2.6	5.2	2.6	5.2
Leucine	4.8	2.9	7.1	4.0	7.7	4.1	8.2	4.0	8.0
Phenylalanine	2.8	1.6	4.0	2.5	4.8	2.4	4.8	2.5	5.0
Tryptophan	1.4	0.8	2.2	0.8	1.5	0.8	1.6	0.7	1.4
(Non-essential amino acids)									
Alanine		2.7	6.7	3.3	6.4	3.2	6.4	3.3	6.6
Arginine		1.1	2.8	3.1	6.0	2.6	5.2	2.9	5.8
Aspartic acid		9.1	22.7	5.3	10.3	5.6	11.2	5.5	11.1
Cysteine		—	—	0.8	1.5	0.7	1.4	0.7	1.4
Glutamic acid		4.0	9.9	7.7	14.9	7.3	14.5	6.0	12.1
Glycine		2.0	5.1	2.6	5.0	2.6	5.2	2.6	5.2
Histidine		0.7	1.7	1.3	2.5	1.2	2.4	1.3	2.6
Proline		—	—	2.1	4.1	2.2	4.4	2.6	5.2
Serine		0.2	0.5	2.5	4.8	2.6	5.2	2.6	5.2
Tyrasine		0.3	0.7	2.3	4.4	2.2	4.4	2.4	4.8

* URAKAMI et al. [26]

Table 8. Lipid composition of *C. lipolytica* cells harvested after 3 days of growth (late logarithmic phase).

Lipid fractions	Percentage of different classes
Phospholipid	21.9
Monoglycerides	9.21
Sterols	7.89
Diglycerides	14.91
Free fatty acids	14.47
Triglycerides	10.09
Sterol ester	21.49

Received April 11, 1985

References

[1] OGATA, K., NISHIKAWA, H., OHSUGI, M.: Agric. Biol. Chem. **33** (1969), 1519.
[2] COONEY, C. L., LEVINE, D. W. — In: Single Cell Protein II. MIT Press, Cambridge, Massachusetts, 1975, 402.
[3] SAHM, H.: Adv. Biochem. Eng. **6** (1977), 77.
[4] KONO, K., OKAMOTO, R., TAKAMATSU, A.: Yeast Cells. (cf. Chem. Abst.) **87** (1977), 20623.
[55] MINAMI, K., YAMAMURA, M., SHIMIZU, S., OGAWA, K., SEKINE, N.: J. Ferment. Technol. **56** (1978), 35.
[6] LOWRY, O. H., ROSEBROUGH, N. J., FARR, A. L., RANDALL, R. J.: J. Biol. Chem. **193** (1951), 265.
[7] BLAUTH, J. O., CHAREZINSKI, M., BERBEC, H.: Anal. Biochem. **6** (1963), 69.
[8] PEDERSEN, T. A.: Acta Chem. Scand. **16** (1962), 374.
[9] FOLCH, J., LEES, M., SLOANE-STANLEY, G. H.: J. Biol. Chem. **226** (1957), 497.
[10] CHENOUDA, M. S., JWANNY, E. W.: J. Gen. Appl. Microbiol. **18** (1972), 181.
[11] MICHAELIS, H. K., HOCKETT, D. P.: Biochem. Biophys. Acta **96** (1965), 272.
[12] DUBOIS, M., GILLES, K. A., HAMILTON, J. K., REBERS, P. A., SMITH, F.: Anal. Chem. **28** (1956), 350.
[13] FRIEDEMANN, T. E., HAUGEN, G. E.: J. Biol. Chem. **226** (1943), 497.
[14] EL-HAWARY, F. M. S., THOMPSON, K. H. S.: Biochem. J. **53** (1953), 340.
[15] ISHERWOOD, F. A.: Nature **173** (1954), 121.
[16] SAHM, H., WAGNER, F.: Arch. Microbiol. **84** (1972), 29.
[17] MIMURA, A., WADA, M., NAKANO, T., HAYAKAWA, S., IGUCHI, T.: J. Ferment. Technol. **56** (1978), 443.
[18] ALLAIS, J. J., BARATTI, J.: J. Ferment. Technol. **61** (1983), 339.
[19] SUZUKI, O., TIGAMI, Y., NAKASATO, S.: Agric. Biol. Chem. **43** (1979), 1343.
[20] FUJII, T., ASADA, Y., TONOMURA, K.: Agric. Biol. Chem. **38** (1974), 1121.
[21] THORNE, R. S. W.: J. Inst. Brew. **55** (1949), 201.
[22] BARTON-WRIGHT, E. C., THORNE, R. S. W.: J. Inst. Brew. **55** (1949), 383.
[23] JONES, M., PRAGNELL, M. J., PIERCE, J. S.: J. Inst. Brew. **75** (1969), 520.
[24] COONEY, C. L., MAKIGUCHI, N.: Biotechnol. Bioeng. Symp. **7** (1977), 65.
[25] RATTRAY, J. B. M., HAMBLETON, J. E.: Can. J. Microbiol. **26** (1980), 190.
[26] URAKAMI, T., TERAO, I., NAGAI, I.: J. Ferment. Technol. **61** (1983), 221.
[27] URAKAMI, T., TAKANO, S.: J. Ferment. Technol. **62** (1984), 111.
[28] Van DIJKEN, J.-P., HARDER, W., BEARDSMORE, A. J., QUAYLE, J. R.: FEMS Microbiol. Lett. **4** (1978), 97.
[29] KATO, K., HIGUCHI, T., SAKAZAWA, C., NISHIZAWA, T., TANI, Y., YAMADA, H.: Biochem. Biophys. Acta **715** (1982), 143.
[30] LEVINE, D. W., COONEY, C. L.: Appl. Microbiol. **26** (1973), 982.
[31] LIOYED, R.: The Arab Congress of Utilization of Nonconventional Food Sources in Feeding Animal. 20—24 Nov., 1982.

Acta Biotechnol. 7 (1987) 1, 39—47

Wirkung von Ölen und Fettsäuren auf Wachstum und Enzymbildung bei Thermoactinomyces vulgaris

VI. Studien zur Wirkungsweise von Öl

LEUCHTENBERGER, A.[1], JAKOVLEVA, M. B.[2], RUTTLOFF, H.[1], LOGINOVA, L. G.[2]

[1] Akademie der Wissenschaften der DDR
Zentralinstitut für Ernährung
Arthur-Scheunert-Allee 114—116, Potsdam-Rehbrücke, 1505 DDR
[2] Akademie der Wissenschaften der UdSSR
117312 Moskau, Prospekt 60-letija Oktjabrja 7a, UdSSR

Summary

Some aspects on the action of vegetable and animal oils and fatty acids on growth and protease synthesis of *T. vulgaris* were studied to interpret the stimulating effect of these substances. It was found that oil does not influence the hydrolysis of the substrate casein by the protease. However, the effect on the gas exchange during fermentation has a substantial importance. Furthermore, the metabolization of lipids shows that they play a role as carbon source.

Einleitung

Mit der Wirkungsweise von Lipiden auf Stoffwechselprozesse bei Mikroorganismen beschäftigen sich verschiedene Autoren. Aus ihren experimentellen Befunden leiten sich Wirkungsmechanismen ab, die man im wesentlichen in vier Gruppen unterteilen kann:

1. Beeinflussung der Enzym-Substrat-Wechselwirkung,
2. Wirkung auf den Stoffaustausch Zelle-Medium,
3. Einfluß auf die Regulation von Protein- bzw. Enzym-Synthese,
4. Nutzung von Öl und Fettsäuren als C-Quelle.

Die bisher vorliegenden Literaturangaben [1] verdeutlichen, daß ein allgemeingültiger Wirkungsmechanismus der Öle und Fettsäuren nicht existiert. Es wird daher angenommen, daß — in Abhängigkeit von den jeweils gegebenen Bedingungen — eine Wirkung auf verschiedene Prozesse im zellulären Stoffwechsel möglich ist.
Befunde über die Wirkungsweise von Ölen und Fettsäuren auf Stoffwechselprozesse bei *T. vulgaris* sind aus der Literatur unseres Wissens nicht bekannt. Nachdem bereits in verschiedenen Mitteilungen über den Effekt derartiger Substanzen auf Proteasebildung und Wachstum dieses Stammes informiert wurde [29, 2—7], beschäftigen sich die nachfolgend mitgeteilten Untersuchungen mit einigen speziellen Aspekten zur Wirkungsweise dieser Substanzen.

Material und Methoden

Stammaterial: *T. vulgaris*, deponiert in der Hinterlegungsstelle für Mikroorganismen im Zentralinstitut für Mikrobiologie und experimentelle Therapie Jena unter Nr. IMET 9515.

Fermentation: Angaben dazu sind aus Mitt. III [5] dieser Veröffentlichungsreihe zu entnehmen.

Oberflächenaktive Substanzen (OAS): Tween 60 und Tween 80, Ferman (Reaktionsprodukt aus Polypropylenglykol und Ethylenoxid, Buna-Werke, DDR), Silikonentschäumer NM 40 (VEB Chemiewerk Nünchritz), Rapsöl (Raffinat, VEB Pflanzenfettkombinat Velten). Zugabe von 0,005%—0,05% OAS jeweils zu Beginn der Fermentation in das Medium der Arbeitskultur.

Wachstum: Proteinbestimmung im Sediment [2]. Darstellung der Ergebnisse abzüglich des t_0-Wertes.

Proteaseaktivität: Modifizierte ANSON-Methode unter Verwendung einer 1%igen Caseinlösung bei 50 °C [8].

Lipaseaktivität: Titrimetrische Methode [9, 10] unter Verwendung einer 10%igen Olivenöl-Gummiarabicum-Emulsion und eines Titrators der Fa. Radiometer Kopenhagen. Die Reaktion erfolgt bei 50 °C und pH 8,0.
α-Amylase- und Glucoamylaseaktivität: Bestimmung nach DNSS (Dinitrosalicylsäure)- und der GOD/POD (Glucoseoxydase/Peroxydase)-Methode bei 50 °C. Aus der Differenz der mit beiden Methoden ermittelten Werte läßt sich auf den α-Amylase- und Glucoamylaseanteil schließen [11].

Zucker (papierchromatographisch): Auf Chromatographiepapier Typ FN 4 werden mit einem Fließmittelgemisch im absteigenden Verfahren die Zucker getrennt und mit einem Sprühmittel sichtbar gemacht. Halbquantitative Auswertung durch Vergleich mit verschiedenen Standardsacchariden [12].

Stärke: Behandlung des Substrates mit Glucoamylase bei 50 °C und nachfolgende Glucosebestimmung mittels GOD/POD [13].

pO_2-Messung: Einsatz einer O_2-Meßsonde (Zentrum für wissenschaftlichen Gerätebau Mytron, Heiligenstadt) und eines O_2-Meßgerätes AM 220 (Forschungsinstitut Meinsberg).

Ergebnisse

Enzym-Substrat-Wechselwirkung

Der Einfluß von Rapsöl auf die proteolytische Wirkung unterschiedlich aktiver Kulturlösungen gegenüber dem Substrat Casein wird untersucht. Dazu wird eine 1%ige Caseinlösung mit 0,1% Rapsöl versetzt und bei 50 °C mit einem Ultra-Turrax ca. 1 min durchmischt. Die Bestimmung der Proteaseaktivität erfolgt jeweils sowohl mit der ölfreien als auch mit der ölhaltigen Caseinlösung. Die Mittelwerte von 2 Versuchsreihen sind in Tab. 1 enthalten. Bei Vergleich der Ergebnisse ist kein signifikanter Unterschied feststellbar.

Tabelle 1. Einfluß von Rapsöl auf die Wechselwirkung Protease—Casein (0,1% Rapsöl, 50 °C)

	Proteolytische Aktivität [TE/ml]			
	Serie I ohne Öl	mit Öl	Serie II ohne Öl	mit Öl
$\bar{x}$ (n = 10)	219	231	582	597
sx	55,65	68,30	142,80	131,15
$s\bar{x}$	17,60	21,60	45,15	41,45
V [%]	25,30	29,55	24,55	21,95
t-Test (ohne Öl → mit Öl)	n. s.		n. s.	

$\bar{x}$ — Mittelwert, sx — Standardabweichung, $s\bar{x}$ — Standardfehler, V — Variabilitätskoeffizient, n. s. — nicht signifikant

O_2-Versorgung und Proteinsynthese

Zunächst wird in Schüttelkultur überprüft, ob Öl auf Grund seiner Oberflächenaktivität die Stoffwechseltätigkeit anregt. Zu diesem Zweck wird der Einfluß verschiedener OAS (Tween 60, Tween 80, Ferman, Silikonöl) in den Konzentrationen von 0,05; 0,01 und 0,005% auf Wachstum und Proteasebildung von *T. vulgaris* in MQW I-Medium untersucht. Nach den erzielten Befunden (Tab. 2) zeigen die getestete OAS — in Abhängigkeit von der Fermentationsdauer und der eingesetzten Konzentration — eine unterschiedlich starke stimulierende oder hemmende Wirkung auf die Proteasebildung.

Tabelle 2. Abhängigkeit der Proteasebildung von der Zusatzmenge an verschiedenen oberflächenaktiven Substanzen (MQW* I-Medium, 500 ml Einstichkolben)

Oberflächenaktive Substanz (OAS)	Zusatzmenge [%]	Aktivitätssteigerung [%] im Vergleich zur Kontrolle						
		Fermentationsdauer [h]						
		16	19	22	40	43	46	64
Kontrolle	0	100	100	100	100	100	100	100
Tween 80	0,05	83	84	90	122	118	119	135
	0,01	80	71	70	73	64	70	74
	0,005	66	89	65	75	86	63	94
Tween 60	0,05	133	94	118	121	128	126	151
	0,01	110	115	102	120	117	108	100
	0,005	88	79	83	96	95	75	99
Ferman	0,05	128	108	70	80	103	67	79
	0,01	150	134	118	146	127	123	127
	0,005	151	115	112	115	130	125	145
Silikonöl	0,05	113	116	69	77	80	72	85
	0,01	129	120	105	102	113	108	125
	0,005	95	119	103	100	108	101	115

* MQW — Maisquellwasser

In 5 l- und 30 l-Fermentoren erfolgen mit MQW V-Medium Untersuchungen zum Einfluß von Rapsöl und inertem Silikonöl auf Wachstum, Proteasebildung und Sauerstoffversorgung der Kultur (Abb. 1—4). Aus Abb. 1 geht hervor, daß beide Öle die Biomassebildung anregen (Rapsöl — 90%, Silikonöl — 60%). Hinsichtlich der Proteasesynthese ergibt sich mit 110% bei Rapsöl und 80% bei Silikonöl ein ähnliches Bild (Abb. 2). Bezogen auf die gebildete Biomassemenge zeigen die verschiedenen Varianten im Verlaufe der Fermentation eine unterschiedliche spezifische Produktbildung (Abb. 3). Sie liegt nach 22 h bei Rapsöl um ca. 10% höher als bei Silikonöl sowie um 40% höher als bei der Kontrolle.
Ein deutlicher Unterschied zwischen den Ölvarianten und der Kontrolle ist bei der pO_2-Konzentration im Kulturmedium zu beobachten. Wie Abb. 4 verdeutlicht, fällt diese während des exponentiellen Wachstums bis zur 5. Stunde bei beiden Ölen auf ca. 60%, bei der Kontrolle nur auf 82% ab.

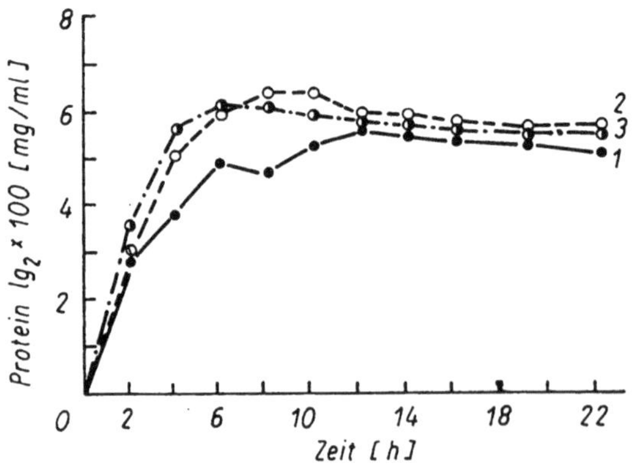

Abb. 1. Wachstumsverlauf während der Fermentation mit und ohne Ölzusatz (MQW V-Medium, 5 l-Fermentor, 0,2% Rapsöl bzw. Silikonöl).

1 — Kontrolle, 2 — Rapsöl, 3 — Silikonöl.

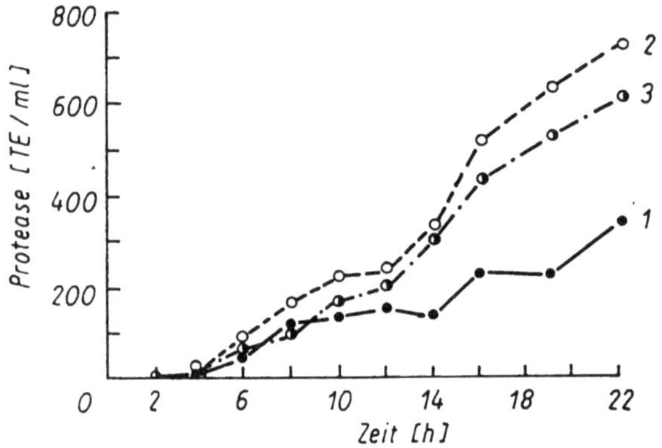

Abb. 2. Proteasebildung während der Fermentation mit und ohne Ölzusatz (MQW V-Medium, 5 l-Fermentor, 0,2% Rapsöl bzw. Silikonöl).

1—3 siehe Abb. 1

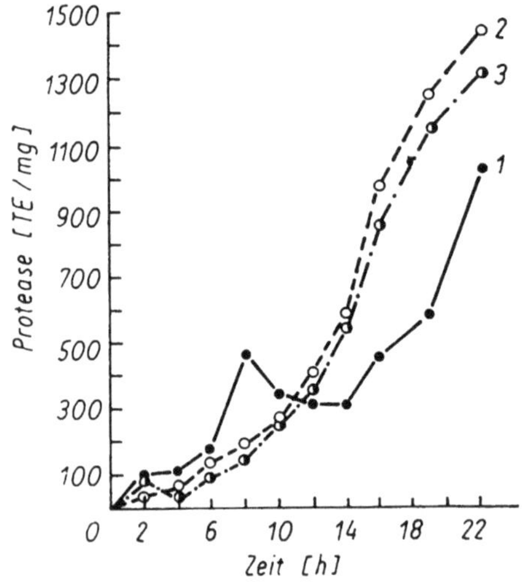

Abb. 3. Proteaseaktivität während der Fermentation mit und ohne Ölzusatz (MQW V-Medium, 5 l-Fermentor, 0,2% Rapsöl bzw. Silikonöl).

1—3 siehe Abb. 1

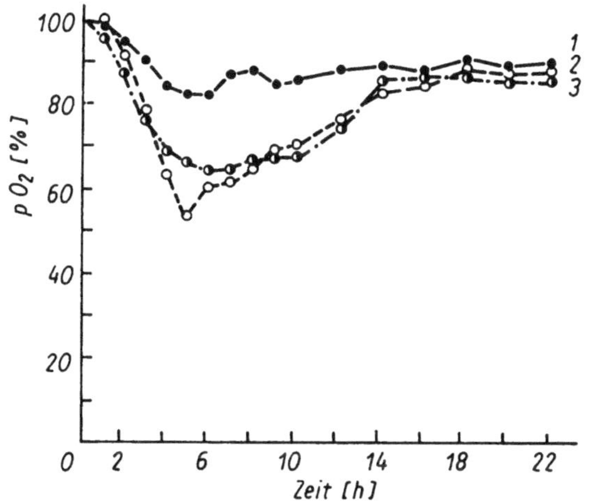

Abb. 4. Veränderung der pO_2-Konzentration im Medium während der Fermentation mit und ohne Ölzusatz (MQW V-Medium, 5 l-Fermentor, 0,2% Rapsöl bzw. Silikonöl). 1—3 siehe Abb. 1

Ölwirkung auf den Kohlenhydratstoffwechsel

Die Beeinflussung des Kohlenhydratstoffwechsels durch Öl wird in 500 ml-Einstichkolben sowie 2 l- und 30 l-Fermentoren unter Verwendung von MQW I- und MQW V-Medium untersucht. Abb. 5 zeigt den Stärkeabbau in MQW V-Medium mit und ohne Ölzusatz im Verlaufe von 22 Stunden.

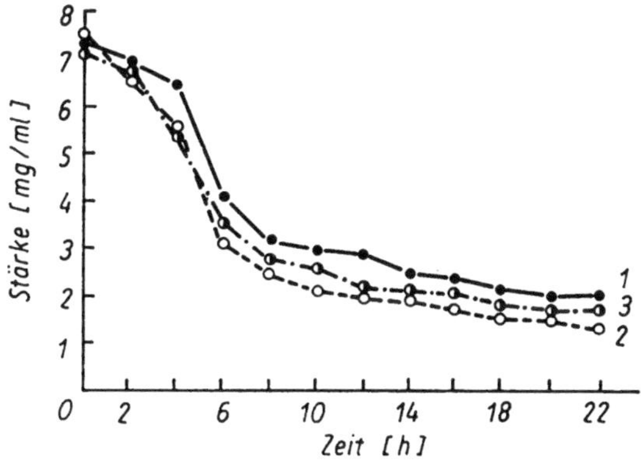

Abb. 5. Stärkeabbau während der Fermentation bei Ab- und Anwesenheit von Öl (MQW V-Medium, 5 l-Fermentor, 0,2% Rapsöl bzw. Silikonöl).
1—3 siehe Abb. 1

Zur Ermittlung der Ursachen des intensiveren Stärkeumsatzes in Gegenwart von Öl wird die Bildung amylolytischer Enzyme analytisch verfolgt. Wie aus Abb. 6 zu entnehmen ist, wird durch beide Ölarten die Synthese der in das Medium ausgeschiedenen α-Amylase angeregt. Sie ist nach 10 h um etwa 100% höher als in der Kontrolle. Glucoamylase wird in nur sehr geringen Mengen gefunden. Die durch die enzymatische Hydrolyse der Stärke freigesetzten Spaltprodukte und deren Metabolisierung lassen sich papierchromatographisch gut nachweisen (Tab. 3).

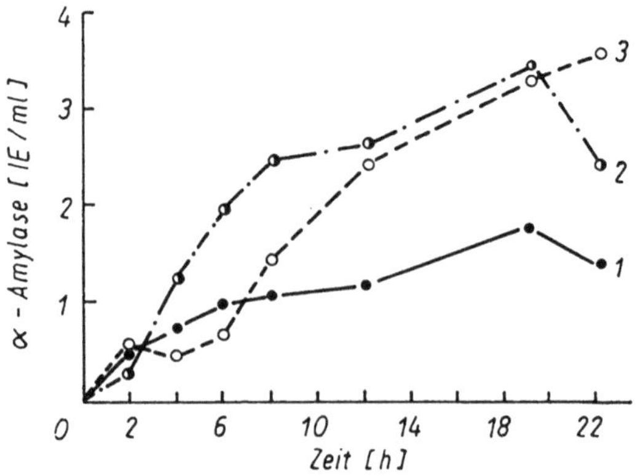

Abb. 6. Bildung von α-Amylase im Verlaufe der Fermentation (MQW V-Medium, 5 l-Fermentor, 0,2% Rapsöl bzw. Silikonöl).
1—3 siehe Abb. 1

Tabelle 3. Papierchromatographische Befunde zum Saccharidumsatz während der Fermentation bei An- und Abwesenheit von Öl (MQW V-Medium, 5 l-Fermentor, 0,2% Raps- bzw. Silikonöl)

Varianten	Kulturdauer [h]	Saccharide [mg/ml]				
		Glucose	Maltose	Maltotriose	Malto-tetraose	Malto-pentaose
Kontrolle	4	0	0,5	0,3	0	0
	8	0	1,3	0,3	0	0
	12	0,1	1,5	0,5	0	0
	16	0	1,1	0,1	0	0
	22	0	1,0	0,1	0	0
Rapsöl 0,2%	4	0	0,8	0,5	0,6	0,1
	8	0	1,7	0,4	0,5	0,1
	12	0	1,8	0,4	0,1	0,1
	16	0,5	1,8	0,2	0,1	0
	22	1,1	2,0	0,1	0	0
Silikonöl 0,2%	4	0	1,0	0,7	0	0
	8	0	0,9	0,7	0,5	0
	12	0,1	1,3	0,2	0,1	0
	16	0,5	1,5	0,1	0	0
	22	0,9	1,2	0	0	0

Man findet hinsichtlich des Saccharidgehaltes der Kulturlösung sowohl qualitative als auch quantitative Unterschiede zwischen der Kontrolle und den Ölvarianten. Im Verlaufe der Fermentation sind in der Variante ohne Ölzusatz keine oder nur Spuren an Glucose sowie Maltotetraose und Maltopentaose nachweisbar. Am stärksten reichert sich die Maltose an (1,5 mg/ml nach 12 h). Im Gegensatz dazu kommt es bei Anwesenheit von Öl zu einem stärkeren Stau von Stärkespaltprodukten (Glucose, Maltose, Maltotriose, Maltotetraose und Maltopentaose).
Im Zusammenhang mit dem Kohlenhydratstoffwechsel ist der Abbau des zugesetzten Öles von Interesse. In Ergänzung zu den in Mitt. IV [6] dargelegten Befunden wird bei einigen Versuchsansätzen im 5 l-Fermentor die Lipaseaktivität gemessen (Tab. 4).

Tabelle 4. Lipasebildung während der Fermentation im 5 l-Fermentor bei An- und Abwesenheit von Öl (LE — Lipaseeinheit, mg — Biomasse (Protein)

Variante	Fermentationsdauer [h]	Lipaseaktivität (LE)			
		intrazellulär		extrazellulär	
		LE/ml	LE/mg	LE/ml	LE/mg
Kontrolle	3	0	0	0	0
	8	0,21	0,05	0	0
	16	0,43	0,11	0	0
	21	0,67	0,17	0	0
Rapsöl	3	0	0	0	0
	8	0,52	0,13	0	0
	16	0,92	0,23	0,21	0,39
	21	2,07	0,52	0,10	0,20
Silikonöl	3	0	0	0	0
	8	0,14	0,04	0	0
	16	0,23	0,06	0,10	0,20
	21	0,30	0,08	0,10	0,22

Diskussion

Gemäß Tab. 1 wird durch Zusatz von 0,1% Rapsöl die Aktivität der Protease gegenüber Casein nicht verändert. Eine Beeinflussung der Enzymstabilität bzw. der Enzym-Substrat-Bindung, wie sie von KONDO [14] sowie von THIELE [15] angegeben werden, liegt demnach nicht vor.

Die zur Überprüfung der Oberflächenwirkung von Ölen durchgeführten Untersuchungen mit OAS weisen in Schüttelkulturversuchen zum Zeitpunkt der höchsten Proteasebildung bei der Kontrolle (nach 22 h) eine Steigerung der Enzymproduktion um ca. 20% aus. Im Vergleich zur aktivitätsfördernden Wirkung verschiedener Öle — sie liegt zu diesem Zeitpunkt in Schüttelkultur zwischen 20 und 60% [6] — ist der Einfluß der OAS jedoch deutlich geringer. Diese Ergebnisse stehen im Gegensatz zu Befunden von KOBAYASHI u. SUZUKI [16], die z. B. bei *Mortierella vinacea* mit verschiedenen OAS keine Steigerung der α-Galaktosidasebildung erzielen; sie stimmen jedoch mit anderen Literaturangaben überein, wonach dieser Effekt bei verschiedenen Mikroorganismen nachgewiesen werden kann [17—19, 4].

Für eine Oberflächenwirkung von Öl sprechen auch andere Befunde. Gemäß Mitt. IV [6] verzögert ein Ölzusatz vor bzw. zu Fermentationsbeginn die Auskeimung der Sporen. Vermutlich hemmt Öl den Stoffaustausch — insbesondere den von Gasen — zwischen Spore und Nährmedium durch Auflagerung eines Ölfilms auf die Zellwand. Erst mit Beginn der Hydrolyse des Öles durch zelleigene Enzyme wird diese Barriere abgebaut.

Eine Schutzwirkung für das Mycel scheint Öl in der Absterbephase desselben zu besitzen. Wie bei der Untersuchung verschiedener Einflußfaktoren in Mitt. III [5] festgestellt wird, steigt mit zunehmender Belüftungsintensität die Mycellyse an. Bei Anwesenheit von Öl ist diese gemindert. Man kann daraus ableiten, daß ein Überangebot von Sauerstoff in der Kulturlösung die Mycellyse durch verstärkte Bildung lytischer Enzyme begünstigt und Öl das Mycel vor diesem Überangebot schützt. Für diese Hypothese sprechen auch Versuchsergebnisse, die im 5 l-Glasfermentor unter Verwendung von MQW V-Medium erzielt worden sind (Abb. 1, 4).

Nach PACA et al. [20] sowie TAKAHASHI u. YOSHIDA [21] führen Zusätze von Entschäumeröl zu einem starken Abfall des O_2-Übergangskoeffizienten (K_La-Wert) in einem Fermentationsmedium. Dieser Effekt ist u. a. von den Eigenschaften des Entschäumeröles abhängig [22]. Entschäumer, denen organische Öle und Fettsäuren als wesentliche Verbindungen zugrundeliegen, führen infolge ihrer Barrierewirkung bei der Gasdiffusion durch die Grenzfläche Gas-Flüssigkeit zu einem Rückgang der K_La-Werte. Dagegen werden nach Ansicht der Autoren Silikonöle andersartig wirksam. Nach Abb. 4 ist bei *T. vulgaris* im Rührfermentor ein gravierender Unterschied der pO_2-Konzentration in Gegenwart von Raps- bzw. Silikonöl nicht erkennbar, so daß man in diesem System auf eine ähnliche Wirkung beider Öle auf die Gasdiffusion schließen muß. Infolge des verminderten O_2-Übergangs von der Gas- in die Flüssigphase kommt es während des exponentiellen Wachstums zu einer stärkeren Abnahme der pO_2-Konzentration im Kulturmedium, woraus vermutlich günstigere Bedingungen für Wachstum und Proteasesynthese resultieren. In Übereinstimmung mit Befunden, die WINKLER [23] unter Einsatz von *Endomycopsis bispora* erzielte, könnte der positive Einfluß von Öl darin bestehen, daß induzierend wirkende Substanzen der Medienkomponenten Hefe oder Maisquellwasser vor einer Inaktivierung durch Oxydationsprozesse geschützt werden. Des weiteren ist auch eine Hemmung der Synthese lytischer Enzyme durch *T. vulgaris* [5] möglich.

Darüberhinaus sind vermutlich noch andere Faktoren wirksam. Dies geht aus dem Vergleich der gebildeten Proteaseaktivitäten bei Zusatz von Rapsöl oder aber von Silikonöl hervor (Abb. 2). So beobachtet man mit Rapsöl im Medium eine um ca. 10% höhere Bildungsrate für die spezifische Enzymaktivität als mit Silikonöl (Abb. 3). Daraus kann abgeleitet werden, daß die Metabolisierung von Rapsöl durch *T. vulgaris* eine zusätzliche Wirkung auf die Proteasesynthese besitzt. Entsprechende Hinweise sind auch von YAMAMOTO et al. [24], KOBAYASHI et al. [16], FRANZKE et al. [25], GALYNKIN et al. [26], PACA et al. [20] sowie EVANS und HALL [22] gegeben worden. Diesbezügliche Untersuchungen lassen vermuten, daß Entschäumeröle im Kohlenhydratstoffwechsel von *T. vulgaris* eine Rolle spielen. Es gibt jedoch Unterschiede zwischen pflanzlichen und tierischen Ölen sowie Silikonöl. Gemäß Mitt. V [7] werden erstere zu etwa 60 bis 70% im Verlaufe der Fermentation umgesetzt, wohingegen Silikonöl nach EVANS u. HALL [22] als inertes Produkt unverändert bleibt.

Die Fähigkeit von *T. vulgaris* zur hydrolytischen Spaltung pflanzlicher Öle wird durch den Nachweis lipolytischer Enzyme in der Kulturlösung bestätigt (Tab. 4). Auf die Fähigkeit zur Lipasebildung bei *T. vulgaris* weisen auch andere Autoren hin. Unter Verwendung von Tributyrin und Tween-80 als Substrate findet FALKOWSKI [27] mittels Diffusionsplattentest nach 120stündiger Bebrütung bei 55 °C sowohl intra- als auch extrazelluläre Lipaseaktivität. ELWAN et al. [28] berichten ebenfalls über die lipolytische Aktivität dieses Stammes. Die Induktion der Lipasebildung erfolgt nach Angaben des Autors durch Butterfett, Maisöl und andere Öle wesentlich intensiver als durch Tributyrin.

Gemäß Abb. 5 ist bei Anwesenheit von Entschäumeröl ein intensiverer Stärkeabbau als bei der Variante ohne Öl zu beobachten. Dies ist auf eine verstärkte α-Amylase-Synthese zurückzuführen (Abb. 6), die vermutlich durch Beeinflussung des Stoffüberganges bei der Zugabe von Entschäumeröl induziert wird. Die in größeren Mengen zur Verfügung stehenden Abbauprodukte der Stärke regen eine intensivere Biomassebildung an. Dabei ist offenbar die Geschwindigkeit der Metabolisierung der Stärkespaltprodukte geringer als deren enzymatische Freisetzung, was insbesondere in einer Anreicherung von Oligosacchariden (Maltose, Maltotriose, Maltotetraose, Maltopentaose) im Kulturmedium zum Ausdruck kommt (Tab. 3). Die bei den Ölvarianten gleichzeitig zu beobachtende Anreicherung von Glucose bleibt unterhalb derjenigen Konzentration, die eine katabolische Repression hervorruft [3]. Der erstmalige Nachweis von Lipase nach 8 h

Fermentation (Tab. 4) sowie das verzögerte Wirksamwerden von Rapsöl auf die Proteasesynthese im synthetischen Medium [3] weisen darauf hin, daß die durch lipolytische Spaltung von pflanzlichen Ölen freiwerdenden Fettsäuren erst am Ende der log-Phase als zusätzliche C-Quelle genutzt werden und somit offenbar für das länger anhaltende Wachstum (Abb. 1) sowie für die höhere Proteaseaktivität (Abb. 2) im Vergleich zur Silikonölvariante mitverantwortlich sind. Zu ähnlichen Befunden gelangen auch YAMAMOTO et al. [24] sowie EVANS et al. [22]. Nach den im synthetischen Medium erhaltenen Resultaten [3, 4] ist es denkbar, daß durch Lipolyse aus Rapsöl freigesetzte Fettsäuren — vor allem Palmitin- und Erucasäure — direkt oder indirekt regulativ auf die Proteasesynthese von *T. vulgaris* Einfluß nehmen.

Eingegangen: 25. 10. 1985

Literatur

[1] LEUCHTENBERGER, A., RUTTLOFF, H.: Z. Allg. Mikrobiol. **19** (1979), 609.
[2] LEUCHTENBERGER, A., SCHAFFNER, B., RUTTLOFF, H.: Z. Allg. Mikrobiol. **21** (1981), 671.
[3] LEUCHTENBERGER, A.: Zur Gewinnung einer thermostabilen Protease mit *Thermoactinomyces vulgaris*. Diss. B, Jena 1982.
[4] YAKOVLEVA, M. B., LEUCHTENBERGER, A., LOGINOVA, L. G., RUTTLOFF, H.: Mikrobiologija **51** (1982), 70.
[5] LEUCHTENBERGER, A., RUTTLOFF, H.: Z. Allg. Mikrobiol. **23** (1983), 635.
[6] LEUCHTENBERGER, A., RUTTLOFF, H.: Acta Biotechnol. (im Druck).
[7] LEUCHTENBERGER, A., RUTTLOFF, H.: Acta Biotechnol. (im Druck).
[8] KLINGENBERG, P., ZICKLER, F., LEUCHTENBERGER, A., RUTTLOFF, H.: Z. Allg. Mikrobiol. **19** (1979), 17.
[9] NÄHER, G. — In: BERGMEYER, H.-U.: Methoden der enzymatischen Analyse. 3. Auflage, Bd. 1. Weinheim: Verlag Chemie, 1974.
[10] SAN CLEMENTE, C. L., VADEHRA, D. V.: Appl. Microbiol. **15** (1967), 110.
[11] RUTTLOFF, H., FRIESE, R., RICHTER, M.: Z. Allg. Mikrobiol. **10** (1970), 341.
[12] RUTTLOFF, H.: Habilitationsschrift, Humboldt-Universität Berlin, 1968.
[13] RUTTLOFF, H., ROTHE, M., FRIESE, R., SCHIERBAUM, F.: Z. Lebensm.-Unters. Forsch. **130** (1966), 201.
[14] KONDO, M.: J. Biochem. [Tokyo] **52** (1962), 279.
[15] THIELE, O. W.: Hippokrates **49** (1978), 178.
[16] KOBAYASHI, H., SUZUKI, H.: J. Ferment. Technol. **50** (1972), 835.
[17] BHUMIBHAMON, O., EKLUND, E. — In: DELLWEG, H.: Abstracts of Papers 5th Intern. Fermentation Sympos. West-Berlin, 1976, 154.
[18] UMESAKI, Y., KAWAI, Y., MUTAI, M.: Appl. Environ. Microbiol. **34** (1977), 115.
[19] REESE, E. T., MAGUIRE, A.: Appl. Microbiol. **17** (1969), 242.
[20] PACA, J., ETTLER, P., GREGR, V.: Kvasny Prumysl. **24** (1978), 181.
[21] TAKAHASHI, H., YOSHIDA, F.: J. Ferment. Technol. **57** (1979), 349.
[22] EVANS, J. I., HALL, M. J.: Process Biochem., (1971) 23.
[23] WINKLER, K.: Untersuchungen zur Glucoamylasebildung des Sproßpilzes *Endomycopsis bispora* in Abhängigkeit von seinen morphologischen Formen. Diss. A, Technische Universität Dresden, 1971.
[24] YAMAMOTO, T., HATTORI, F., TAKATSU, H.: Bull. Inst. Chem. Res. Kyoto **42** (1964), 252.
[25] FRANZKE, C., GÖBEL, R., KROLL, J., LIEBS, P., RITTER, G., SCHULTZE, M.: Nahrung **23** (1979), 283.
[26] GALYNKIN, V. A., EFIMOVA, T. P., GREKOVA, V. K., KOROLEVA, T. A.: Khim-Farm. Zh. **12**, (1978), 116.
[27] FALKOWSKI, J.: Zbl. Bakteriol., Hyg. I. Orig. B, **168** (1979), 361.
[28] ELWAN, S. H., MOSTAFA, S. A., KHODAIR, A. A., ALI, O.: Zbl. Bakteriol. II., **133** (1978), 706.
[29] LEUCHTENBERGER, A., KLINGENBERG, P., RUTTLOFF, H.: Z. Allg. Mikrobiol. **19** (1979) 27.

Acta Biotechnol. 7 (1987) 1, 48

Book Review

K. Maramorosch (Ed.)

Advances in Cell Culture, Vol. 4

Orlando, San Diego, New York, London, Toronto, Montreal, Sydney, Tokyo: Academic Press, Inc. 1985. 268 S. 49.50 $, 43.50 £

Volume 4 of Advances in Cell Culture continues serial publication of progress in foundamental knowledge and in vitro methodology. First article illustrates morphological, cytochemical and biochemical changes of muscle culture systems that are presently used to study pathogenesis of hereditary and acquired human myopathies, and normal muscle regeneration. Following papers are devoted to basic aspects such as the biology of malignant colonic cells for development of more effective treatment modes, cell shape and growth control. Two further reviews are directed toward the use of in vertebrate cells in spermiogenesis studies and mycotoxin action. A survey on recent advances in knowledge of growth, nutrition, and metabolic reguirements of cancer cells from human solid tumors lend hope that cell lines can be established from all carcinomas with regard to the development of effective cancer therapy. Important practical highlights are the chapters on hybridoma technology as well as structure and application of microcarriers in cell culture. Thus, authors made the book particular interesting by stimulating the reader with valuable practical hints, fresh approaches, and new concepts for solving clinical, biochemical and biotechnological problems in the diverse field of in vitro research. In conclusion, this volume presents profound perceptions till 1984 in the current contributions and should be therefore in the hand of those who intend to work in the forefront of cell culture.

H.-D. Grimmecke

Acta Biotechnol. 7 (1987) 1, 49–53

Einfluß einiger Faktoren in Abhängigkeit von den Methoden bei der Immobilisierung eines bakteriellen Enzympräparats mit Milchkoagulationswirkung

DOBREVA, E.

Bulgarische Akademie der Wissenschaften
Institut für Mikrobiologie
1113 Sofia, Akad. G. Bontshevstr. 26

Summary

The influence of different factors (binding time, pH, temperature, and enzyme/carrier ratio) on the efficiency of immobilization of enzyme preparation with milk clotting activity from *Bacillus mesentericus* M strain by use of different carriers and methods of binding was investigated. It has been established that the dependence of immobilization from pH is most pronounced and is mainly influenced by the nature of the carrier. The binding time is of essential importance only by immobilization through adsorption. The temperature has not any influence on the efficiency of immobilization of the indicate preparation. Saturation of many of the used carriers by the different methods for immobilization was achieved at enzyme/carrier ratio 1 : 10, with exception of the cases where glutaraldehyde was used as binding reagent. In this case then the ratio was 1 : 4.

Einleitung

Bei der Auswahl von passenden Trägern und Methoden zur Enzymimmobilisierung sind die Durchführungsbedingungen von großer Bedeutung [1–4]. In Abhängigkeit von der zur Immobilisierung verwendeten Methode beeinflussen bestimmte Bedingungen die Wirksamkeit der Enzymfixierung. Die Erkennung dieser Abhängigkeit für ein bestimmtes Enzym erleichtert die Optimierung der Bedingungen seiner Immobilisierung.
Das Ziel des vorliegenden Aufsatzes ist, den Grad des Einflusses einiger Bedingungen in Abhängigkeit von den verwendeten Methoden zur Immobilisierung des Enzympräparats mit Milchkoagulationswirkung aus *Bac. mesentericus* Stamm M zu untersuchen.

Material und Methoden

Die Untersuchungen wurden mit dem Milchkoagulationsenzympräparat aus *Bac. mesentericus* Stamm M durchgeführt, das eine Neutralproteinase mit einer Aktivität 150000 PUN/g darstellt. Letztere wurde nach verschiedenen Methoden immobilisiert, die Tabelle 1 angibt.
Die angegebenen Methoden sind in unseren früheren Arbeiten detailliert beschrieben [5, 6].
Bei der nach diesen Methoden durchgeführten Immobilisierung des Enzympräparats wurde der Einfluß folgender Faktoren verfolgt:

Bindungszeit — 1,4 und 20 Stunden,
Temperatur — 4 und 25 °C,
pH — von 5,0 bis 12,0; in Abhängigkeit von der verwendeten Methode,

Verhältnis Enzym/Träger — 5,10, 25 und 50 mg Enzym je 100 mg Träger.

Tag. 1. Methoden der Immobilisierung der Neutralproteinase.

Methode	Träger	Immobilisierungsweise
1	Amberlite IRC 50	Adsorption
2	DEAE-Cellulose	Adsorption
3	CM-Cellulose mit Benzidin modifiziert	durch Glutaraldehyd
4	DEAE-Cellulose	durch Glutaraldehyd
5	CM-Cellulose	nach Aktivierung mit Thionylchlorid
6	Bentonit	nach Aktivierung mit Thionylchlorid
7	CM-Cellulose mit Benzidin modifiziert	nach Diazotierung

Die Wirksamkeit der Immobilisierung wurde durch die Bestimmung der proteolytischen Aktivität der immobilisierten Proben gegenüber dem Substrat Kasein nach der Methode von Hagihara verfolgt [7]. Zum besseren Vergleich wurden die Angaben der höchsten Aktivität in Prozent für einen bestimmten Träger in Abhängigkeit von der Bedingung, bei der die Bindung erfolgte, vorgestellt. (Die höchste Aktivität wird als 100% bezeichnet).

Ergebnisse und Diskussion

Die Ergebnisse der Prüfung des Einflusses der Bindungszeit auf die Wirksamkeit der Immobilisierung des Enzympräparats nach verschiedenen Methoden werden in Abb. 1 vorgestellt. Beim Vergleich der vorliegenden Diagramme ist ersichtlich, daß auf Grund des Faktors „Bindungszeit" die geprüften Methoden auf folgende Weise eingeordnet werden können:

I. Methoden 1, 2 und 4 — Bei ihnen nimmt bei Steigerung der Bindungszeit die Aktivität der immobilisierten Proben ab. In diesem Fall handelt es sich offensichtlich um die

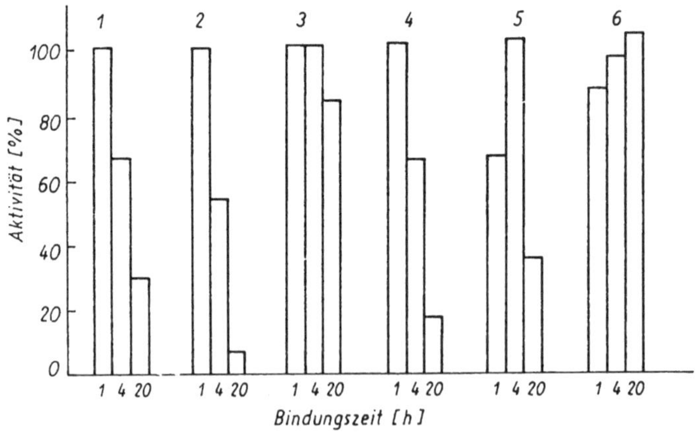

Abb. 1. Grad der Immobilisierung des Enzympräparates bei verschiedenen Methoden in Abhängigkeit von der Bindungszeit.

Immobilisierung durch Adsorption und um eine Abnahme der Aktivität wegen der natürlichen Desorption des Enzyms und dessen Übergehen in die Lösung. Ähnliche Beobachtungen machten auch Messing [8, 9] bei der Immobilisierung von Papain und RNA-se an poröses Glas und Balcon [3] bei der Immobilisierung von Katalase an DEAE Cellulose.

II. Methoden 3 und 6 — Bei ihnen hat die Bindungszeit keinen wesentlichen Einfluß. Die Immobilisierung ist dabei kovalent und wird sehr schnell realisiert, in der Zeit danach wird keine wesentliche Änderung der schon fixierten Enzymmenge beobachtet. Untersuchungen von Monsan [4] bei Bindung der Invertase an mit Cyanurchlorid aktiviertem Bentonit bestätigen dies.

III. Methode 5 — Bei dieser ist eine Zeit von 4 Stunden optimal für die Bindung. Hier handelt es sich offensichtlich um eine chemische Reaktion zwischen dem Säurechlorid und dem Enzym, für die dies eine optimale Zeit ist.

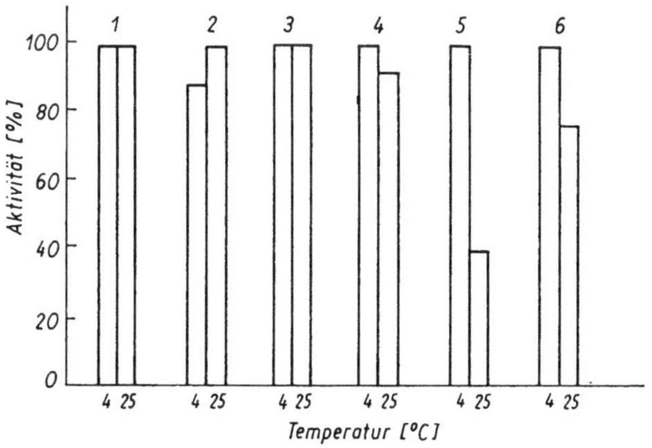

Abb. 2. Grad der Immobilisierung des Enzympräparates bei verschiedenen Methoden in Abhängigkeit von der Temperatur.

Der Einfluß der Temperatur auf die Immobilisierung des Enzympräparats nach verschiedenen Methoden wird in Abb. 2 dargestellt. Aus den auf dieser Abbildung gegebenen Angaben kann man schließen, daß die Bindung des entsprechenden Enzyms nicht wesentlich von der Temperatur abhängt, zumindest in dem Bereich, in dem die Untersuchung durchgeführt wurde. Die einzige Ausnahme stellt die Methode 5 dar, bei der die Aktivität des bei 4 °C gewonnenen Präparats wesentlich höher ist. Der gleiche Effekt, aber viel schwächer ausgedrückt, wird bei der Methode 6 beobachtet, bei der deutlich wird, daß die Reaktion zwischen dem Säurechlorid und dem Enzym bei niedriger Temperatur besser verläuft.

Die Resultate der Untersuchung des Einflusses des pH-Wertes bei der Immobilisierung des Enzympräparats nach den o. a. Methoden werden in Abb. 3 und Abb. 4 graphisch dargestellt. Daraus ist ersichtlich, daß die Abhängigkeit „Wirksamkeit der Immobilisierung/pH" nicht wesentlich von der Bindungsweise, sondern viel mehr von der Natur des Trägers beeinflußt wird. So ist bei dem Kationaustauschträger Amberlit IRC 50 die Immobilisierung bei pH 6,0—7,0 am besten (Methode 1) und bei dem Anionaustauschträger DEAE Cellulose bei pH 10,0 (Methode 2). An CM-Cellulose ist die Immobilisierung optimal bei pH 8,0 ohne Rücksicht auf ihre Modifikation (mit Benzidin — Methoden 3 und 7 oder mit Thionylchlorid — Methode 5) und auf ihre Bindungsweise (durch Glutaraldehyd — Methode 3 oder nach Diazotierung — Methode 7) (Abb. 4).

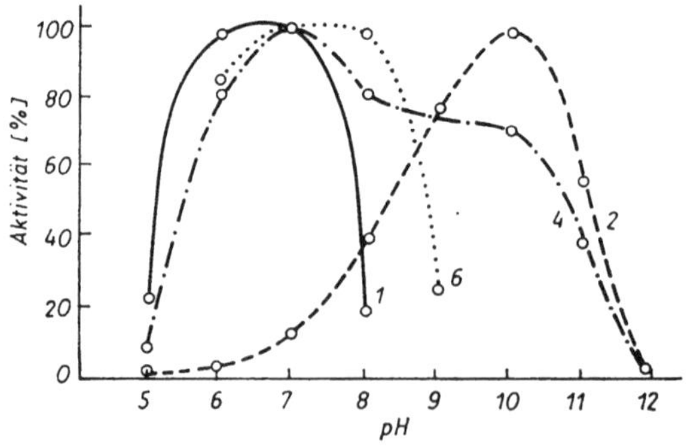

Abb. 3. Grad der Immobilisierung des Enzympräparats bei verschiedenen Methoden in Abhängigkeit vom pH.

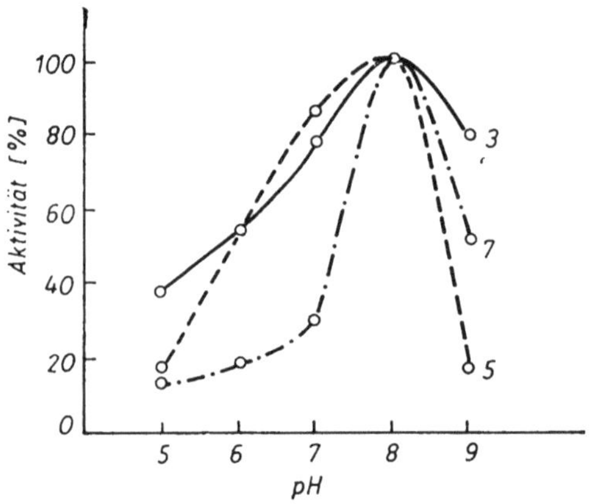

Abb. 4. Grad der Immobilisierung des Enzympräparats auf modifizierte CM-Cellulose in Abhängigkeit vom pH.

In diesem Fall besteht der Unterschied nur in der Breite des pH-Bereichs, in dem die Immobilisierung an CM-Cellulose möglich ist. Interessant ist das Resultat bei Methode 4 (Abb. 3), d. h. der Enzymimmobilisierung an DEAE-Cellulose durch Glutaraldehyd. In diesem Fall ist die Bindung bei pH 7,0 am besten. Es handelt sich hierbei deutlich um eine Änderung des Bindungsmechanismus im Vergleich mit dem bei Methode 2. Vermutlich reagiert der Glutaraldehyd mit den Enzymaminogruppen, wodurch in die Enzymmoleküle neue, stark polare funktionelle Gruppen (—CHO) eingeführt werden. Diese liefern feste Bindungen mit dem Träger bei niedrigeren pH-Werten. Bei höherem pH-Wert tritt dagegen auch die für DEAE-Cellulose charakteristische Adsorption ein, was durch das Vorhandensein eines zweiten, niedrigeren Optimums bei pH 10,0 auf Kurve 4 der Abb. 3 bestätigt wird.

In Abb. 5 werden die Resultate der Untersuchung der Abhängigkeit der Immobilisierung bei verschiedenen Methoden von dem Verhältnis Enzym/Träger vergleichend dargestellt. Die angegebenen Graphiken widerspiegeln deutlich den Einfluß der Enzymmenge auf die Immobilisierungswirksamkeit, die die Leistung eines gegebenen Trägers bestimmt. Man sieht, daß bei der Mehrzahl der verwendeten Methoden die Trägersaturierung schon bei 10 mg Enzym je 100 mg Träger auftritt.

Bei den Methoden 5, 6 und 7 führt die Erhöhung dieser Enzymquantität zu keiner wesentlichen Änderung der Aktivität des immobilisierten Präparats. Die Immobilisierung durch Adsorption an DEAE-Cellulose und besonders an Amberlit IRC 50 (Kurven 1 und 2) führt zu einer deutlichen Abnahme der Aktivität bei Zunahme der Enzymmenge auf über 10 mg je 100 mg Träger. Diese Tatsache ist mit der größeren Enzymmenge, die auf einer kleinen Fläche angehäuft wird, zu erklären.

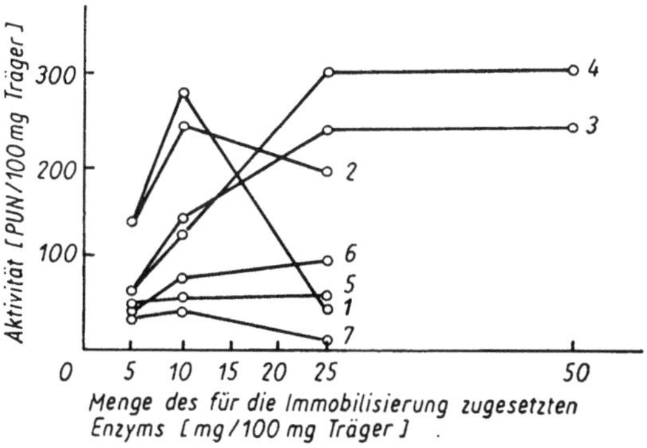

Abb. 5. Grad der Immobilisierung des Enzympräparats bei verschiedenen Methoden in Abhängigkeit von dem Verhältnis Enzym/Träger.

Nach Poltorak [10] erleichtert dies die Bildung von Enzymassoziaten und führt zu der Abnahme der Aktivität. Bei der Immobilisierung des Enzympräparats mittels Glutaraldehyd an die mit Benzidin modifizierte CM-Cellulose bzw. DEAE-Cellulose wird der Träger bei Anwendung von 25 mg Enzym je 100 mg Träger gesättigt.
Aus Abb. 5 sieht man auch, daß die höchste Aktivität der immobilisierten Präparate bei Methoden 1, 2, 3 und 4 erhalten wurde.

Schlußfolgerung

Bei der Untersuchung des Einflusses verschiedener Bedingungen in Abhängigkeit von den Methoden bei der Immobilisierung des Enzympräparats mit Milchkoagulationswirkung aus *Bac. mesentericus* Stamm M wurde folgendes festgestellt:
Der pH-Einfluß auf die Immobilisierungswirksamkeit des oben erwähnten Präparats wird hauptsächlich von der Natur des Trägers bestimmt.
Die Bindungszeit ist nur bei der Immobilisierung des Enzympräparats durch Adsorption von wesentlicher Bedeutung.
Die Temperatur beeinflußt nicht wesentlich die Immobilisierungseffektivität des Enzympräparats.
Die Saturierung der meisten der verwendeten Träger bei den verschiedenen Methoden zur Immobilisierung des Präparats erfolgt bei einem Verhältnis Enzym/Träger 1 : 10. Bei den Methoden unter Anwendung von Glutaraldehyd als Bindungsreagens ist dieses Verhältnis 1 : 4.

Eingegangen: 7. 5. 1985

Literatur

[1] Мурель А., Федосеев В., Панк Марет, Кирет О.: Изв. АН Эст. ССР, Химия, Геол. **26** (1977) 3, 235.
[2] Федосеев В., Панк-Марет, Хейло Х., Мурель А., Кирет О.: Изв. АН Эст. ССР, Химия, Геол. **26** (1977) 4, 326.
[3] Balcon, J., Foulkes, P., Olson, N. F., Richardson, T.: Process Biochem. **6** (1971) 8, 42.
[4] Monsan, P.: These d'Universite „P. Sabatier", Toulouse, France, 1971.
[5] Dobreva, E.: Acta Microbiol. Bulg. (1979) 3, 14.
[6[Добрева Е.: IV Конгресс по микробиологии, София **3** (1978), 23.
[7] Hagihara, B., Matsubara, H., Nakai, M., Okunuki, K.: J. Biochem. **45** (1958) 3, 185.
[8] Messing, R. A.: Enzymologija **38** (1970), 38.
[9] Messing, R. A.: Enzymologija **38** (1970), 370.
[10] Полторак О. М., Пряхин А. Н., Чухрай Е. С.: Вестн. Моск. у-та, Химия **18** (1977) 2, 125.

Acta Biotechnol. 7 (1987) 1, 55—59

Abhängigkeit der dimensionslosen Mischgeschwindigkeit (n/K) von der Re-Zahl bei einem dreistufigen Rührersystem

WOLF, K.-H., STIEBITZ, O., VENUS, J.

Technische Universität Dresden
Sektion Verarbeitungs- und Verfahrenstechnik
Mommsenstraße 13, Dresden, 8027, DDR

Summary

The mixing-time characteristic in the three-stage stirrer system is analyzed for the range of $6 \cdot 10^4 \leqq \mathrm{Re} \leqq 3.2 \cdot 10^5$ using the pH-method. Different from the one-stage systems, for the baffled reactor (BW = 0.8) it was estimated that the homogenized number increases with $c_\mathrm{H} \sim \mathrm{Re}^\mathrm{m}$ (m = 0.77). Obviously, between two stirrers it exists a kernel sphere with a delay mass transfer in consequence of low axial current components.
The measurements can be described with high significance by the model for dimensionless "mixing-rate number" of KHANG and LEVENSPIEL.

Problem- und Zielstellung

Aus ökonomischen und verfahrenstechnischen Gründen werden zunehmend mehretagige Rührersysteme zum Stoffvereinigen eingesetzt. Für eine Reihe fermentativer Prozesse werden Bioreaktoren mit drei Rührelementen auf einer gemeinsamen Rührwelle bestückt, da bei hohen Reaktoren ($h_0/h_1 > 1$) dadurch eine gleichmäßigere Turbulenzverteilung erzielt wird.
Während für die verschiedensten apparativen Varianten einfacher Rührwerke (ein Rührelement) ein breites Literaturspektrum [1] vorliegt, gibt es über 3-stufige Rührersysteme nur wenige Publikationen. Eine Auswertung einiger dieser Arbeiten [2—6] zeigt, daß wesentliche geometrische und Betriebsparameter einen veränderten Einfluß auf die Hydrodynamik haben, als er durch einstufige vergleichbare Systeme bekannt ist. Für die kinetische Modellierung mikrobieller Prozesse in Rührfermentoren und zur Maßstabsübertragung werden Angaben über die Mischzeit benötigt. Für das bewehrte dreistufige System (Abb. 1) stehen zur Zeit keine entsprechenden Berechnungsgrundlagen zur Verfügung. Für die Maßstabsübertragung empfiehlt sich das auf Grundlage einer begründeten Vorstellung (Rührkesselkaskade idealer Mischer mit Rückführung) aufgestellte Modell von KHANG und LEVENSPIEL [7].

$$\left(\frac{n}{K}\right) \cdot \left(\frac{d_2}{d_1}\right)^\alpha = f(\mathrm{Re}) \tag{1}$$

wobei

$$A = 2 \exp\left(-K t_\mathrm{H}\right) = \frac{c_\mathrm{max} - c_\mathrm{min}}{|c_\infty - c_0|} \tag{2}$$

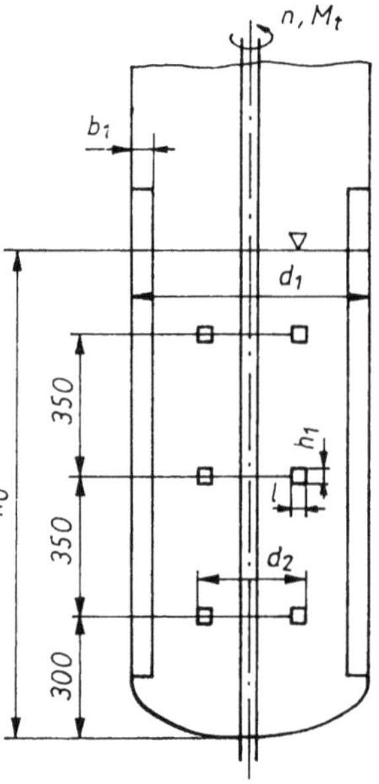

Abb. 1. Abmessungen des Modellrührwerkes.
$d_1 = 600$ mm; $\frac{d_2}{d_1} = 0{,}33$; $h_0 = 1240$ mm
Bewehrung: 4 Wandstrombrecher $BW = 0{,}8$, $b_1 = 60$ mm

und

$$t_H = -\frac{\ln(A/2)}{K} \tag{3}$$

Da für technische Prozesse in Rührfermentoren beim Einsatz von Schaufelrührern Re $> 10^4$, soll nur die Mischzeitcharakteristik im vollturbulenten Bereich untersucht und möglichst durch das Modell Gl. (1) dargestellt werden.

Lösungsweg

Für den dreistufigen Rührapparat mit folgenden geometrischen Kennwerten

$V = 0{,}35$ m³ $\quad h_{2m} = 0{,}3$ m $\quad a_1 = 0$
$d_1 = 0{,}6$ m $\quad b_1 = 0{,}060$ m
$h_3 = 1{,}23$ m

$\frac{d_2}{d_1}$	$\frac{h_0}{d_1}$	$\frac{h_1}{d_2}$	$\frac{h_{2m}}{d_1}$	BW
0,33	1,2	0,2	0,5	0,8

wurden im Bereich

$$1{,}66 \leqq n \leqq 8{,}33$$
$$n \text{ in } s^{-1}$$

mit dem System Wasser die Mischzeiten t_H nach der pH-Meßmethode [8] ermittelt. Dem Rührmedium wurde eine Menge von $1/10^4$ (36 ml) 4 N NaOH bzw. 4 N HCl mit einer Spritze in das Medium in $t = 0{,}5$ s injiziert. Der von EINSELE [8] empfohlene Bereich wurde eingehalten, $3 < \text{pH} < 6$ bzw. $\text{pH} > 8$. Der Verlauf $c = f(t)$ wurde auf einem Schreiber erfaßt.

Jede Einstellung wurde viermal wiederholt und der endgültige Wert daraus gemittelt. Die Bestimmung der Mischzeit t_H erfolgte nach dem 5% Kriterium (Restschwankung), (Abb. 2).
Die Auswertung erfolgte zunächst graphisch in doppeltlogarithmischer Darstellung und numerisch durch lineare Regression.

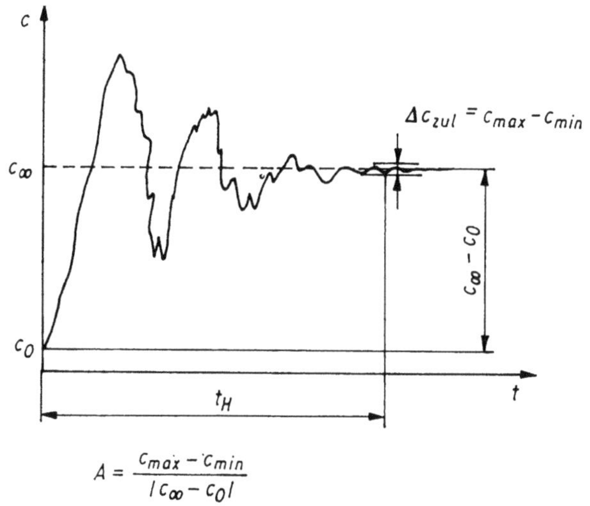

Abb. 2. Zeitlicher Verlauf des Abbaues von Konzentrationsunterschieden in der Rührmaschine.

Ergebnisse

Abb. 3 zeigt den Verlauf der Homogenisierbeiwerte über der Re-Zahl. Es ist festzustellen, daß für $6 \cdot 10^4 < \mathrm{Re} < 3{,}2 \cdot 10^5$ der Homogenisierbeiwert mit der Re-Zahl nach $c_H \sim \mathrm{Re}^m$ wächst. Als Exponent läßt sich für $m = 0{,}77$ ermitteln. Dieser Wert liegt in der Größenordnung mit $c_p \sim \mathrm{Re}^{-p}$ ($p = 0{,}125$). Diese Übereinstimmung begründet sich aus

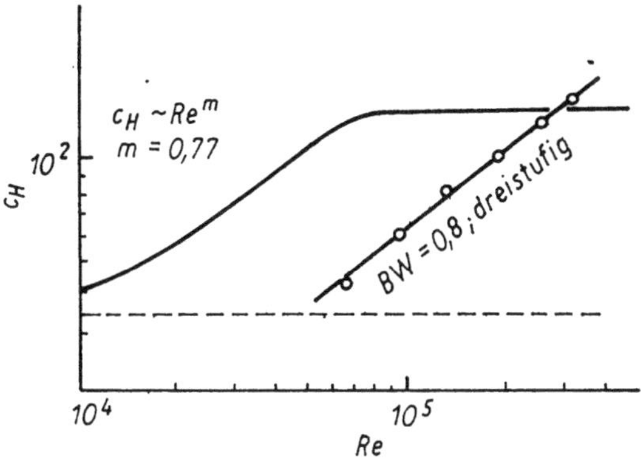

Abb. 3. Homogenisierbeiwerte (○) des Schaufelrührers.
$d_2/d_1 = 0{,}33$, dreistufig, $h_0/d_1 = 2$; $d_1 = 0{,}6$ m; $h_{2m}/d_1 = 0{,}5$; $BW = 0{,}8$
——— ohne Einbauten; einstufig; $d_2/d_1 = 0{,}4$; $h_0/d_1 = 1$ [1, S. 92]
\- - - - 4 WSB; einstufig; $d_2/d_1 = 1$ [1, S. 88]

der Analogie zwischen Impuls- und Stofftransport. Das Ergebnis ist insofern überraschend, da der gefundene Verlauf der Homogenisierbeiwerte für unbewehrte einstufige Systeme gilt, der dreistufige Rührreaktor jedoch eine ausreichende Bewehrung $BW = 0{,}8$ besitzt. Für den vorgenannten Re-Bereich muß das dreistufige System mit $BW = 0{,}8$ offensichtlich als unbewehrtes System gewertet werden. Die Erscheinung läßt sich aus dem Sekundärströmungsfeld zwischen den Rührern erklären, in der eine Zone verschleppter Entmischung existieren muß. Durch die starken radialen und tangentialen Strömungsanteile zwischen zwei Rührern ist die axiale Strömungskomponente zu gering, um ein Kerngebiet verminderten Stoffaustausches zu vermeiden.
Die Zusammenfassung der Meßwerte durch das Modell von KHANG und LEVENSPIEL [7] ergab nachfolgende Beziehung

$$\frac{n}{K} = -\frac{nt_{\mathrm{H}}}{\ln (A/2)} = 2{,}3 \cdot 10^{-3}\,\mathrm{Re}^{0{,}77} \tag{4}$$

$$B^* = 0{,}99$$
$$s_{\mathrm{R}} = 0{,}02$$
$$v = 1{,}74\%$$

Der statistisch hoch abgesicherte Zusammenhang gilt für dreietagige Rührsysteme. Weitere Arbeiten sollten sich auf hydrodynamische Untersuchungen im Bereich

$$10^4 \leqq \mathrm{Re} \leqq 6 \cdot 10^4 \quad \text{und} \quad 2 \cdot 10^5 \leqq \mathrm{Re} \leqq 10^6$$

und $BW = 0{,}8$ beziehen.

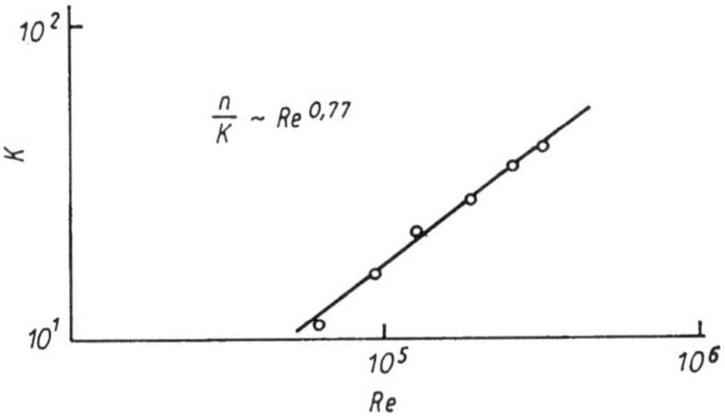

Abb. 4. Abhängigkeit der dimensionslosen Mischgeschwindigkeit.
$N_{\mathrm{mix}} = n/K$ von Re für ein dreistufiges Rührersystem;
$BW = 0{,}8$; $d_1 = 0{,}6$; $d_2 = 0{,}196$ m

Symbolverzeichnis

$BW = c_{\mathrm{W}} N_{\mathrm{s}} \frac{b_1}{d_1} \frac{h_3}{d_1}$	Bewehrungskennzahl [1]	—
b_1	Breite der Strombrecher	m
a_1	Spaltbreite zwischen Behälterwand und Strombrecher	m
c	Konzentration	kgm^{-3}
c_{W}	Widerstandsbeiwert (bei Strombrechern $c_{\mathrm{W}} = 1$)	—
d_1	Behälterinnendurchmesser	m
d_2	Rührerdurchmesser	m
h_1	Rührerblatthöhe	m

h_{2m}	Rührereinbauhöhe, bezogen auf Rührermitte	m
h_3	Höhe der Strombrecher oder anderer Einbauten	m
K	Amplituden-Abklinggeschwindigkeitskonstante, Gl. (2)	s^{-1}
n	Rührerdrehzahl	s^{-1}
$\mathrm{Re} = \frac{n d_2^2}{\nu}$	REYNOLD-Zahl	—
t_H	Homogenisierzeit, technische Mischzeit	s
α	Exponent in Gl. (1)	—
ν	kinematische Viskosität	m^2s^{-1}

Eingegangen: 26. 2. 1986

Literatur

[1] LIEPE, F., MÖCKEL, H.-O.: Verfahrenstechnische Berechnungsmethoden. Teil 4/1. Stoffvereinigen in fluiden Phasen. Leipzig: VEB Deutscher Verlag für Grundstoffindustrie, 1978.
[2] AIBA, S., HUMPHREY, A., MILLIS, N. F.: Biochemical Engineering. New York: Academic Press, 1965.
[3] MORITZ, V. S. u. a.: Ferment. Technol. **52** (1974), 127.
[4] STEIFF, A.: Untersuchungen zum Wärme- und Impulsaustausch in ungerührten und gerührten Gas-Flüssigkeits-Reaktoren. Diss. Iniversität Dortmund, 1976.
[5] NIENOW, A. W., LILLY, M. D.: Biotechnol. Bioeng. **21** (1979), 2431.
[6] MÖCKEL, H.-O.: Chem. Tech. **35** (1983), 344.
[7] KHANG, S. J., LEVENSPIEL, O.: Chem. Eng. Sci. **31** (1976), 569.
[8] EINSELE, A., FIECHTER, A.: Chem. Rundsch. **29** (1976), 53.

Acta Biotechnol. 7 (1987) 1, 60

Book Review

Howard M. SHAPIRO

Practical Flow Cytometry

New York: Alan R. Liss Inc., 1985 (Second Printing), 37.50 £

From the beginning of flow cytometry H. M. SHAPIRO has been engaged in methodological work in this field. Founded on such experience the author wrote and illustrated his book "Practical Flow Cytometry" in an unconventional manner using a computer himself. This book may attract the interested outsiders to work in flow cytometry and stimulate professionals who have been dealing with flow cytometry for many years. The reader will be completely informed about all that is necessary for active work in flow cytometry. The description of SHAPIRO's flow cytometric device is given in detail and combined with optical and electronic instructions for explanation so that one might be encouraged to start immedately with the building of a self-made flow cytometer. Besides this instruction the specific performances of commercial flow cytometers are discussed and the potential user gets the knowledge to balance commercial flow cytometers against others as well as against a self-made one, depending on the circumstances. One third of the book is concerned with the use of flow cytometry in the diverse fields of medicine, biology and biotechnology. By a fundamental consideration the results in cell fixation and fluorescence probing are reviewed and the trends are stressed. Cell sorting and data analysis, both of which need much expensive electronics, are presented with a discussion of how much is needed to solve a particular problem. Similar advice is given in many other chapters. The 623 references, the extended tables of contents and a carefully compiled index correspond well to the attribute "practical" in the book's title. A splendid book: up-to-date, interesting and convincing, competent and, if you could look at SHAPIRO's Law's of Cytometry, not without some humour.

A. Schmidt

Acta Biotechnol. 7 (1987) 1, 61–69

Verfahren und Analysenautomat zur Bestimmung von Nährlösungskomponenten und Stoffwechselprodukten

WEISE, H.[1], KREIBICH, G.[1], SCHELLER, F.[2]

[1] VE Forschungszentrum Biotechnologie im Pharmazeutischen Werk Dresden GERMED
Alt-Stralau 62, Berlin, 1017 DDR
[2] Akademie der Wissenschaften der DDR
Zentralinstitut für Molekularbiologie
Bereich Angewandte Enzymologie
DDR-1115 Berlin, Rösslestraße 10

Summary

A continuous flow system with an enzyme electrode and with a new type of measuring instrument, the BIOXY-Meter, is described as an example for the determination of glucose. The BIOXY-Meter combined with various biosensors may be used in biotechnology and in other fields for the solution of a number of analytical problems. Measuring principle is the reduction of oxygen. The peak of first derivation of the current-time-curve is proportional to oxygen consumption and to decrease of substrate concentration. Enzyme fixation with common adhesive and a device for automated saturation of measuring samples with air in the sampler are also described. With the apparatus 60 samples per hour are analyzed for glucose. The linear measuring range is 0.1 mMol to 1.5 mMol with a coefficient of variation of $V_k = 0.93$ per cent. The described type of enzyme electrode is stable for 2 to 12 weeks or 1,000 to 5,000 measurements.

Bei der Untersuchung von Rohstoffen sowie bei der Bewertung von Fermentationsprozessen gibt es auch heute noch Probleme bei der Bestimmung der Substrate und Enzymaktivitäten. Insbesondere bei der Bestimmung von Kohlenhydraten, wie Glukose, Saccharose oder Stärke, wird oft aus Gründen der einfachen Handhabung die Bestimmung der reduzierenden Substanzen als Maß für den Glukosegehalt oder, nach ihrer vorherigen Spaltung, als Maß für den Saccharose- oder Stärkegehalt verwendet. Ähnliche Verhältnisse bestehen bei der Bestimmung von Enzymaktivitäten, speziell solcher Enzyme, die Glukose aus den Substraten freisetzen. Mit einer spezifischen und für Routineanalytik geeigneten Glukose-Bestimmungsmethode könnten somit viele der Probleme gelöst werden.

Bei der Suche nach einer einfachen spezifischen Bestimmungsmethode rückte in den letzten Jahren immer mehr die enzymatische Bestimmung in den Vordergrund. Das am häufigsten angewendete Verfahren ist dabei die Bestimmung der Glukose mit Hilfe von Glukoseoxydase (GOD) und Peroxydase (POD) unter Verwendung einer farbbildenden Substanz [1]. Bei der Untersuchung komplexer Medien, wie beispielsweise Fermentationslösungen, ist eine problemlose Anwendung dieser Bestimmungsmethode jedoch nicht möglich, da insbesondere die Reinheit der verwendeten Enzyme als auch mögliche Störsubstanzen im Untersuchungsmedium zu Fehlereinflüssen führen.

Diese Probleme sowie die hohen Enzymkosten und der große Zeitaufwand, die die Einführung dieser Bestimmungsmethode in die Routineanalytik der Fermentationskontrolle bisher verhindert haben, werden durch den Einsatz von Enzymelektroden ausgeschaltet. Bei dieser Bestimmungsmethode wird durch die Kopplung von enzymatischer

Reaktion mit Elektrodenprozessen die Spezifität der enzymatischen Bestimmungen und eine vorhandene, gut entwickelte Meßtechnik sinnvoll miteinander verknüpft [2].
Die Zahl der Veröffentlichungen über die Entwicklung von Enzymelektroden hat sich gerade in den letzten Jahren sehr erhöht, wobei zur Meßwerterfassung sowohl amperometrische als auch potentiometrische Meßwertgeber Anwendung finden. So reicht die Palette der Sensoren von pO_2-Elektroden [3] über Thermistoren [4], pH-Elektroden [5] bis hin zu speziellen ionenselektiven Elektroden [6] bzw. ionensensitiven Feldeffekttransistoren (ISFET) [7].
Für die enzymatische Bestimmung der Glukose in wäßriger Lösung ergeben sich zwei Möglichkeiten.

1. Nach der sogenannten Doppelenzym-Methode, bei der unter Einwirkung von Hexokinase (HK) die Glukose mit Adenosintriphosphat (ATP) reagiert, wobei Glukose-6-phosphat (G-6-P) und Adenosindiphosphat (ADP) entstehen.

$$\text{Glukose} + \text{ATP} \xrightleftharpoons{\text{HK}} \text{G-6-P} + \text{ADP}$$

In Gegenwart des Koenzyms Nicotinsäureamid-adenin-dinucleotid-phosphat (NADP) und unter der katalytischen Mitwirkung von Glukose-6-phosphat-dehydrogenase (G-6-PDH) wird G-6-P zu 6-Phosphoglucono-δ-lacton(6-PG) oxydiert, während NADP reduziert wird.

$$\text{G-6-P} + \text{NADP}^+ \xrightleftharpoons{\text{G-6-PDH}} \text{6-PG} + \text{NADPH} + \text{H}^+$$

2. Wesentlich weiter verbreitet ist die Bestimmung mit Glukoseoxydase (GOD), die folgende Reaktion katalysiert (8):

$$\text{Glukose} + H_2O + O_2 \xrightarrow{\text{GOD}} \text{Glukonsäure} + H_2O_2$$

Aus der Reaktionsgleichung ergeben sich folgende Möglichkeiten für die Meßwerterfassung:

— Registrierung des Reaktionsproduktes H_2O_2.
— Nachweis des O_2-Verbrauchs. Voraussetzung ist der Einsatz luftgesättigter Lösungen oder die Anwendung einer Differenzmessung [9, 10]. Bei anderen Autoren [11, 12] wird unter Ausschluß von Sauerstoff gearbeitet, wobei als Akzeptoren organische Redoxsysteme eingesetzt werden, deren reduzierte Form stabil und quantitativ elektrochemisch oxidierbar ist.

$$\text{Glukose} + A_{ox} + H_2O \xrightarrow{\text{GOD}} \text{Glukonsäure} + A_{red} + 2H^+$$

Bei Anwendung von Ferrozen kann die Reaktion auch in Gegenwart von Sauerstoff durchgeführt werden [13].
— Messung der in einer zweiten Enzymreaktion umgewandelten Form des Substrats (z. B. mit jodidsensiblen Elektroden [6]);
— Messung der Glukonsäure durch pH-Messung [5].

Wir stellten uns die Aufgabe, die Glukosebestimmung unter Verwendung einer Enzymelektrode durchzuführen, deren Grundsensor eine Sauerstoffelektrode ist. Das Meßsystem sollte zu einer Analysenstrecke zusammengefaßt werden und eine schnelle und exakte Bestimmung von Nährlösungskomponenten und Stoffwechselprodukten ermöglichen.
Bei der Untersuchung von Fermentationslösungen ergibt sich als allgemeines Problem, daß sich in der Probe die gesamte Mikroorganismenflora befindet, die sich weiter vermehrt. Der Stoffwechsel der Mikroorganismen ist ein generelles Problem bei der Analytik, da hierdurch alle Analysenwerte eine fortlaufende Veränderung erfahren. Es besteht

somit die Aufgabe, die Mikroorganismen aus der Untersuchungslösung zu entfernen bzw. ihre Aktivität zu unterdrücken. Problemlos ergibt sich das durch ein Abzentrifugieren, wodurch gleichzeitig mögliche unlösliche Substanzen entfernt werden. Der Einsatz eines kontinuierlich arbeitenden Filtrationsbausteines [22] wird z. Z. untersucht.
Als spezielles Problem für den Einsatz einer Enzymelektrode auf der Basis einer Sauerstoffelektrode ergibt sich, daß diese Lösungen stark sauerstoffverarmt sind und somit eine direkte Messung unmöglich machen.
Um dieses Problem auszuschalten, kann die Messung beispielsweise mit einer Sauerstoffelektrode und einer Enzymelektrode durchgeführt werden, wobei die Differenz beider Werte als Meßgröße dient [9, 10]. Von CLELAND und ENFORS [14] wird eine sogenannte sauerstoffstabilisierte Enzymelektrode beschrieben, bei der unmittelbar vor der Elektrode Sauerstoff erzeugt wird.
Bei dem von uns angestrebten Meßsystem sollte die zu untersuchende Lösung unmittelbar vor der Messung luftgesättigt werden.

Herstellung der Enzymelektrode

Da bei den von uns durchgeführten Untersuchungen der Sauerstoffverbrauch gemessen werden sollte, wurde eine Stabmeßzelle zur pO_2-Messung vom VEB Metra Meß- und Frequenztechnik (Dresden-Radebeul) als elektrochemischer Sensor verwendet. Der Sauerstoffpartialdruck (pO_2) in einer Flüssigkeit wird bei der Stabmeßzelle dadurch bestimmt, daß ein Teil des durch die Membranfolie diffundierenden Sauerstoffs an der Katode elektrochemisch reduziert wird und dabei einen meßbaren Strom verursacht. Das Elektrodensystem wird bei dieser Ausführung durch nur für Gase durchlässige Polyethylenfolie vor störenden Bestandteilen der Meßflüssigkeit geschützt. Um die eigentliche Enzymelektrode zu erhalten, wird auf die Polyethylenfolie trägerfixierte GOD angeordnet, über die eine die Enzymschicht abdeckende semipermeable Membran gespannt wird. Die Enzymelektrode besteht somit aus einer äußeren Folie, die für Sauerstoff und Glukose, aber nicht für Glukoseoxydase durchlässig ist, und einer inneren Folie, die für Sauerstoff, aber nicht für Glukose und Glukoseoxydase durchlässig ist. Die Anwendung von immobilisiertem Enzym hat den Vorteil, daß wesentlich weniger Enzym verbraucht wird. Im Institut für Technische Mikrobiologie wurde ein Verfahren zur Herstellung von enzymbeschichteter Folie entwickelt, bei dem unter Verwendung eines handelsüblichen Klebstoffes (z. B. Epasol-Kontakt) wasserlösliche Glukoseoxydase auf der zu verwendenden Polyethylenfolie fixiert wird [15]. Hierbei werden für eine Enzymfolie 0,5 mg GOD verwendet. Diese Enzymfolien sind bei trockener Lagerung im Kühlschrank über 1 Jahr haltbar. Die präparierten Enzymelektroden sind ca. 14 Tage bis 3 Monate einsetzbar und können für mindestens 1000 Messungen genutzt werden, wobei Maximalwerte von 5000 Messungen erreicht werden.

Meßgerät

Das Meßsignal kann mit dem Sauerstoff-Meßverstärker M 65 F vom VEB Metra Meß- und Frequenztechnik registriert werden. Die Glukosebestimmung kann dabei über die Bestimmung des stationären Endwertes erfolgen, d. h. Messung des restlichen an der Elektrode herandiffundierenden O_2-Stromes im stationären Zustand.
Wesentlich günstiger ist, wenn man den zeitlichen Verlauf der Stromänderung als Maß für die Glukosekonzentration verwendet. Hierbei benutzt man als Meßsignal den Maximalwert der ersten Ableitung der Strom-Zeit-Kurve. Nach diesem als kinetische Methode bezeichneten Meßprinzip arbeiten beispielsweise das Analog-Glukometer [16, 17] und das Meßgerät GKM 01 [18].

In Zusammenarbeit mit dem Zentrum für wissenschaftlichen Gerätebau der AdW der DDR wurde ein neues Meßgerät, das BIOXY-Meter, entwickelt, das speziell für Untersuchungen von Fermentationslösungen und für den Einsatz von Biosensoren geeignet ist. Bei diesem Gerät wurde ein neues Verfahren angewendet, bei dem die kinetische Bestimmung des Sauerstoffverbrauchs weitgehend unabhängig wird von der Anfangshöhe des Sauerstoffdiffusionsstromes [20]. Das wurde dadurch erreicht, daß eine durch unterschiedliche Größe des Sauerstoffreduktionsstromes bedingte unterschiedliche Größe des Basissignals, unmittelbar vor der Messung durch Veränderung der Verstärkung, auf eine vorgegebene konstante Basissignalhöhe automatisch eingestellt wird und in diesem Kalibrierungszustand verharrt. Durch die Regelung der Verstärkung des Strom-Spannungs-Wandlers wird erreicht, daß Änderungen des Basisstromes infolge Schwankens oder Triftens des an die Elektrode diffundierenden Sauerstoffs ohne Einfluß auf die Größe des Meßwertes bleiben.
Weiterhin sind in der Auswerteeinrichtung ein verzögert geregelter Strom-Spannungs-Wandler, ein Differentiator, ein Schwellwertschalter und eine Vorrichtung zum Speichern des Zustandes des geregelten Strom-Spannungs-Wandlers vorhanden.
Diffundiert beispielsweise Glukose an die GOD-Enzymelektrode, so kommt es an der Elektrode zur Enzymreaktion, verbunden mit einer Verminderung des Sauerstoffreduktionsstromes. Ein dem Strom-Spannungs-Wandler nachgeschalteter Differentiator signalisiert diese Stromabnahme und erwirkt, indem er einen Schwellwertschalter betätigt, die Unterbrechung der Verstärkungsregelung für den Strom-Spannungs-Wandler, wobei dessen Verstärkungsgrad, wie er vor der Glukosereaktion eingeregelt wurde, vermittels eines geeigneten Speichers erhalten bleibt. Die Verzögerung bei der Verstärkungsregelung des Strom-Spannungs-Wandlers verhindert dabei, daß sich das Ausgangsbasissignal innerhalb der Zeitspanne zwischen Zugabe der Glukose und Unterbrechung der Verstärkungsregelung vermindert.

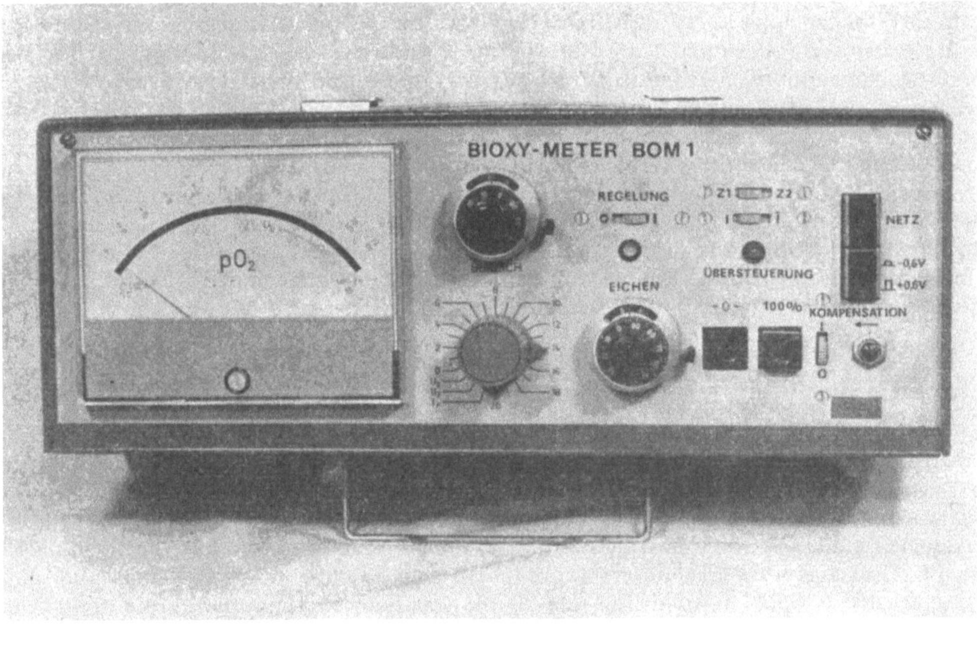

Abb. 1. Meßgerät BIOXY-Meter.

Die automatische Einstellung des Basissignals unmittelbar vor der Messung auf eine konstante Basissignalhöhe bringt gegenüber den bisherigen Meßgeräten folgende Vorteile:

— Die kinetische Bestimmung des Sauerstoffverbrauches mittels biospezifischer Sensoren kann weitgehend unabhängig von der Anfangshöhe des Sauerstoffdiffusionsstromes erfolgen, indem unmittelbar vor der Messung automatisch eine Kalibrierung der Auswerteeinrichtung auf einen vorgegebenen Sollwert erfolgt und diese Kalibrierung bei einer durch die Meßprobenzugabe hervorgerufenen Änderung der Sauerstoffkonzentration automatisch beibehalten wird.
— Durch die Möglichkeit der Messung vor Einstellung eines stationären Wertes kann die Zahl der Messungen pro Stunde wesentlich erhöht werden.
— Durch die gleiche Höhe des Basissignals bei allen Messungen werden die nachfolgenden kinetischen Messungen der Strom-Zeit-Kurve wesentlich genauer.

Durchflußzelle

Bei den ersten Arbeiten zur Glukosemessung von Nährlösungskomponenten und Stoffwechselprodukten erfolgte die Messung diskontinuierlich. Es wurde eine zylindrische Meßkammer verwendet, in welche die Elektrode hineinragte [16, 17]. Die Proben wurden mittels einer Wasserstrahlpumpe abgesaugt. Die Rührung konnte sowohl über eine Laborkleinrührwerk als auch über einen Magnetrührer erfolgen.
BERTRAM et al. [19] beschrieben die Messung von Blutglukose unter Verwendung der vom VEB Metra angebotenen Durchflußzelle. Die Anwendung dieser Durchflußzellen bei Fermentationslösungen brachte keinen Erfolg. Es zeigte sich, daß die Enzymelektroden bereits nach ca. 1—2 Tagen ihre Aktivität verloren. Die genaue Untersuchung ergab, daß die Meßkammer aus Messing besteht, das vernickelt ist. An der vernickelten Oberfläche zeigten sich insbesondere im Bereich der Verschraubungen bald Risse, so daß dort Kupferionen in Lösung gehen konnten und die Ursache für eine Inaktivierung des Enzyms bildeten.
Die selbst entwickelten Durchflußzellen wurden aus Piacryl gefertigt und wiesen folgende Vorteile auf:

— einfache Herstellung
— geringe Abmessungen, Reaktionsraum $< 10\ \mu l$
— der Flüssigkeitsstrom kann beobachtet werden
— durch ein direktes Anspülen der Elektrode unter einem Winkel von 150° bis 160°, wobei die Ableitung vertikal erfolgt, wird ein schnelles Freispülen der Membran und ein problemloser Durchgang von Luftsegmenten erreicht.

Automatischer Probenehmer

Zum Aufbau eines Analysenautomaten ist der Einsatz eines automatischen Probenehmers notwendig. Es wurde auf den Einsatz des automatischen Probenspeichers APS-4 vom VEB MLW Labortechnik Ilmenau orientiert. Bei diesem Gerätetyp befinden sich jeweils 11 Probenbehälter in Magazinen und werden der Reihe nach abgearbeitet. Dazu taucht eine Kanüle in den betreffenden Probenbehälter und unter Verwendung einer Pumpe wird die Probe abgesaugt. Nach dem Absaugen taucht die Kanüle in einen Behälter mit Pufferlösung. Die Dauer der Probenentnahme und der Entnahme von Pufferlösung sind auf Kontaktscheiben programmiert.
Durch Auswechseln der Kontaktscheiben läßt sich die Zahl der Proben/Stunde verändern.

Um das Ziel, die Luftsättigung der Probe unmittelbar vor der Messung zu erreichen, sollte der Probenspeicher die notwendigen Funktionen für ein Einleiten von Luft übernehmen.
Der Probenspeicher APS-4, der nur einen Taucharm besitzt und somit auch nur eine Funktion ausübt, wurde in der Weise umgebaut, daß zu dem vorhandenen Arm (zur Probenahme) im Winkel von 90° ein zweiter Arm (zum Einblasen von Luft) angeordnet wurde. Es ergibt sich somit die Möglichkeit, die vorherige Begasung der Proben, d. h. deren Luftsättigung, unmittelbar vor der Probenahme durchzuführen [21].
Zur Realisierung dieser Aufgabe wurde der Vorschub des Probenmagazins der Zeit des Umschwenkens des Probenehmers angepaßt, wozu der Tauchhebereinschub mit dem Magazinvorschub zeitlich abgeglichen wurde.

Abb. 2. Probenspeicher APS-4 in Stellung Pufferentnahme und Luftsättigung der Proben.

Dosierpumpe

Für das Dosieren der Lösungen und der Luft wurde die Dosierpumpe DP 2-2 vom VEB MLW Labortechnik Ilmenau verwendet. Es handelt sich hierbei um eine nach dem Schlauchabquetschprinzip arbeitende Schlauchpumpe mit 2×8 Kanälen. Über einen Rollenkäfig ist das Schlauchspann- und -halterungssystem angebracht. Das Abquetschen der Schläuche bewirkt ein Drucklager, welches durch eine Spannvorrichtung elastisch auf den Rollenkäfig gedrückt wird. Durch Einlegen von Schläuchen mit verschiedenem Innendurchmesser, durch umschaltbare Gänge und durch Parallelschalten von entsprechenden Schläuchen ergibt sich eine große Variation der Fördermenge.
Bei dieser Punpe ergibt sich jedoch wie bei allen nach dem Abquetschprinzip arbeitenden Pumpen ein prinzipbedingtes Pulsieren des Förderstromes. Um den negativen Einfluß des Pulsierens zu unterdrücken, wurden die auf einer Welle angeordneten zwei Rollenkäfige so verändert, daß die Rollen zueinander versetzt sind, wobei über diese jeweils ein Schlauch gleichen Durchmessers geführt wird, dessen Ansaug und/oder Druckseiten miteinander verbunden sind, wodurch die Förderleistung im Minimum des einen Schlauches jeweils im Maximum des anderen Schlauches liegt.

Arbeitsweise des Analysenautomaten

Der Analysenautomat besteht aus einem automatischen Probenehmer I, einer Durchflußzelle II mit Enzymelektrode IIa, dem Meßgerät BIOXY-Meter III sowie einer Dosierpumpe IV. Wahlweise ist der Anschluß eines Schreibers V möglich (Abb. 3).

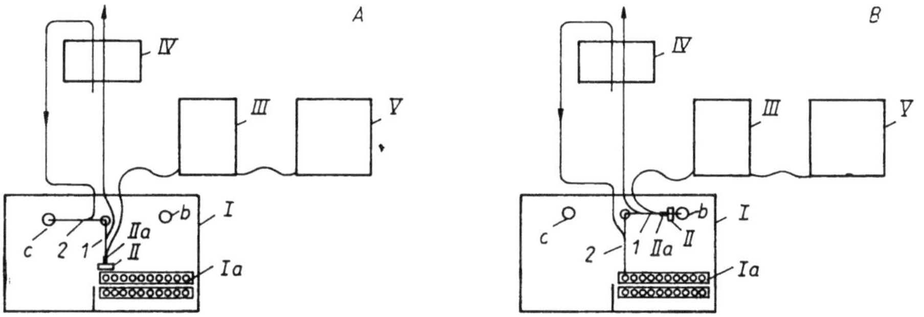

Abb. 3. Gerätekombination.

I — automatischer Probenehmer, II — Durchflußzelle, IIa — Enzymelektrode, III — BIOXY-Meter, IV — Dosierpumpe, V — Laborbandschreiber, Ia — Probebecher, 1 — Taucharm, 2 — Taucharm, b — Puffer-Vorratsgefäß, c — Spulgefäß.

Abb. 3A. Probenehmer in der Stellung Probenahme.

Abb. 3B. Probenehmer in der Stellung Ansaugen von Phosphatpuffer bei gleichzeitiger Luftsättigung der Probe.

In den Probebechern Ia des automatischen Probenspeichers I befinden sich Phosphatpuffer und die zu untersuchende Lösung. Der Probenspeicher I ist durch das Vorhandensein von zwei im Winkel von 90° angebrachten Taucharmen gekennzeichnet. Auf dem Taucharm 1 befindet sich die Durchflußzelle II mit der Enzymelektrode IIa. Der Taucharm 2 wird zum Einblasen von Luft verwendet.
In Stellung 1 (Abb. 3B) steht der Taucharm 1 über dem Vorratsgefäß b und mit der Dosierpumpe IV wird durch die an der Durchflußzelle sich befindende Kanüle Phosphatpuffer, an der Enzymelektrode IIa vorbei, gepumpt und es stellt sich ein stationärer Wert entsprechend dem Sauerstoffdiffusionsstrom ein. In dieser Zeit wird Luft mit der gleichen Dosierpumpe IV über eine am Taucharm 2 sich befindende Kanüle in die zu analysierende Lösung gedrückt. Das Eindrücken von Luft erfolgt bis unmittelbar vor Probenahme, d. h. es endet mit Umschwenken des Tauchhebers und der nachfolgenden Probenahme entsprechend Stellung 2 (Abb. 3A). In Stellung 2 (Abb. 3A) wird unter Verwendung des Taucharmes 1 mit der Pumpe IV Meßflüssigkeit durch die Durchflußzelle II, an der Enzymelektrode IIa vorbei, gepumpt, während die am Taucharm 2 sich befindende Kanüle in das Spülgefäß c taucht. Nach dem Kontakt der Meßflüssigkeit mit der Enzymelektrode verändert sich sofort der Diffusionsstrom. Am BIOXY-Meter III kann wahlweise der Verlauf der Stromänderung oder der elektronisch ermittelte Anstieg der Strom-Zeit-Kurve abgelesen werden. Durch Verwendung eines Schreibers V kann sowohl die Strom-Zeit-Kurve als auch der Anstieg der Kurve aufgezeichnet werden. Die Dauer eines Meßzyklus, d. h. die Umschaltung von Phosphatpufferdurchsatz auf Durchsatz von Meßflüssigkeit, ist abhängig von dem Ansprechverhalten der Sauerstoffelektrode, der Geschwindigkeit der Enzymreaktion und der Diffusionsgeschwindigkeit. Da auf eine kinetische Messung orientiert wird, liegt die Kontaktzeit für die Meßflüssigkeit bei 5 bis 10 Sekunden. Für den Phosphatpufferdurchsatz zur

Einstellung des Diffusionsgleichgewichtes ergibt sich eine Zeit weniger als 60 Sekunden. Damit ist die Möglichkeit gegeben, die Proben im Minutenrhythmus zu analysieren, d. h. 60 Proben pro Stunde.
Der lineare Meßbereich für die Glukosebestimmung liegt zwischen 0,1 mMol und 1,5 mMol. Der ermittelte serielle Variationskoeffizient V_k liegt bei 0,93%.

Abb. 4. Analysenautomat.

Anwendungsmöglichkeiten für den Analysenautomaten

Beim Einsatz von Enzymelektroden zur Messung von Fermentationslösungen unterscheidet man zwischen dem Einsatz bei der Substratbestimmung und der Bestimmung der Enzymaktivität. Das von uns entwickelte Verfahren, das den Einsatz einer Glukoseoxydase-Enzymelektrode vorsieht, ist sowohl zur Substratbestimmung als auch Enzymaktivitätsbestimmung geeignet. Hierbei erfolgt die Bestimmung direkt oder nach vorheriger Spaltung der Substrate und der nachfolgenden Messung des Spaltproduktes Glukose. Bei der Bestimmung glukosespaltender Enzyme ergibt sich als Meßgröße für die Enzymaktivität die unter definierten Bedingungen pro Zeiteinheit gebildete Glukose.
Für die Substratbestimmung existieren z. Z. folgende Analysenvorschriften:

1. Glukose
2. Saccharose $\xrightarrow{+\text{ Invertase}}$ Glukosebestimmung
3. Stärke $\xrightarrow[+\text{ Glukoamylase}]{\text{Dispergieren m. NaOH}}$ Glukosebestimmung
4. Glukosinolat $\xrightarrow{+\text{ Myrosinase}}$ Glukosebestimmung

Wenn nur Saccharose bzw. nur Glukosinolat vorhanden ist, kann die Bestimmung durch eine Bienzym-Elektrode erfolgen. Hierbei werden zwei Enzyme, Invertase-Glukoseoxydase bzw. Myrosinase-Glukoseoxydase, vor der Elektrode fixiert. Liegen diese Substrate jedoch im Gemisch mit Glukose vor, so muß die Bestimmung in einem 2-Schritt-Verfahren erfolgen, d. h. es erfolgt eine Bestimmung der Anfangsglukose und der Gesamtglukose. Der Substratgehalt wird rechnerisch aus beiden Werten ermittelt. Bei der Stärkebestimmung ergibt sich das Problem, daß es sich hierbei um kein einheitliches Produkt handelt und deshalb eine kinetische Messung ähnlich der Saccharosebestimmung nicht möglich ist. Die Messung über die Glukosebestimmung erfolgt deshalb generell nach dem Dispergieren und nach der Spaltung durch Glukoamylase.

Der entwickelte Analysenautomat mit dem BIOXY-Meter bietet die Voraussetzung für eine weitere breite Anwendung von Enzymelektroden, insbesondere von Oxydasen, beispielsweise zur Bestimmung von Aminosäuren, Alkohol, Ascorbinsäure und Milchsäure.
Weitere Anwendungsmöglichkeiten ergeben sich durch den Einsatz von mikrobiologischen Sensoren, d. h. von Elektroden mit vorgeschalteten trägerfixierten Mikroorganismen.

Eingegangen: 20. 12. 1985

Literatur

[1] Weise, H.: Diss. Humboldt-Universität zu Berlin, 1974.
[2] Scheller, F., Pfeiffer, F.: Z. Chem. **18** (1978), 50.
[3] Romette, J.-L., Froment, B., Thomas, D.: Clin. Chim. Acta **95** (1979), 259.
[4] Danielsson, B., Gadd, K., Mattiasson, B., Mosbach, K.: Clin. Chim. Acta **81** (1977), 163.
[5] Nilsson, H., Akerlund, A.-Ch., Mosbach, K.: Biochim. Biophys. Acta **320** (1973), 529.
[6] Nagy, G., von Storp, L., Guilbault, G.: Anal. Chim. Acta **66** (1973), 443.
[7] Pinkerton, T., Lawson, B.: Clin. Chim. **28** (1982), 1946.
[8] Keilin, D., Hartree, E. F.: Biochem. J. **50** (1952), 331.
[9] Bessman, S.: Diabetes 25 (1976), 81.
[10] Clark jun., Leland, C.: DE-AS 1598285.
[11] Mindt, W., Racine, Ph., Schläpfer, P.: DD-AS 100556.
[12] Schläpfer, P., Mindt, W., Racine, Ph.: Clin. Chim. Acta **57** (1974), 283.
[13] Cass, A., Davis, G., Francis, G., Hill, H., Aston, W., Higgins, J., Plotkin, E., Scott, L., Turner, A.: Anal. Chem. **56** (1984), 667.
[14] Cleland, N., Enfors, S.-O.: Eur. J. Appl. Microbiol. Biotechnol. **18** (1983), 141.
[15] Weise, H. u. a.: DD-WP 150760.
[16] Scheller, F. u. a.: Acta biol. med. germ. **39** (1980), 671.
[17] Weise, H., Scheller, F., Kreibich, G., Pfeiffer, D.: Lebensmittelindustrie **28** (1981), 491.
[18] Jänchen, M., Scheller, F., Pfeiffer, D., Wiegand, P., Nentwich, J.: Z. med. Labor.-Diagn. **23** (1982), 39.
[19] Bertermann, K., Scheller, F., Pfeiffer, D., Jänchen, M., Lutter, J.: Z. med. Labor.-Diagn. **22** (1981), 83.
[20] Weise, H. u. a.: WP G 01 N/272048.
[21] Weise, H. u. a.: WP G 01 N/272049.
[22] Geppert, G., Thielemann, H.: Chem. Tech. **35** (1983), 517.

Acta Biotechnol. 7 (1987) 1, 71–77

Fluorimetrische und photometrische Schnellbestimmung von L-Lysin in Fermentationslösungen

HESSE, G., RÖMER, W., MIOSGA, N., WACHTEL, E.

Akademie der Wissenschaften der DDR
Zentralinstitut für Mikrobiologie und experimentelle Therapie
Beutenbergstr. 11, Jena, 6900 DDR

Summary

In a methodical paper two analytical-chemical procedures are described which concern the control of L-lysine formation during fermentation processes. On one hand lysine is determined fluorimetrically with o-phthalaldehyde at defined pH, on the other spectrophotometrically with ninhydrine in acidic medium. The analytical conditions are discussed with respect to specifity, interferences, accuracy and reproducibility. A comparison of values with Autoanalyzer technique and two other methods for lysine determination offers the possibility to decide in what manner and with which aim of desired accuracy, economy and rapidity a given specimen has to be analyzed.

Einleitung

Zur spezifischen Bestimmung der basischen Aminosäure Lysin in Hydrolysaten und Fermentationslösungen sind kaum vortrennungsfreie chemische [1], aber einzelne Methoden auf enzymatischer Basis (L-Lysin-α-aminooxidase; L-Lysin-decarboxylase) mit spektrophotometrischer bzw. amperometrischer Endpunktanzeige ohne Vortrennung bekannt [2, 3], hingegen eine größere Zahl arbeitsaufwendigerer Verfahren, die Lysin nach dessen Isolierung (z. B. an Ionenaustauschern) molekülspezifisch erfassen [z. B. 4]. Bestimmungsmethode der Wahl dürfte die Kombination enzymatischer und elektrochemischer Methoden in Form von Messungen mit Enzymelektroden (CO_2-Messungen mit Hilfe immobilisierter L-Lysindecarboxylase aus *E. coli* über pH-Wert-Änderungen einer Bikarbonatschicht) für Fermentationslösungen sein [z. B. 5].
Lysinreiche Fermentationslösungen gestatten aber auch wegen ihres vergleichsweise hohen Lysingehaltes den Einsatz weniger spezifischer Methoden, wenn deren Leistungsfähigkeit, Grenzen und der Einfluß variabler Größen untersucht sind. Es werden in diesem Zusammenhang zwei schnelle, ökonomische und hinsichtlich ihrer Spezifität charakterisierte Methoden zur Lysinbestimmung in Fermentationslösungen mitgeteilt, die auf fluorimetrischer bzw. photometrischer Endpunktanzeige beruhen.

Material und Methoden

Fluorimetrische Bestimmung des Lysingehaltes mit o-Phthalaldehyd (OPA)

ROTH [6] setzte Aminosäuren zwischen pH 8 und 11 mit dem fluorogenen Reagenz OPA in Gegenwart des starken Reduktionsmittels 2-Mercaptoethanol um und erhielt blau fluoreszierende Reaktionsprodukte. Lysin fällt als einzige Aminosäure aus diesen Reaktionsbedingungen heraus und zeigt ein Maximum an Fluoreszenz bei pH 6. Wir benutzen dieses Verhalten des Lysins zu einer Lysinbestimmung bei pH 5,4 und arbeiteten dabei von MENDEZ und GAVRILANES [7] mitgeteilte Modifikationen in die ROTHsche Technik ein.

Reagenzien und Geräte

M/15 Phosphatpuffer, pH 5,4; 0,05% 2-Mercaptoethanol enthaltend (PP). Der Puffer wird wöchentlich neu angesetzt, 2-Mercaptoethanol einer Arbeitspuffermenge täglich frisch zugesetzt.
OPA: 0,3 mg/ml, wäßrige Lösung, täglich neu herstellen.
Lysinstandard: ~ 2,7 mg Lys · HCl (2 h bei 105 °C getrocknet) in 10 ml H_2O lösen, 0,2 ml dieser Lösung entsprechen dem höchsten Standardwert, der mit durch die Bestimmung geführt wird.
SPEKOL 11 oder SPEKOL 10 mit „ZV", Fluoreszenzmeßansatz „FK" und Photomultiplier PhoM 3/19 (alles VEB Carl Zeiss Jena), Sekundärfilter 743 Nr. 86520 (GA 43).

Durchführung

Vor Beginn der Messungen werden die Fluorimeter mit dem höchsten Lysinstandard geeicht.
SPEKOL 11: Hg-Lampe Hg E/3, Anregungswellenlänge 365 nm, Filter einsetzen, Nullabgleich mit wassergefüllter Küvette, höchsten Standardwert auf z. B. Anzeige 1000 bringen (s. Gebrauchsanleitung SPEKOL 11).
SPEKOL 10: Nitraleuchte, Anregungswellenlänge 340 nm, Filter einsetzen, Nullabgleich mit wassergefüllter Küvette, Galvanometer auf Fast-Vollausschlag (z. B. 95 Skalenteile) unter Benutzung der nötigen Verstärkungsstufen (z. B. „ZV" 200; PhoM 5) bringen.
Kulturlösungen (KL) zentrifugieren, von Überstand (KF) je nach zu erwartendem Lysingehalt mit H_2O wie folgt verdünnen: ~ 6 mg/ml 1 : 25; ~ 15 mg/ml 1 : 50; ~30 mg/ml 1 : 100, usw.
Meßlösung: Von den Verdünnungen (VL) in der Reihenfolge der Aufführung in Reagenzgläser pipettieren: 0,2 ml VL + 1,4 ml PP + 0,4 ml OPA; 35 min bei 22 °C stehenlassen, Fluoreszenz der Serie innerhalb von 15 min messen. Blindwert und drei Standards mit durch die Bestimmung führen, Auswertung über eine täglich überprüfte Eichkurve.

Photometrische Bestimmung mit Ninhydrin

Schon 1952 hatte CHINARD [8] festgestellt, daß Lysin bei pH 1 relativ spezifisch mit Ninhydrin eine gelbbraune, photometrisch auswertbare Farbe entwickelt. Dieses Verfahren und andere aus ihm abgeleitete schreiben ein Eisessig/H_3PO_4 — Reaktionsgemisch vor.
Wir änderten die Reaktionsbedingungen ab auf das Arbeiten in 1 N schwefelsaurer wäßriger Lösung in Gegenwart eines geeigneten hochsiedenden Lösungsmittels, das mit Wasser mischbar ist.

Reagenzien und Geräte

Ninhydrin-DMSO-Stammlösung: 5%ige (w/v) Lösung von Ninhydrin, p. a., in DMSO oder peroxidfreiem Ethylenglykolmonomethylether (Methylcellosolve). Tagesbedarf-Reagenzlösung (RL) : 5 ml der Stammlösung mit 25 ml DMSO oder Methylcellosolve verdünnen. 1 N H_2SO_4; 60%iges wäßriges EtOH (60 Vol. Tle EtOH)
Geeignetes Spektralphotometer (z. B. SPEKOL 10 oder 11 oder VSU — 2 P, alles VEB Carl Zeiss Jena)
Wasserbad, Laborzentrifuge

Durchführung

KL zentrifugieren, vom KF Verdünnungen mit 1 N H_2SO_4 etwa wie folgt vornehmen: ~ 6 mg/ml 1 : 50; ~ 15 mg/ml 1 : 100; ~ 30 mg/ml 1 : 200, usw.
Meßlösung; 1 ml verdünnte Probe + 0,5 ml 1 N H_2SO_4 + 1,2 ml RL in Reagenzgläser pipettieren, Gläser verschließen (Parafilm oder Plaststopfen), 60 min bei 100 °C halten, in Eiswasser auf Zimmertemperatur abkühlen, 2 ml 60%iges EtOH zugeben, Extinktion bei 458 nm gegen Wasser messen ($d = 1$ cm).
Einen Blindwert und zwei Standardproben durch die Bestimmung führen, Extinktion des Blindwertes ($\leq 0,030$) rechnerisch berücksichtigen, zur Auswertung Eichkurve benutzen. Diese umfaßt den Konzentrationsbereich von 40 μg Lys · HCl/4,7 ml Meßlösung bis 240 μg Lys · HCl/4,7 Meßlösung ($0,075 < E < 0,500$). Die Eichkurve wird mit den entsprechenden Konzentrationen von Lys · HCl in 1 N H_2SO_4 aufgestellt.

Ergebnisse

Statistische Auswertung der Analysenergebnisse

OPA-Methode

Lys · HCl: 30 μg/2 ml eingesetzt (mittlerer Eichkurvenwert): $n = 10$; $\bar{x} = 29,8$ μg/2 ml; $s = 0,7$ μg/2 ml; Streubereich [9] $\Delta x = t_{95\%} \cdot s = 1,6$ μg/2 ml. Wiederfindung: 99,3%.
KF von ~ 30 mg/ml Lys · HCl: $n = 10$; $\bar{x} = 31,9$ mg/ml; $s = 0,6$ mg/ml; Streubereich $\Delta x = 1,4$ mg/ml; Richtigkeit (accuracy): Probe eines KF bestimmt mit 20,8 mg/ml Lys · HCl, zugewogen 16,1 mg/ml Lys · HCl; theoretisches Gesamt-Lys · HCl 36,9 mg/ml; gefundenes Lys · HCl 36,6 mg/ml ≙ 98% der Zuwaage.

Ninhydrinmethode

Lys · HCl: 103 μg/4,7 ml eingesetzt (mittlerer Eichkurvenwert): $n = 10$; $\bar{x} = 105$ µg/ 4,7 ml; $s = 4$ µg/4,7 ml; Streubereich $\Delta x = 9$ µg/4,7 ml. Wiederfindung: 101,9%.
KF von ~ 30 mg/ml Lys · HCl: $n = 10$; $\bar{x} = 31,6$ mg/ml; $s = 0,6$ mg/ml; $\Delta x = 1,4$ mg/ ml. Richtigkeit: Probe eines KF bestimmt mit 22,5 mg/ml Lys · HCl; zugewogen 25,3 mg/ ml. Lys · HCl; theoretisches Gesamt-Lys · HCl 47,8 mg/ml; gefundenes Lys · HCl 49,0 mg/ml ≙ 105% der Zuwaage.

Spektrale Verläufe der Signale der OPA- und Ninhydrinmethode

In der Abb. 1 sind die normierten Anregungs- bzw. Emissionsspektren der OPA-Methode und der wellenlängenabhängige Extinktionsverlauf der Ninhydrinmethode aufgezeichnet.

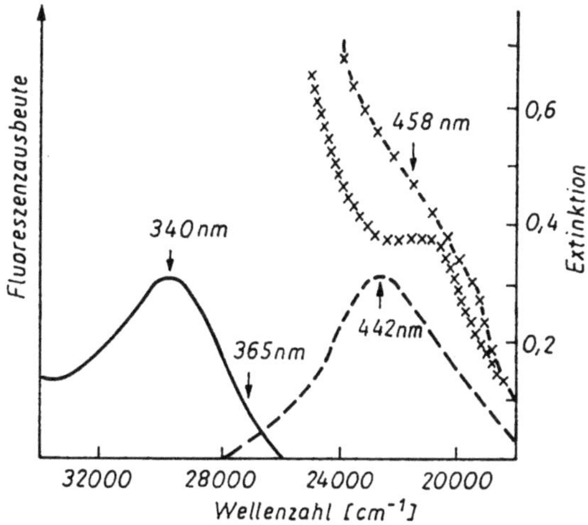

Abb. 1. Normierte Anregungs- und Emissionsspektren der OPA-Methode; Extinktionsverläufe der Ninhydrinfärbung

—— Anregungsspektrum OPA
---- Emissionsspektrum OPA
×—× Absorptionsspektrum Ninhydrin (DMSO)
××× Absorptionsspektrum Ninhydrin (Methylcellosolve).

Beide Ninhydrinspektren mit 200 μg Lys · HCl/4,7 ml Meßlösung.

Signalkonstanz, Erhitzungsdauer, Aziditäts-, Salz- und Lösungsmitteleinflüsse

Die Extinktionen der Meßlösungen der Ninhydrinmethode sind mindestens 1 h konstant. Dagegen ändern sich die Fluoreszenzausbeuten der OPA-Meßlösungen in Abhängigkeit von ihrer Standzeit (Abb. 2). Es ist bei Serienmessungen zu bedenken, daß alle Proben innerhalb 40 min bis 55 min nach der OPA-Zugabe vermessen werden müssen. Die bekannte Temperaturabhängigkeit der Fluoreszenz spielt keine Rolle, weil die mitgeführten Standardproben Temperaturschwankungen ebenfalls unterworfen sind. Der Einfluß der Erhitzungsdauer auf die Extinktionen der Ninhydrinmethode ist aus Abb. 3 ersichtlich. Danach gewinnt man nach 60 min bei 100 °C kaum noch an Farbtiefe; die Blindwerte bleiben über die ganze Zeit konstant (Lösungsmittel DMSO).

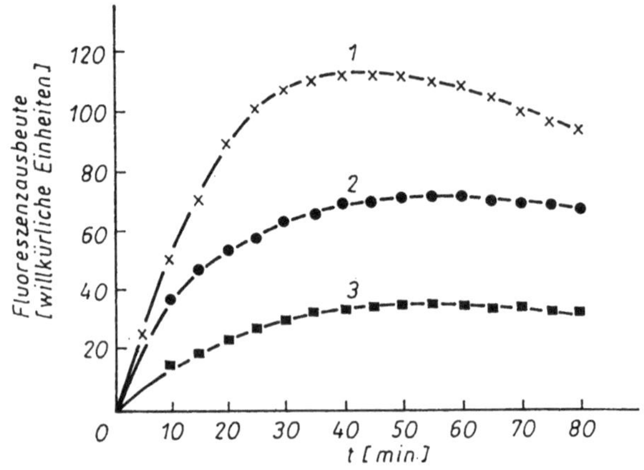

Abb. 2. Zeitabhängigkeit der Fluoreszenzausbeute des OPA-Lysin-Reaktionsproduktes.
1 — Standardlösung ~ 40 μg Lys · HCl/2 ml Meßlösung, Anregung 340 nm;
2 — Verdünntes KF, Anregung 340 nm;
3 — Standardlösung ~ 7 μg Lys · HCl/2 ml Meßlösung, Anregung 365 nm

Die Abhängigkeit der Meßextinktion der Ninhydrinmethode von der Azidität der Probe (nicht Meßlösung) ist in Abb. 4 dargestellt. Das schmale Plateau zwischen pH 0,8 und 1,0 verbietet die Einbringung neutraler oder alkalischer Probelösungen. Die OPA-Methode wird von der pH-Konstanz des verwendeten PP beeinflußt. Steigende (alter Puffer) oder sinkende (oxydiertes 2-Mercaptoethanol) pH-Werte der Reaktionslösung ergeben erhöhte oder verminderte Fluoreszenzausbeuten [6].

Salze wie NaCl, $(NH_4)_2SO_4$, KH_2PO_4, Na_2HPO_4, $MgCl_2$ und Glukose oder Saccharose in Konzentrationen, in denen sie in Fermentationslösungen vorkommen, interferieren bei beiden Methoden prinzipiell oder wegen der nötigen hohen Vorverdünnung nicht.

Für die Ninhydrinmethode wurden folgende mit Wasser mischbare Lösungsmittel zur Stabilisierung der erhaltenen Farbe in einer Phase erprobt: DMSO, DMF, Triethylenglykol, Monomethylethylenglykol (Cellosolve), 1,4-Dioxan, Glycerin und Glykol. An der von der Eigenabsorption des Ninhydrins unbeeinflußten geeigneten Meßwellenlänge bei ~ 460 nm sieht man im Spektrum eine Absorptionsschulter mit Methylcellosolve ($\lambda = 458$ nm; $\varepsilon_{mol} = 1400$ l/mol), bei Verwendung von DMSO oder Triethylenglykol muß bei 458 nm an der abfallenden Flanke gemessen werden (DMSO: $\varepsilon_{mol} = 1750$ l/mol; Triethylenglykol: $\varepsilon_{mol} = 1650$ l/mol). Alle anderen Lösungsmittel zeigen entweder zu hohe Blindwerte bei zu geringen Meßextinktionen (Glycerin, Glykol) oder Trübung (durch Verdampfen des Lösungsmittels) nach der Erhitzung (Dioxan) oder sind über eine Stunde bei 100 °C instabil (DMF).

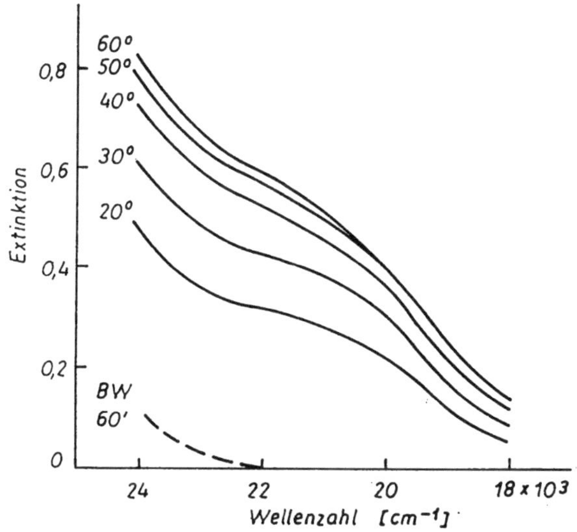

Abb. 3. Abhängigkeit der Extinktion von der Erhitzungsdauer (Ninhydrinmethode)
230 μg Lys · HCl/4,7 ml Meßlösung; Lösungsmittel DMSO.

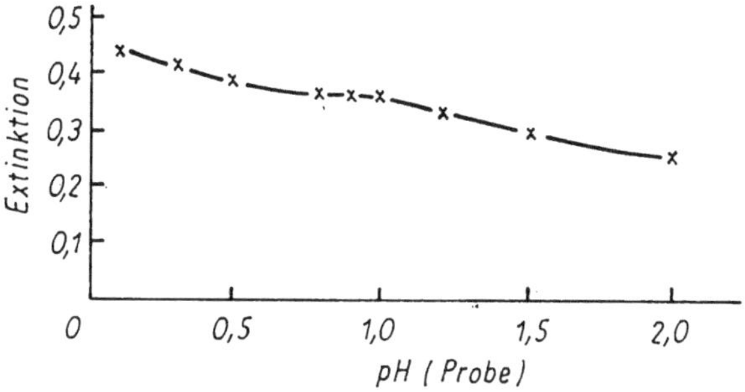

Abb. 4. Einfluß des pH-Wertes der Probelösung auf die Extinktion der Meßlösungen (Ninhydrinmethode)
150 μg Lys · HCl/4,7 ml Meßlösung; Lösungsmittel DMSO.

Beiträge anderer Aminosäuren zur Fluoreszenz (OPA) oder Extinktion (Ninhydrin)

Die Fluoreszenzausbeute bzw. die Extinktion des Lysins wurde in beiden Schnellmethoden für eine gewählte Konzentration als 100% angenommen. Andere Aminosäuren (ferner das NH_4^+-Kation und Ornithin) zeigen dann, in äquimolaren Massen in die Bestimmungen eingeführt, die in Tab. 1 aufgeführten Störungen.

Methodenvergleich OPA-Methode, Ninhydrinmethode, Furfuralmethode [1] und Aminosäureanalysator (AA)

Abb. 5 enthält die Verfolgung der Produktbildung mit den beiden beschriebenen Schnellmethoden, außerdem Werte nach PATTHY und PATTHY [1], die Lysin ohne dessen Vortrennung mit Furfural (Analogie Furfural + Anilinsalze: rotviolette Verbindungen) bestimmen. Ferner sind Lysingehalte dargestellt, die nach Vortrennung des Lysins (zu-

Tabelle 1. Interferenzen durch andere Aminosäuren
Ninhydrinmethode: $c = 2{,}4 \cdot 10^{-4}$ mol/l Lys · HCl in der Meßlösung; Meßwellenlänge 458 nm.
OPA-Methode: $c = 9{,}6 \cdot 10^{-5}$ mol/l Lys · HCl in der Meßlösung; Anregungswellenlänge 340;
1) α, ε-Diaminopimelinsäure

AS	Äquimolares Signal				AS
	OPA-Methode (%)	Ninhydrin Methode (%)	OPA-Methode (%)	Ninhydrin Methode (%)	
Lys · HCl	100	100	100	100	Lys · HCl
His	278	65	78	1	Arg
Pro	1	37	28	0	Ser
Leu	19	0	26	0	Thr
Ileu	27	0	30	1	Phe
Val	26	0	31	7	Tyr
Glu	11	0	6	3	Asp
Met	40	1	2	51	CySH
Ala	17	0	77	137	1)
Gly	71	0	117	51	Lys-NH_2
Try	33	47	0	0	NH_4^+

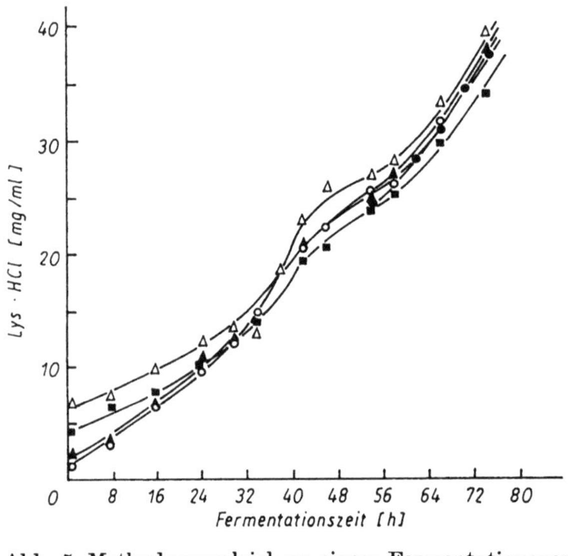

Abb. 5. Methodenvergleich an einem Fermentationsverlauf.
●— Aminosäureanalysator; △— OPA-Methode; ■— Ninhydrinmethode; ▲— Furfuralmethode [1] direkt; ○—Furfuralmethode [1] nach Säulenvortrennung

sammen mit Arg und His) an dem auf pH 5,8 äquilibrierten makroretikulären Wofatit-Carboxyl-Kationenaustauscher CA 20 und nachfolgender Furfuralreaktion erhalten wurden. Für das Ende der Fermentation (42 h bis 72 h) wurden dieselben Proben einem AA zugeführt; diese Werte werden als wahre Gehalte angenommen (Tab. 2).

Tabelle 2. Prozentuale Abweichungen der bestimmten Lysingehalte von AA-Werten ab 24,5 mg Lys · HCl/ml (Auswertung der Abb. 5.)

Lys · HCL [mg/ml] (AA-Werte)	Furfural [%]	Furfural (Säule) [%]	OPA [%]	Ninhydrin [%]
24,5	+4,1	+4,1	+9,3	−2,8
26,0	+4,6	+3,1	+7,4	−3,2
29,5	+3,9	+1,0	+7,2	−3,6
34,5	+2,6	0	+4,7	−6,7
37,5	+0,5	−0,8	+4,2	−7,4

Diskussion

Die exakte quantitative Erfassung des tatsächlich im Untersuchungsmaterial vorliegenden Lysins konnte im gegebenen Fall nur durch mehrfachen Methodenvergleich [9] gesichert werden.

Bei Annahme der Richtigkeit der durch AA erhaltenen Werte geben beide Furfuralkurven die tatsächlichen Lysingehalte am besten wieder. Das geht für das Ende der Fermentation aus der guten Übereinstimmung mit dem AA hervor; die Werte der nullten bis achten Stunde sind ebenfalls allein nach Reaktion mit Furfural real. In Tab. 2 sind die prozentualen Abweichungen aller Methoden von 24,5 mg Lys · HCl/ml an aufwärts als Vergleich zum AA aufgeführt.

Die Schulter in den Kurvenzügen der Fermentation um 46 h erklärt sich durch den Übergang der Kultur in die stationäre Wachstumsphase und den damit verbundenen Inbalancen in den Metabolitflüssen.

Die Fehler der OPA- bzw. Ninhydrinmethode überschreiten nicht die für schnelle betriebliche Verlaufskontrollen in Kauf zu nehmenden Beträge (Tab. 2). Die Minuswerte der Ninhydrinmethode erklären sich aus den geringeren Beiträgen anderer präsenter Aminosäuren zur Farbentwicklung (Tab. 1). Es soll allerdings darauf hingeweisen werden, daß in den heterogenen Kulturen bei Direktbestimmungen (ohne Vortrennung) weder für Furfural noch für o-Phthalaldehyd oder Ninhydrin Aminosäuren allein für Fluoreszenz- oder Farbausbeuten in Frage kommen werden und daß die Störungen von Fermentation zu Fermentation nicht als konstant angenommen werden können.

Herrn Dr. STEINERT, AdL der DDR, ACUB Halle, danken wir für die Ausführung der Lysinanalysen am Aminosäure-Analysator.

Eingegangen: 20. 8. 1985

Literatur

[1] PATTHY, M., PATTHY, A.: Acta Biochim. Biophys. Acad. Sci. Hung. **10** (1975), 277.
[2] KUSAKABE, H., KODAMA, K., KUNINAKA, A., YOSHINO, H., SODA, K.: Agric. Biol. Chem. **43** (1979), 1749.
[3] MACHOLAN, L.: Collect. Czech. Chem. Comm. **43** (1978), 1811.
[4] KOFRÁNYI, H.: Hoppe-Seyler's Z. Physiol. Chem. **305** (1956), 61.
[5] WHITE, W. C., GUILBAULT, G. G.: Anal. Chem. **50** (1978), 1481.
[6] ROTH, M.: Anal. Chem. **43** (1971), 880.
[7] MENDEZ, E., GAVRILANES, J. G.: Anal. Biochem. **72** (1976), 473.
[8] CHINARD, F. P.: J. Biol. Chem. **199** (1952), 91.
[9] DOERFFEL, K.: Statistik in der analytischen Chemie. Leipzig: VEB Deutscher Verlag für Grundstoffindustrie, 1982.

Acta Biotechnol. 7 (1987) 1, 78

Book Review

DECHEMA Monographien Bd. 102
Grundlagen von Elektrodenreaktionen

Vorträge von der Tagung der Fachgruppe Angewandte Elektrochemie der GDCh vom 2. bis 4. Oktober 1985
Weinheim; Deerfield Beach, Florida; Basel; VCH-Verlagsgesellschaft, 1986, 628 S.

Der vorliegende Band enthält 43 Beiträge zu aktuellen Problemen, neuen Ideen und Theorien, interessanten Systemen und Methoden, gegliedert in folgende fünf Abschnitte:
Die Phasengrenze fest/flüssig; Kinetik von Elektrodenreaktionen; Elektrokatalyse; Photoelektrochemie; Festkörperaspekte und neue Materialien. Selektive Elektroden spielen in der Biotechnologie eine bedeutende Rolle, nicht nur zum Nachweis bestimmter Kationen und Gase, sondern vor allem in Kombination mit Enzymen und Mikroorganismen als Biosensoren. Wenn auch in der Monographie auf die Biotechnologie nicht unmittelbar eingegangen wird, so sind die Elektrodenvorgänge doch oft von unmittelbarem Interesse für die Entwicklung von Sensoren und für das nähere Verständnis ihrer Arbeitsweise. Dem trägt die vorwiegend theoretische Behandlung der erhaltenen experimentellen Ergebnisse Rechnung.
Besondere Beachtung verdienen Beiträge zu den Grundlagen der Elektrokatalyse, zu chemisch modifizierten Elektroden, insbesondere Elektrodenreaktionen an Polypyrrol-Elektroden, zur heterogenen Redoxkatalyse an Oxidelektroden, zu photoelektrochemischen Systemen, unter anderem CO_2-Reduktion an Elektroden, zu Hexacyanometallat-modifizierten Elektroden für Kationenselektivität und schließlich zur Bestimmung von Tensiden mit ionen-spezifischen Elektroden. Dem Entwickler und Anwender elektrochemischer Sensoren werden eine Menge Anregungen gegeben, die dem neuesten Stand entsprechen.

G. Geppert

Acta Biotechnol. 7 (1987) 1, 79–85

Der neue Hemmhofmeßautomat HZAA-3 des Autoselect-Systems

Mühlig, P., Bauer, E., Schicht, G., Günther, H., Harz, D.

Akademie der Wissenschaften der DDR
Zentralinstitut für Mikrobiologie und experimentelle Therapie
Beutenbergstraße 11, Jena, 6900 DDR

Summary

For the automatical measuring of inhibition zones on bioassay plates a new zone reader of the Autoselect-System has been developed. This third type is described. It combines all advantages of the preceding types as, for instance, the special principle of measuring and the transport of the plates. However, details of the function, the service, and the design have been improved.

Einleitung

Aktivitätsbestimmungen von Antibiotikabildnern auf der Grundlage des Agarlochplatten-Diffusionstestes erfordern möglichst genaue Messungen der Hemmhofdurchmesser auf den Testplatten. Ursprünglich mußten diese Messungen visuell und manuell durch Anlegen eines Lineals an die einzelnen Hemmzonen auf der Unterseite der Platte vorgenommen werden. Bemühungen zur Rationalisierung dieser Arbeiten durch Umbau von Lesegeräten oder Trichinoskopen sowie durch Entwicklung von mechanisch-optischen Geräten bis hin zu Automaten verfolgten das Ziel, diese Messungen nicht nur zu erleichtern, sondern auch zu beschleunigen und zu objektivieren. Da die Genauigkeit der Meßergebnisse aber von der Güte des Gerätes beeinflußt wird, gibt man — vor allem bei umfangreichen Meßreihen — automatischen Geräten mit objektiver Messung den Vorzug gegenüber einfachen Geräten mit subjektiver Erfassung der Meßergebnisse. Unter den automatischen Geräten lassen sich auf Grund des Meßprinzips zwei Gruppen unterscheiden:

1. Geräte, die auf dem Scanningverfahren beruhen und universell zu verwenden sind. Hierher gehören z. B. der Antibiotic Zone Reader von Fisher Scientific Co. [1], der Zonenanalysator AR-140 von Millipore [2, 3] und der Automatic Zone Reader von Atago, Optical Works Co. Ltd. Auch Bildanalysatoren, welche nach dem genannten Prinzip arbeiten, können bei Vorhandensein bestimmter Zusatzeinrichtungen für Hemmhofmessungen eingesetzt werden, z. B. das Quantimet 720 P von Imanco [4, 5].
2. Spezielle Photometer, die als Hemmhofmeßgeräte konzipiert wurden, z. B. das automatische Meßgerät Automatic Microbial Assay Equipment von Autodata, der Autobiocoder von Saxby, der AS-4A Zone Computer von Yamato sowie der Hemmhofmeßautomat HMA-2 des Autoselect-Systems [6], der nunmehr als verbessertes drittes

Modell vorliegt. Er war wie die hier genannten automatischen speziellen Meßgeräte schon seit dem ersten Modell [7] für die Messung großer quadratischer Testplatten mit 64 Löchern konzipiert [8]. Um Übereinstimmung der Positionen der Stanzlöcher beim Herstellen und beim Messen der Hemmhöfe zu gewährleisten, war beim ersten Gerät ein Stanzer eingebaut.
Beim zweiten Modell unseres Hemmhofmeßautomaten wurde die Funktion des Stanzens nicht mehr im Meßgerät integriert, sondern es wurde ein Achtfachstanzautomat [9] konstruiert, welcher in nur 8 Arbeitsschritten 64 Löcher in den Agar einer Testplatte stanzt. Dadurch konnte eine beträchtliche Zeiteinsparung erzielt werden, und infolge der präzisen Fertigung beider Geräte blieb die hohe Genauigkeit erhalten.
Aus dem Hemmhofmeßautomaten HMA-2 entwickelten wir mit Unterstützung des VEB Carl Zeiss JENA eine verbesserte Variante, den Hemmhofauswerteautomaten HZAA-3. Dieser Automatentyp wird hier beschrieben.

Äußeres

An der Vorderfront des kastenförmigen Meßgerätes (Abb. 1) mit den Abmessungen von $825 \times 645 \times 675$ mm befinden sich

— eine Schiebetür zum Verschließen der Öffnung für das Einbringen und Herausnehmen der Testplatte;

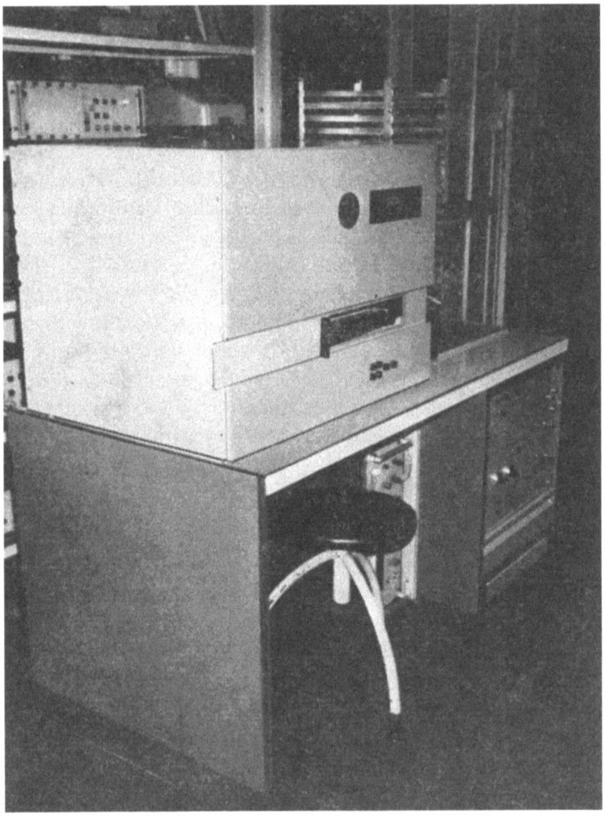

Abb. 1. Der Hemmzonenauswerteautomat HZAA-3

— Drucktasten für die Steuerung des Testplattentransportes;
— eine Digitalanzeige, welche die Lage des eben gemessenen Hemmhofs auf der Testplatte und dessen Durchmesser angibt;
— eine Mattscheibe, auf welcher der jeweils im Strahlengang befindliche Hemmhof abgebildet wird.

Anschlußmöglichkeiten für einen Drucker, einen Locher und einen Computer wurden an der Rückseite des Meßgerätes untergebracht.
Das Meßgerät selbst steht auf einem Tisch, welcher in 3 Einschüben den Hauptteil der Elektronik enthält.

— An dem oberen Einschub befinden sich der Zündknopf für die 8 Leuchtstoffröhren der Beleuchtungseinheit des Meßgerätes sowie 8 Dioden, an denen sich erkennen läßt, ob alle 8 Leuchtstoffröhren brennen.
— Der mittlere Einschub besitzt 2 Potentiometer, mit denen die auf einer Eichplatte vorgegebenen Meßwerte kalibriert werden können (Eichanpassung). Die Vorderfront läßt sich durch Lösen der 2 rechten Schrauben aufklappen, wodurch 4 Potentiometer zum elektrischen Abgleich der Helligkeitswerte zugänglich werden.
— Am untersten Einschub befinden sich die Bedienungsknöpfe zum Ein- und Ausschalten des Gerätes.

Das Meßprinzip

Das Meßprinzip des als spezielles Photometer zu betrachtenden HZAA-3 besteht darin, daß die Helligkeitswerte von 3 Bildsegmenten gemessen werden (Abb. 2):

L — Gesamthelligkeitswert über dem Hemmhofmeßfeld (spaltförmige Segmente)
L_u — Helligkeitswert des ständig unbewachsenen Bereiches in unmittelbarer Nachbarschaft des Stanzloches (ringförmiges Segment)
L_b — Helligkeitswert des bewachsenen Bereichs, d. h. außerhalb des Hemmhofs (ringförmige Segmente).

Da L_u und L_b plattenspezifische Konstanten sind und an objektfreier Stelle der Abgleich $L = L_u = L_b$ erfolgt, erhält man für L folgende Gleichung:

$$L = \alpha \cdot L_u + (1 - \alpha) L_b \quad \text{mit} \quad 0 \leqq \alpha \leqq 1. \tag{1}$$

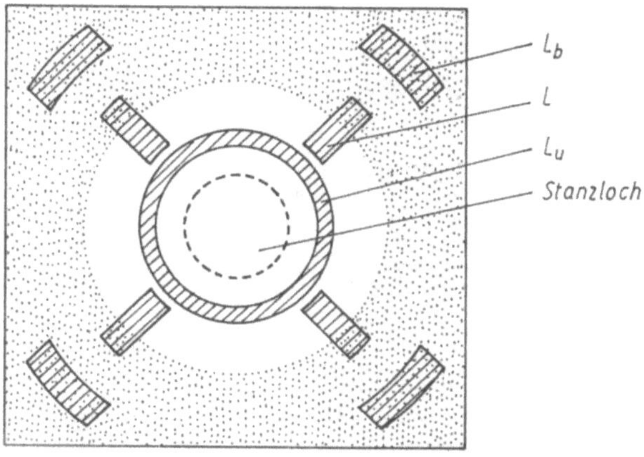

Abb. 2. Die Meßbereiche in Relation zum Hemmhof (nicht punktiert).
L_b — bewachsener Bereich, L — Meßbereich des Hemmhofs, L_u — unbewachsener Bereich

Demzufolge gilt:

$\alpha = 0$ für $L = L_b$, d. h. bei Fehlen eines Hemmhofes und

$\alpha = 1$ für $L = L_u$, d. h. bei einem Hemmhof, der die obere Meßgrenze überschreitet. Die technisch bedingten Meßgrenzen betragen beim HZAA-3:

$\alpha = 0$ für Hemmhofdurchmesser $< 17{,}5$ mm

$\alpha = 1$ für Hemmhofdurchmesser $> 29{,}5$ mm.

Auf der Bestimmung von α basiert das Meßverfahren für den Hemmhofdurchmesser D

$$\text{gemäß } \alpha = \frac{L - L_b}{L_u - L_b}. \tag{2}$$

Durch die Meßblendengestalt für L (Abb. 3) erreicht man die lineare Abhängigkeit

$$D(\alpha) = a \cdot \alpha + b. \tag{3}$$

Die Werte der Konstanten a und b dienen der Eichanpassung der Hemmzonenwerte an mitgelieferte Standards.
Aus (2) und (3) ergibt sich somit

$$D(\alpha) = a \frac{L - L_b}{L_u - L_b} + b. \tag{4}$$

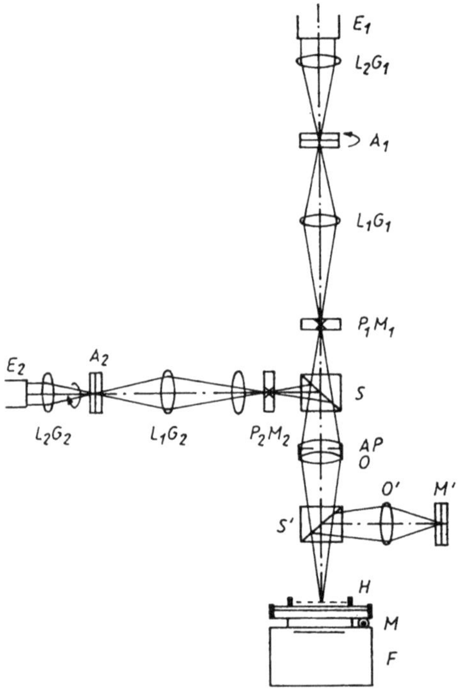

Abb. 4. Prinzipskizze des Strahlenganges.
F — Flächenstrahler (8 Leuchtstoffröhren); M — Mattglas; H — Hemmhof; S' — Strahlenteiler; O' — Objektiv; M' — Mattscheibe; O — Objektiv; AP — Austrittspupille; S — Strahlenteiler; P_1M_1, P_2M_2 — Polarisationsfilter und Masken; L_1G_1, L_1G_2 — Linsengruppen; A_1, A_2 — Analysatoren; L_2G_1, L_2G_2 — Projektive; E_1, E_2 — Empfänger (SEV)

Die Beziehung (4) ist für den Meßbereich

$$17{,}5 \text{ mm} < D < 29{,}5 \text{ mm}$$

am HZAA-3 kalibriert. Außerhalb dieses Meßbereichs werden folgende Hemmhofwerte angezeigt:

$D < 17{,}5$ mm der Wert 0,

$29{,}5 \text{ mm} < D \leqq 44{,}0$ mm der Wert 29,5 und

$D > 44{,}0$ mm der Wert 0.

Der optische Strahlengang

Eine Prinzipskizze des Strahlenganges zeigt Abb. 4. Von einem Flächenstrahler F, bestehend aus 8 nebeneinander angeordneten Leuchtstoffröhren und einem darüber befindlichen Mattglas M, wird der Hemmhof H gleichmäßig ausgeleuchtet.
Mit dem Objektiv O und dem Strahlenteiler S werden von H zwei Bilder in den Ebenen P_1M_1 und P_2M_2 erzeugt. Die dort angebrachten Bildfilter bestehen aus je einem Polarisator mit Maske (Abb. 3). Über die Linsengruppe L_1G_1 und L_1G_2 wird ein zweites Zwischenbild in der Ebene der Analysatoren A_1 und A_2 erzeugt.

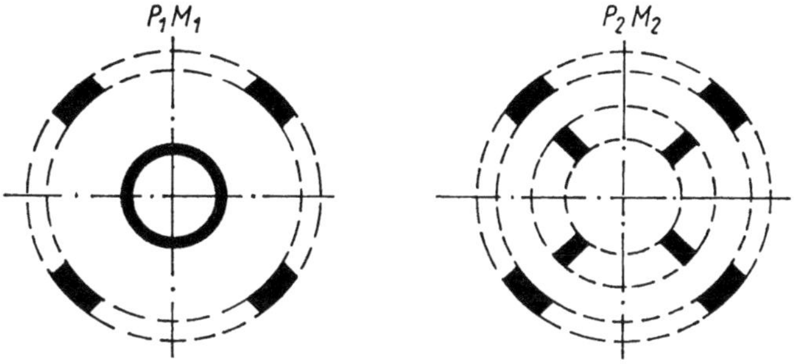

Abb. 3. Bildmasken der beiden Strahlengänge (schwarz die Durchtrittsstellen für das Licht).

Alternierend werden an dieser Stelle die Polarisationsrichtungen so geändert, daß zeitlich aufeinanderfolgend die in den Ebenen P_1M_1 und P_2M_2 gefilterten und unterschiedlich polarisierten Bildanteile des Hemmhofes durchgelassen werden und über die Projektive L_2G_1 und L_2G_2 auf die Empfänger E_1 und E_2 treffen. Dabei gelangt das Licht aus den verschiedenen Regionen des Hemmhofs und seiner Umgebung (L, L_b, L_u) stets auf die gleiche Fläche des Targets des Sekundärelektronenvervielfachers, d. h. die Projektive bilden die Austrittspupille AP des Objektivs O auf E_1 bzw. E_2 und H ins Unendliche ab.
Der Strahlenteiler S' führt dem Objektiv O' 20% des durch H gehenden Lichts zu. Dadurch wird auf der Mattscheibe M' ein Bild des Hemmhofs erzeugt, dessen Lagekontrolle durch 3 konzentrische Ringe auf der Mattscheibe erleichtert wird. Der innere entspricht dem Stanzloch, der mittlere der unteren und der äußere der oberen Meßgrenze.

Die elektronische Meßanordnung

Die Sekundärelektronenvervielfacher liefern eine Spannung U proportional zur Lichtintensität (Abb. 5). Jeder Empfänger gibt periodisch nacheinander zwei verschiedene Signalspannungsimpulse auf Analogspeicher (Sample-Hold). Schaltimpulse zur Zeit des Maximums der Impulse öffnen diese Speicher und verriegeln sie danach, so daß die Gleichspannungen nach Glättung durch steile Tiefpaßfilter (Butterworth-Filter 2. Ordnung) auf Differenzverstärker gegeben werden können, die aus den Einzelsignalen die Differenzen $U_L - U_{L_b}$ und $U_{L_u} - U_{L_b}$ bilden. Gemäß Gleichung (4) wird in einer Analog-Dividierstufe der Quotient gebildet.

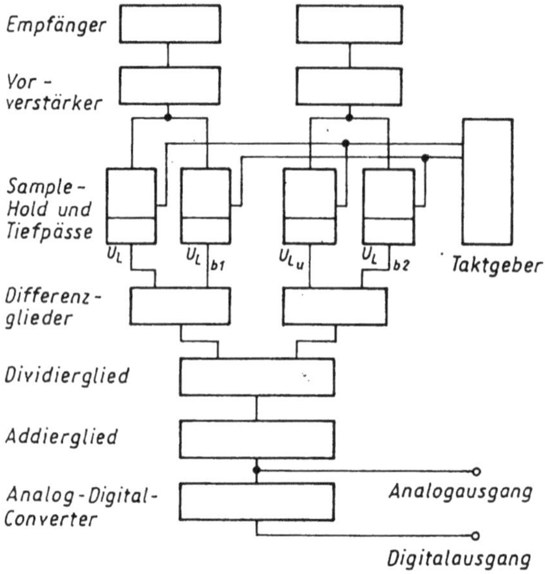

Abb. 5. Die elektronische Meßanordnung.
U_L. U_{L_b}, U_{L_u} = Spannungssignale für den Hemmhofbereich, den bewachsenen und unbewachsenen Bereich

Zur Anpassung der Meßwerte an reale Durchmesserwerte der Hemmzonen dienen zwei Eichhemmhöfe vom Durchmesser 20 und 26 mm. Die Eichung wird so vorgenommen, daß über 2 Potentiometer an der Vorderfront des Elektroniktisches (mittlerer Einschub) die Konstanten a und b (Gleichung 4) so bestimmt werden, daß den beiden Hemmhofwerten die Meßspannungen 2,0 bzw. 2,6 V entsprechen. Zur Überprüfung der Linearität dienen weitere 4 Hemmhöfe vom Durchmesser 18 bis 28 mm auf derselben Eichplatte.

Arbeitsweise

Voraussetzungen zur Inbetriebnahme: Nach dem Nivellieren des Gerätes entsprechend der Bedienungsvorschrift wird es an das Netz angeschlossen, eingeschaltet und die Eichplatte mit Modellhemmhöfen von 18, 20, 22, 24, 26 und 28 mm aus transparenter Folie auf den Plattenwagen des Gerätes gebracht. Sollte sich die Notwendigkeit ergeben, den Modellhof nach dem Mattscheibenbild zu zentrieren, so genügt ein einmaliges Korrigieren mit 2 Rändelschrauben.
Messen der Testplatten: Mit der Tastatur an der Vorderfront des Meßgerätes werden der automatische mäanderförmige Transport, die Hemmhofmessung und die Meßwertan-

zeige ausgelöst. Rationelles Arbeiten erfordert den Anschluß eines Druckers oder Lochers bzw. eines Kleinrechners.
Da das Meßgerät für die Geschwindigkeit von 4 min pro Platte konzipiert wurde, lassen sich unter Einbeziehung der Zeit für das Wechseln der Platten in einer Stunde mindestens 12 Platten mit je 64 Hemmhöfen messen. Eine Verdoppelung der Durchlaufgeschwindigkeit ist grundsätzlich möglich. Soll ein bestimmter Hemmhof nochmals gemessen werden oder befinden sich nur wenige Hemmhöfe auf der Platte, so können diese mit Hilfe der Tastatur direkt angesteuert werden.

Diskussion

Der HZAA-3 weist gegenüber dem zweiten Modell unseres Meßautomaten ein moderneres Design mit übersichtlicher Anordnung der Bedien- und Kontrollelemente auf.
Die Elektronik ist in den Tischeinschüben untergebracht und verbessert worden. Stromversorgungsgeräte und Anzeigeperipherie sind in das Gerät einbezogen.
Das bewährte optische System wurde beibehalten, im Detail aber verbessert. So wurden z. B. zur homogenen Ausleuchtung des Hemmhofes nicht mehr nur eine Glühlampe mit Kollimator, sondern 8 Leuchtstoffröhren mit parabolischen Reflektoren eingesetzt. Die Stromversorgung für die Leuchtstoffröhren weist Besonderheiten auf, die frühzeitige Alterung der Lichtquellen weitgehend vermeiden und stabile Betriebsbedingungen erwarten lassen.
Der mechanische Aufbau des Plattenwagens zeichnet sich durch eine höhere Stabilität und eine noch bessere Positioniergenauigkeit der Hemmhöfe aus. Der Plattendurchlauf wurde durch den automatischen Rücklauf in die Ausgangsstellung ergänzt. Gegenüber dem bisherigen Modell wurde der HZAA-3 selbst mit einer Digitalanzeige des Hemmhofdurchmessers ausgestattet, die es erlaubt, schon während der Messung die Meßwerte zu verfolgen. Zusätzlich wird links neben dem Meßwert auch die Position des betreffenden Hemmhofs durch Angabe der Spalte und Reihe des Ortes auf der Testplatte angezeigt.
Nach wie vor weist der HZAA-3 die schon beim HMA-2 vorhandenen Vorteile auf, z. B.

— hohe Meßgenauigkeit (besser als 0,1 mm),
— Bestimmung eines effektiven Durchmessers bei nicht völlig kreisförmigen oder anderweitig beeinträchtigten Hemmhöfen,
— objektive Messung durch Ermitteln der Hemmhofgrenze am 50%-Wert des Dichtegradienten zur Umgebung,
— dadurch Möglichkeit der Messung von Hemmzonen mit diffuser Begrenzung.

Eingegangen: 31. 7. 1985

Literatur

[1] SEYFARTH, H., EWALD, O. P.: Pharm. Ind. **34** (1972), 40.
[2] WALLHÄUSSER, K. H.: Pharm. Ind. **34** (1972), 127.
[3] BARTOSCHEWITSCH, Yu. E., EGOROV, A. A., GOLDSTEIN, V. L., JONES, I. J., LUNI, P. A.: Antibiotiki XXI, **2** (1976), 122.
[4] BARCLAY, G. A.: Abstracts of 2nd Int. Symp. on Rapid Methods and Automation on Microbiology, Cambridge 1976, 63.
[5] SHARPE, J. E., LIGHTBOWN, J. W., JONES, A.: Abstracts of 2nd Int. Symp. on Rapid Methods and Automation on Microbiology, Cambridge 1976, 63.
[6] MÜHLIG, P., BAUER, E., KNÖLL, H., BRADLER, G., SCHICHT, G., STEININGER, H., GÜNTHER, H.: Acta Biotechnol. **2** (1982) 2, 193.
[7] BRADLER, G., KNÖLL, H.: Zbl. Pharm. **113** (1974) 6, 563.
[8] KNÖLL, H., BRADLER, G., SCHICHT, G., FORBERG, W.: Acta Biotechnol. **1** (1981), 167.
[9] SCHICHT, G., STEININGER, H., GROSSE, P., KNÖLL, H.: Acta Biotechnol. **1** (1981), 300.

Acta Biotechnol. 7 (1987) 1, 87–92

Short Communications

Effects of Temperature on Microbial Ethanol Production
I. The Temperature Profile Curve of Ethanol Production of the Yeast Strain Saccharomyces cerevisiae Sc 5

RICHTER, K., BECKER, U.

Academy of Sciences of the G.D.R.
Institute of Biotechnology, Leipzig
Permoserstraße 15, Leipzig, 7050 DDR

Summary

The temperature-profile curve of ethanol production of the yeast *Saccharomyces cerevisiae* Sc 5 is shown. Accordingly the biokinetic sphere of ethanol formation consists of 5 ranges. The maximum specific ethanol formation rate ν_0 is reached within the temperature limits of $32\,°C \leqq T \leqq 36\,°C$, and the maximum temperature for ethanol formation amounts to $T''_{max} = 57\,°C$. Within the first suboptimum temperature range ethanol formation is not retarded thermally. Using a modified ARRHENIUS equation the activation energy of ethanol formation was calculated to be $\Delta H^{A}_{EtOH} = (78.5 \pm 2.2)$ kJ/mol.

Introduction

Microbial ethanol production as well as cell growth is always the result of sequences of biochemical processes catalyzed by enzymes, the reaction rates of which are significantly dependent on temperature. In principle, the fundamental relationships which are applied to the single steps in this connection also define the characteristics of the resulting metabolic activities of the microorganisms.
Regarding the temperature dependence of the specific growth rate of ethanol producing yeasts a great quantity of experimental data and theoretical considerations can be found in the literature [1–10]. Thus it is known that cell growth can only occur within a so-called biokinetic temperature range limited by a minimum and a maximum temperature for growth (T'_{min}, T'_{max}). The ARRHENIUS equation is valid for the suboptimum temperature range where an increase of temperature is followed by an exclusive activation of cell growth [11–13]:

$$\frac{d\ln \mu}{dT} = \frac{\Delta H^{A}}{RT^2}. \tag{1}$$

But within the superoptimum temperature range an increase of temperature always causes a decrease of the specific growth rate, since degradations and denaturation reactions dominate the reactions of synthesis on the molecular level. Additionally the death rate can increase, too. Such an associated thermal growth profile was found for *Saccharomyces cerevisiae* cells [1, 3, 4].
The following considerations are to give an answer to the question how the specific ethanol formation rate is affected by increasing temperatures.

Materials and Methods

The microorganism used was the yeast strain *Saccharomyces cerevisiae* Sc 5. The precultivation of cells and the discontinuous fermentation experiments were carried out in the same manner as described previously [14—16]. However, the CO_2 formed during the process was conveyed through an effective condenser system to avoid the escape of ethanol with the gas output, especially at higher temperatures.
The medium used for batch fermentation was a mineral salt medium enriched with yeast extract:

Sucrose:	100···200 g/l
yeast extract:	5 g/l
$(NH_4)_2SO_4$:	0.5 g/l
$MgSO_4 \cdot 7H_2O$:	0.4 g/l
K_2SO_4:	0.4 g/l
H_3PO_4 (85%):	0.4 ml/l
$ZnSO_4 \cdot 7H_2O$:	4.4 mg/l
$FeSO_4 \cdot 7H_2O$:	13.0 mg/l
$CuSO_4 \cdot 5H_2O$:	0.8 mg/l
$MnSO_4 \cdot 5H_2O$:	4.4 mg/l

The pH-value was adjusted and regulated by means of a diluted ammonia solution at pH = 4.5. The initial biomass concentration amounted to 5—20 g/l and the temperature was varied between 10 °C and 60 °C.
The biomass was determined gravimetrically and for ethanol estimation a gas chromatograph (GCH F 18.3, VEB Chromatron Berlin; 10% PEG on porolith, acetone as an internal standard, column length: 3 m) was used. The actual specific ethanol formation rate was calculated by means of the method presented in [14, 15]. Its maximum possible values ν_0' (referring to a determinate temperature in each case) could be obtained by extrapolation of the actual course of the function $\nu = f(t)_T$ or $\nu = f(P)_T$, to the point $P = 0$ using a regression computer program.

Results and Conclusions

The course of the function $\nu_0' = f(T)$ is shown in Fig. 1. Besides, some numerical values of it are presented in Table 1. It can be clearly noticed that the biokinetic sphere of ethanol formation is divided into 5 different ranges by 6 characteristic temperatures (Table 2).

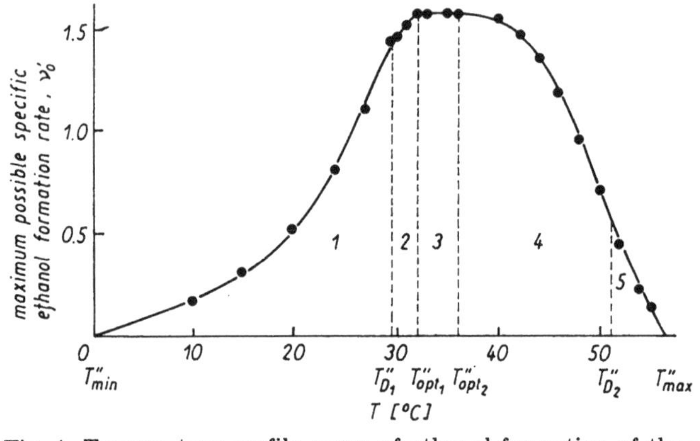

Fig. 1. Temperature profile curve of ethanol formation of the strain *Saccharomyces cerevisiae* Sc 5.

● ● experimentally found values
—— calculated curve

Table 1. The maximum possible values ν_0' of the specific ethanol formation rate of the strain *Saccharomyces cerevisiae* Sc 5 at different temperatures

T [°C]	ν_0' [g/g · h]	T [°C]	ν_0' [g/g · h]
10	0.17	36	1.58
15	0.30	40	1.55
20	0.52	42	1.48
24	0.81	44	1.36
27	1.11	46	1.18
29.5	1.435	48	0.96
30	1.465	50	0.71
31	1.525	52	0.45
32	1.58	54	0.22
33	1.58	55	0.14
35	1.58	57	0

Table 2. The different ranges of the temperature profile curve of ethanol formation (strain *Saccharomyces cerevisiae* Sc 5)

No.	range	limits	characteristics
1	range of thermally non-retarded ethanol formation in the suboptimum temperature interval	$T''_{\min} \leqq T < T''_{D_1}$ $0\,°C \leqq T < 29.5\,°C$	increase of temperature causes an activation of ethanol formation
2	transitional range within the suboptimum temperature interval	$T''_{D_1} \leqq T < T''_{\text{opt.}_1}$ $29.5\,°C \leqq T < 32\,°C$	the reversible deactivation of ethanol formation begins at T''_{D_1}, but the thermal activation still dominates
3	optimum temperature range of ethanol formation	$T''_{\text{opt}_1} \leqq T \leqq T''_{\text{opt}_2}$ $32\,°C \leqq T \leqq 36\,°C$	balance between thermal activation and reversible deactivation, the maximum value ν_0 can be reached by the cells
4	first range of thermal deactivation within the superoptimum temperature interval	$T''_{\text{opt}_2} < T < T''_{D_2}$ $36\,°C < T < 51\,°C$	an increase of temperature is followed by a decrease of the specific ethanol formation rate
5	second range of thermal deactivation within the superoptimum temperature interval	$T''_{D_2} \leqq T < T''_{\max}$ $51\,°C \leqq T < 57\,°C$	the irreversible deactivation of ethanol formation begins at T''_{D_2}, ethanol formation ceases completely at the temperature $T''_{\max}$.

The first suboptimum temperature range is characterized by thermally non-retarded ethanol formation. A marked increase in temperature causes an activation of glycolysis. Thus the ARRHENIUS equation (1) modified for the gross ethanol formation process is applicable in this case:

$$\frac{\mathrm{d}\ln \nu}{\mathrm{d}T} = \frac{\Delta H^{\mathrm{A}}_{\mathrm{EtOH}}}{RT^2} \quad [17]. \tag{2}$$

The factor ΔH^{A}_{EtOH} can be interpreted as the activation energy of ethanol formation. Using the integrated form of equation (2)

$$\Delta H^{A}_{EtOH} = \frac{R \ln (\nu_1/\nu_2)}{\left(\frac{1}{T^2} - \frac{1}{T_1}\right)} \quad (3)$$

the activation energy of ethanol formation was calculated from the experimentally found ν_0'-values. The absolute error could be evaluated by means of the equation

$$\Delta\Delta H^{A}_{EtOH} = \frac{R}{\left(\frac{1}{T_2} - \frac{1}{T_1}\right)} \cdot \left(\frac{\Delta\nu_1}{\nu_1} + \frac{\Delta\nu_2}{\nu_2}\right) \quad (4)$$

assuming an error of $\Delta\nu_1 = \Delta\nu_2 = \pm 0.005$ g/g · h for the specific ethanol formation rates ν_0'.

The results of this estimate are shown in Table 3. Accordingly the activation energy of ethanol formation amounts to $\Delta H^{A}_{EtOH} = (78.5 \pm 2.2)$ kJ/mol for the strain Sc 5.

In comparison with this finding the activation energy of ethanol formation was calculated in the same way for a few other strains on the basis of literature data. The corresponding logarithmic plots (Fig. 2) illustrate that the strain Sc 5 is comparable with other microorganisms in regard of the thermal activation of ethanol formation.

The presented temperature profile curve of ethanol formation shows a suboptimum transitional range, where alcohol formation is mainly determined by thermal activation, but also by the beginning thermal deactivation. The maximum specific ethanol formation rate is realized within the temperature limits of 32 °C $\leqq T \leqq$ 36 °C. Above this interval

Table 3. The activation energy of ethanol formation calculated from experimentally found ν_0' — values of the yeast strain *Saccharomyces cerevisiae* Sc 5

temperature intervall T_1 [K]	T_2 [K]	experimentally found ν_0'-values ν'_{01} [g/g · h]	ν'_{02} [g/g · h]	ΔH^{A}_{EtOH} calculated [J/mol]	$\Delta\Delta H^{A}_{EtOH}$ calculated [J/mol]
283.16	288.16	0.17	0.30	77109.52	6255.6
283.16	293.16	0.17	0.52	77208.76	2695.1
283.16	297.16	0.17	0.81	78061.89	1779.2
283.16	300.16	0.17	1.11	78040.21	1410.7
283.16	302.56	0.17	1.43	78238.85	1208.9
288.16	293.16	0.30	0.52	77311.51	3694.1
288.16	297.16	0.30	0.81	78617.15	1807.8
288.16	300.16	0.30	1.11	78451.29	1269.5
288.16	302.56	0.30	1.43	78657.84	1015.6
293.16	297.16	0.52	0.81	80300.17	2860.5
293.16	300.16	0.52	1.11	79299.33	1476.6
293.16	302.56	0.52	1.43	79409.79	1029.3
297.16	300.16	0.81	1.11	77933.02	2641.0
297.16	302.56	0.81	1.43	78729.09	1339.3
300.16	302.56	1.11	1.43	79742.26	2518.7
			average:	78474.04	2200.1

$\Delta H^{A}_{EtOH} = (78.5 \pm 2.2)$ kJ/mol

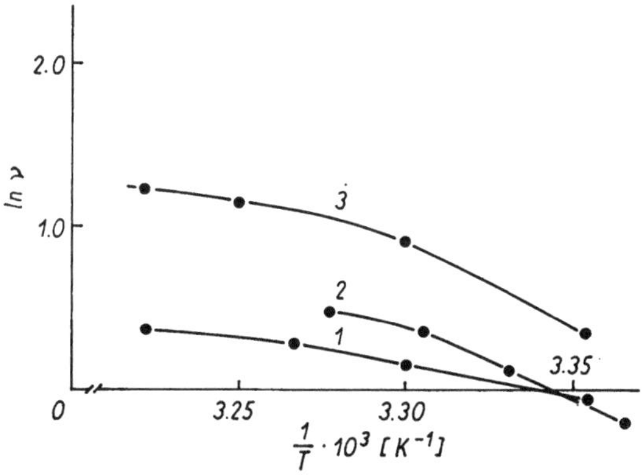

Fig. 2. The *Arrhenius* plots of different ethanol producing microorganisms.
1 — *Saccharomyces uvarum* AtCC 26602 [11]
2 — *Saccharomyces cerevisiae* Sc 5
3 — *Zymomonas mobilis* ZM1 [18].

the thermal deactivation of ethanol production dominates and consequently the specific ethanol formation rate decreases with increasing temperatures. When the maximum temperature T''_{max} is reached, the specific ethanol formation rate becomes nought. The superoptimum temperature interval consists of 2 ranges. In the first one the thermal deactivation of ethanol formation is reversible, but in the second range a further term of thermal deactivation supervenes, which seems to have an irreversible character.
A thermodynamic interpretation of these results will be attempted in our next paper.

Symbols

μ	specific growth rate	h^{-1}
ν	specific ethanol formation rate	g/g · h, mol/g · h
ν_0'	maximum possible specific ethanol formation rate obtained in the absence of ethanol at the temperature T	g/g · h, mol/g · h
$\Delta\nu_1$, $\Delta\nu_2$	error made in the estimation of ν_0'	g/g · h, mol/g · h
ν_0	maximum specific ethanol formation rate obtained in the absence of ethanol at the optimum temperature	g/g · h, mol/g · h
T	temperature	°C, K
T''_{min}	minimum temperature for ethanol formation	°C, K
T''_{D_1}	temperature at which the reversible deactivation of ethanol formation begins	°C, K
T''_{opt_1}	optimum temperature (lower limit of the optimum temperature range)	°C, K
T''_{opt_2}	optimum temperature (upper limit of the optimum temperature range	°C, K
T''_{D_2}	temperature at which the irreversible deactivation of ethanol formation begins	°C, K
T''_{max}	maximum temperature for ethanol formation	°C, K
P	ethanol concentration	g/l, mol/l
ΔH^{A}	activation energy of growth	kJ/mol
ΔH^{A}_{EtOH}	activation energy of ethanol formation	kJ/mol
$\Delta\Delta H^{A}_{EtOH}$	absolute error of ΔH^{A}_{EtOH}	kJ/mol
R	gas constant	8.3191 J/mol · deg.

Received December 3, 1985

References

[1] VAN UDEN, N., MADEIRA-LOPEZ, A.: Z. Allg. Mikrobiol. **10** (1970), 515.
[2] OLIVEIRO-BAPTISTA, A., VAN UDEN, N.: Z. Allg. Mikrobiol. **11** (1971), 59.
[3] VAN UDEN, N., MADEIRA-LOPEZ, A.: Arch. Microbiol. **104** (1975), 23.
[4] VAN UDEN, N., MADEIRA-LOPEZ, A.: Biotechnol. Bioeng., **18** (1976), 791.
[5] SIMOES-MENDES, B., MADEIRA-LOPEZ, A., VAN UDEN, N.: Z. Allg. Mikrobiol. **18** (1978), 275.
[6] WALTON, E. T., PRINGLE, I. R.: Arch. Microbiol. **124** (1980), 285.
[7] BROWN, S. W., OLIVER, S. G.: Biotechnol. Lett. **4** (1982), 262.
[8] TÖRÖK, T., REICHART, O.: Eur. J. Appl. Microbiol. Biotechnol. **17** (1983), 191.
[9] MAVRINA, L., POZMOGOVA, I., SCHADE, W.: Z. Allg. Mikrobiol. **23** (1983), 571.
[10] MAVRINA, L., POZMOGOVA, I., RABOTNOVA, I., SCHADE, W.: Z. Allgem. Mikrobiol., **23** (1983), 581.
[11] LEE, J. H., WILLIAMSON, D., ROGERS, P. L.: Biotechnol. Lett. **2** (1980), 141.
[12] VAN UDEN, N., DA CRUZ DUARTE, H.: Z. Allgem. Mikrobiol. **21** (1981), 743.
[13] CHISTYAKOVA, T. A., MINKEVICH, I. G., EROSHIN, V. K.: Eur. J. Appl. Microbiol. Biotechnol. **18** (1983), 25.
[14] RICHTER, K., BECKER, U.: 3rd Symposium of Socialist Countries on Biotechnology, Bratislava, 25.–29. 4. 1983.
[15] RICHTER, K., BECKER, U.: Acta Biotechnol. **5** (1985), 9.
[16] RICHTER, K., BECKER, U.: Acta Biotechnol. **5** (1985), 145.
[17] FRANZ, B.: Die Nahrung **5** (1961), 457.
[18] LEE, K. J., SKOTNICKI, M. L., TRIBE, D. E., ROGERS, P. L.: Biotechnol. Lett. **3** (1981), 291.

Acta Biotechnol. 7 (1987) 1, 93–96

Study on the Enzyme Hydrolysis of Waste Photoemulsions

KRYSTEVA, M. A.[1], PEEV, G. A.[2], YOTOVA, L. K.[1]

[1] Higher Institute of Chemical Technology
Department of Biotechnology
1156 Sofia, Bulgaria
[2] Higher Institute of Chemical Technology
Department of Chemical Engineering
1156 Sofia, Bulgaria

Summary

Conditions for silver recovery from waste photoemulsions by means of the enzymes trypsin and alkaline protease hydrolysis of the gelatine and the mechanism of process were investigated. The degree of silver regeneration is depending on time during the first minutes. Temperature has a decisive significance only during the initial minutes of the process.
The sedimentation of silver chloride from the enzyme-degraded photoemulsions depends strongly on the valency of the anions present in the solution.

The high price of silver justifies the interest in the methods for its regeneration from waste photoemulsions. This may be achieved by chemical or enzyme hydrolysis of the gelatine to which the silver halogenide is bound. Chemical hydrolysis involves most frequently heating at high temperature in the presence of acids. Therefore its substitution by enzyme treatment has been the subject of a number of studies [1, 2, 3]. Almost complete regeneration of the silver halogenide has been obtained when waste photoemulsions are treated with enzyme preparations at 45–50 °C for 1 to 3 h. In the case of alkaline protease use the enzyme hydrolysis has been followed by the reduction of pH to 4.0, inactivating by this way the enzyme and encouraging the settling of the silver particles, the separation of which occured during a period of 4–6 h [3].
The purpose of the present study is to find suitable conditions for silver regeneration from waste photoemulsions by means of enzyme hydrolysis of gelatine in them as well as to investigate the mechanism of processes taking place in the procedure.

Materials and Methods

The experiments were carried out on waste photoemulsion containing 17.8 g/kg silver chloride and 231 g/kg gelatine. The alkaline protease from *Bac. subtilis*, strain DY, trypsin from bovine pancreas obtained after the release of chymotrypsin and used without additional purification, are industrial products available in Bulgaria. The flocculating agent polyacrylamide was supplied from the Metallurgical Works in Kaluga (USSR).
The proteolytic activity of the enzymes used was determined by BERGMEYR's method [4], using caseine as substrate. Their esterase activity was monitored according to MARKWARDT and RICHTER [5] by means of N-benzoylarginine-p-nitroanilide · HCl. The silver content was determined titrimetrically by VOLGARD method [6]. The enzyme hydrolysis was performed in a vessel, 3 cm in diameter, at temperatures of 20 and 40 °C. The emulsion was vigorously stirred using an electro-

magnetic stirrer, magnet diameter 2 cm at 600 rpm; 4 ml of the emulsion. Were mixed with 4 ml 0.1 M buffer, pH 8.0 to which an appropriate amount of enzyme was added, depending on its proteolytic activity. This amount was 0.25 mg/ml for trypsin and 1 mg/ml for the alkaline protease. The hydrolysis was interrupted by adding 1 ml 0.1 M HCl in the case of kinetic measurements. The hydrolysates were filtered or centrifuged, and the amount of silver chloride in the sediments was determined after washing. Disc polyacrylamide electrophoresis was performed according to [7]. The electrode buffer used was with pH 8.3 and the gel concentration was 7.5 per cent. The rheological properties of the emulsion and of the hydrolysates were determined by means of a viscosimeter with coaxial cylinders "Rheotest RV-2".

Results and Discussion

The degree of silver regeneration depending on time, when the photoemulsions are treated with trypsin (A) and with alkaline protease (B) at 20 and 40 °C are demonstrated in Fig. 1. As it is to be seen from the figure, the rate of the process is very high and the main amount of silver is released in the very first minutes (82 per cent at 20 °C and 88 per cent at 40 °C in the case when trypsin was used). The last 10—15 per cent of the metal are released relatively more slowly, but at the end of the 60th min the silver is almost completely recovered. Temperature has a decisive significance only during the initial minutes of the process, but it has no essential effect on silver yields after 30th min. The results obtained using the two enzymes are similar.
In the course of the experiments it became apparent that the sedimentation of the silver chloride from the enzyme-degraded photoemulsions depends strongly on the valency

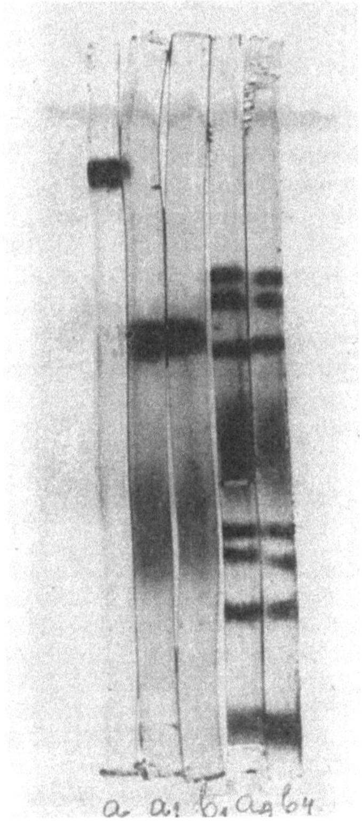

Fig. 2. Disc polyacrylamide electrophoresis of photoemulsion enzyme hydrolysates.
a — untreated gelatine, a_1 and a_4 — photoemulsions treated with trypsin for 1 and 4 h, respectively, at 20 °C, b_1 and b_4 — photoemulsions treated with alkaline protease for 1 and 4 h, respectively, at 20 °C

of the anions present in the solution. Quantitative release of silver chloride was obtained in the presence of phosphate and borate anions. However, there was no sedimentation of silver chloride when the enzyme hydrolysis was performed in the presence of nitrate, hydroxylic, bicarbonate and carbonate anions (Table 1). Addition of polyacrylamide to the hydrolysates obtained resulted in easily filtred AgCl sediment with quantitative yield.

In order to elucidate the mechanism of silver separation from the photoemulsions by means of enzyme hydrolysis, we applied disc polyacrylamide electrophoresis of the enzyme hydrolysates obtained after 1 and 4 h treatment with alkaline protease and trypsin. As is to be seen from Fig. 2, the enzyme hydrolysates obtained after one h treatment manifest a rather weak fragmentation of the gelatine molecule, whereas after 4 h the number of the fragments is considerably greater. There is almost complete accordance between the number and the mobility of the gelatine fragments, obtained by means of trypsin and alkaline protease.

Table 1. Effect of different anions on the amount of silver released after enzyme treatment of photoemulsions for 1 h at 20 °C

Enzyme used	Specific proteolytic activity [U/mg]	Specific esterase activity [U/mg]	Amount of silver released [%]			
			PO_4^{-3}	$B_3O_7^{-3}$	OH^-, NO_3^- HCO_3^- CO_3^{-2}	OH^-, NO_3^-, CO_3^{-2}, HCO_3^- with addition of polyacrylamide
Trypsin	2.0	919	98.64	98.10	no sediment	99.30
Alkaline protease	0.55	271	97.80	98.0	no sediment	99.10

The rheological study of the hydrolysates after one-hour treatment with alkaline protease has shown that they behave as Newtonian fluids with viscosity of 1.3×10^{-3} Pas at 40 °C and 1.95×10^{-3} Pas at 20 °C. In view of the thixotropic character of the initial emulsion and the decomposition taking place, it is possible to define reliably only the hydrodynamic conditions at the end of the process which correspond to turbulent mixing (REYNOLDS number at 20 and 40 °C — 2050 and 3080, respectively).

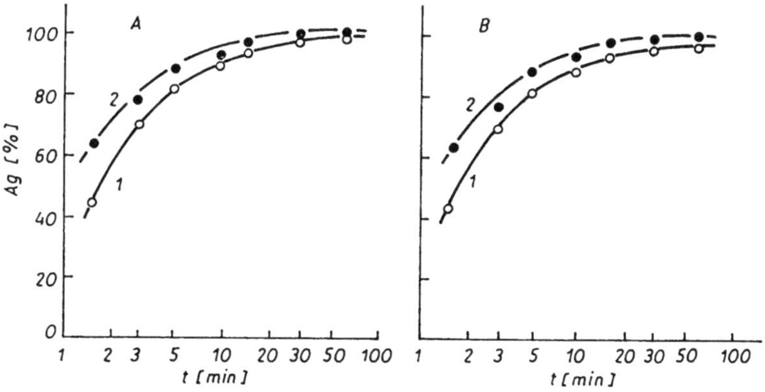

Fig. 1. Degree of silver recovery (per cent) in dependence on time at 20 °C (curves 1) and at 40 °C (curves 2).

A — photoemulsion treated with trypsin, *B* — photoemulsion treated with alkaline protease

The results obtained suggest that limited proteolysis of gelatine is sufficient for complete silver recovery from photoemulsion. However, the gelatine fragments obtained probably play the role of protective coloid and AgCl cannot be separated directly in the form of sediment. Its coagulation reguires the presence of higher members of the anionic series established [8], such as PO_4^{-3} and $B_3O_7^{-3}$, or of the polyelectrolyte polyacrylamide, used as flocculating agent in the purification of effluent waters [8].
The fact that alkaline protease from *Bac. subtilis*, strain DY, does not differ in its effect on the gelatine compared with trypsin gives grounds for its successful use on an industrial scale, because of its low price. On the one hand owing to the short duration of enzyme effect on the gelatine, the enzymes used do not undergo noticeable changes in their activity. This permits the process of silver regeneration to be carried out with recycling of the hydrolysates. On the other hand, the partially decomposed gelatine, after suitable ultrafiltration could possibly be used in the cosmetic industry.

Acknowledgement

The authors would like to express their gratitude to Dipl. Eng. E. Petrusenko for the kindly provided waste photoemulsion.

Received October 15, 1985

References

[1] Tsyperovich, A. S., Zlatopolski, A. I., Mishunin, I. F.: Inventor's certificate No. 240690 (1976). U.S.S.R.
[2] Mishunin, I. F., Tsyperovich, A. S.: Kino- und Fernsehtechnik (1977) 11, 50 (in Russian).
[3] Godfrey, T., Reichelt, J.: Industrial Enzymology. The Nature Press, 1983, 370.
[4] Bergmeyr, H. U.: Methoden der Enzymatischen Analyse, Bd. II. 1970, 974.
[5] Markwardt, M., Richter, M.: Die Pharmazie **24** (1969), 620.
[6] Volgard, J.: Pract. Chem. **117** (1874), 217.
[7] Davis, D.: Ann. N. Y. Acad. Sci. **121** (1964), 404.
[8] Proskuryakov, V. A., Schmidt, L.: Reinigung von Abwässern in der chemischen Industrie. Leningrad: Chimiya, 1977, 98 (in Russian).

Acta Biotechnol. 7 (1987) 1, 97—99

Ethanol Concentration Determination Using Solvent Extraction and Refractometry

LAZAROVA, G., GENOVA, L., KOSTOV, V.

Bulgarian Academy of Sciences
Institute of Microbiology
Sofia 1113, Bulgaria

Summary

The feasibility to avoid the distillation in the case of refractometric determination of the ethanol concentration in fermentation samples is examined. The results obtained show that the ethanol concentration in the range 0 to 100 $g \cdot l^{-1}$ can be easily determined by means of benzyl alcohol extraction and refractometry.

Despite the availability of several different methods for ethanol concentration determination (gas chromatography [1], enzyme [2—4] and chemical [5, 6], methods, the refractometry finds wide application to the control of alcohol production [7—9]. This method is used after the sample is distilled and the ethanol concentration is determined by measuring the distillate refraction.
Distillation is an obligatory step of treatment by most of the ethanol assaying methods and it is considerably time-consuming. VARMA et al. propose the use of an alternative ethanol separation method — solvent extraction, followed by gas chromatography analysis [10].
The objective of our investigation was by using solvent extraction in combination with refractometry to devise a rapid and easily applicable method for determination of ethanol concentration in fermentation broths.

Materials and Methods

Organic Solvents

The following solvents were used for the extraction: n-butanol, toluol, and benzyl alcohol, pfa. For the preparation of standard solutions absolute ethyl alcohol, pfa, was used.

Fermentable Sugars

In order to investigate the influence of the presented sugars on the results for the ethanol concentration in the samples, standard solutions of the following sugars were examined: glucose, lactose, sucrose, pfa. Solutions of substrates used in the industrial alcohol production (molasses with 46.7% fermentable sugar content and dried whey ultrafiltrate with 99% lactose content) were studied as well.

Sample Treatment

Two ml of the organic solvent were added to an equal volume of the sample. The samples were extracted 1 h at 30 °C using an rotary shaker with 300 rpm. Separation of the two layers was achieved by sample centrifugation (2500 rpm, 5 min). The position of the organic phase (up/down) depends on the solvent specific weight and on the sugar — ethanol concentration of the sample. 0.2 ml from the organic phase were used for the refractometry.

Refractometry

Throughout the experiments an ABBE refractometer, Zeiss, Jena, was used. It was thermostated at 30 °C.

Results and Discussion

The refractometry application to ethanol concentration determination after solvent extraction of the sample poses a specific requirement to the organic solvent used: its solutions with the component to be determined must posses a linear and steep enough relation between its refraction coefficient and the analysed component concentration, because the accuracy of the measurement is determined by the calibration line slope [11].

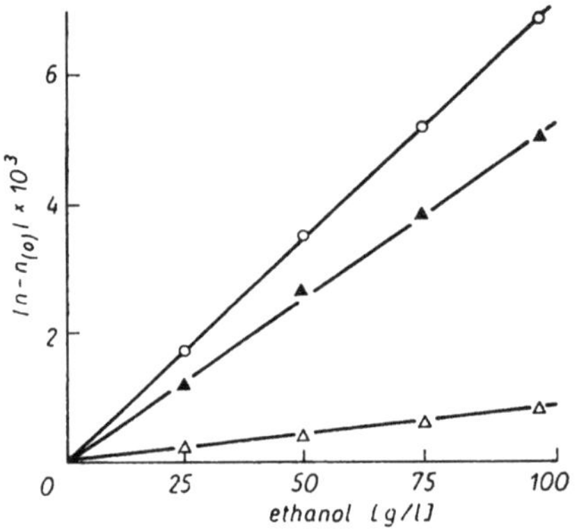

Fig. 1. Relationship between the ethanol concentration in the sample and the refraction coefficient of the distillate of the organic phase after extraction, respectively. $n_{(O)}$ denotes the refraction coefficient at nil ethanol concentration.
○ — distillate, n = 1.3316; ▲ — extract with butanol, n = 1.3874; △ — extract with toluol, n = 1.4913.

For the conventional ethanol assaying method refractometry is applied to a distillate, i.e. to a bicomponent system water — ethanol. That is why the results obtained by extraction with the different solvents have been compared with the refractometric data for a distillate, where the refraction coefficient plot vs. the ethanol concentration has a slope of 6.8×10^{-5} l · g^{-1} (Fig. 1).
Our investigations of the n-butanol applicability as an extractant in the combination extraction—refractometry in the case of ethanol concentration analysis have shown that it is unsuitable because the slope of the refraction coefficient plot vs. the initial sample ethanol concentration is low: 5.0×10^{-5} l · g^{-1} (Fig. 1).
The toluol proved to be unsuitable either, the value of the slope in this case is only 0.8×10^{-5} l · g^{-1} (Fig. 1).
The best results were obtained by applying the benzyl alcohol as the extractant. The plot of the organic phase refraction coefficient vs. the initial ethanol concentration in the water phase is shown in Fig. 2. The linear regression on analysis of the results obtained with standard ethanol—water solutions has revealed a line slope of 14.2×10^{-5} l · g^{-1} with a correlation coefficient 0.999. This slope is two times greater than that in the case of the distillate and the refractometric determination of the ethanol concentration

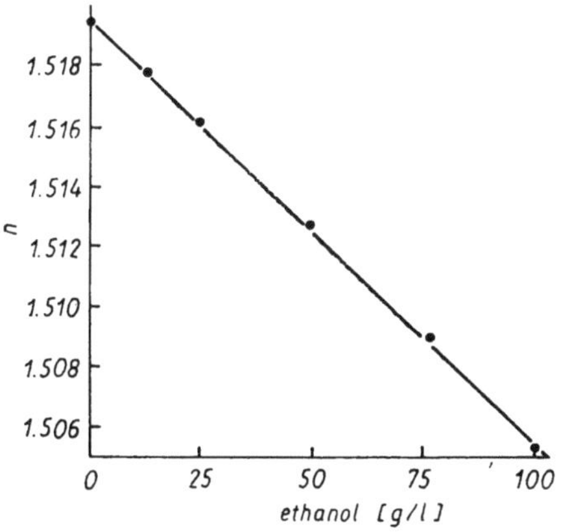

Fig. 2. Standard plot of the organic phase refraction coefficient versus the initial ethanol concentration in the water phase in the case of benzyl alcohol extraction.

after benzyl alcohol extraction proves to be more accurate in comparison with the distillation.

To check the influence of the sugar content in the samples on the refraction coefficient of the organic phase, experiments with water solutions of sugars (0 to 100 g $\cdot$ l^{-1}) and ethanol (0 to 50 g $\cdot$ l^{-1}) were carried out. The results obtained have shown that the sugar presence in the investigated concentration range does not effect the accuracy of the ethanol concentration determination, this being true for molasses in 240 g $\cdot$ l^{-1} concentration as well.

To evaluate the error of the method proposed, 3 separate sets of experiments with 6 replications were carried out with samples being taken from fermentation broths. The statistical analysis revealed a relative error of 1.86% with 0.36% standard deviation.

Conclusion

From the organic solvents tested the benzyl alcohol proved to be the most suitable as an ethanol extractant. The results presented show that the ethanol concentration in a fermentation sample can be easily determined by means of benzyl alcohol extraction and refractometry. The method is convenient particularly in the case when a great number of samples is to be analysed, because up to 50 samples can be measured in 2 hours.

Received February 26, 1985

References

[1] COMBERBACH, M. D., SCHARER, J. M., MOO-YOUNG, M.: Biotechnol. Lett. 6 (1984) 2, 91.
[2] BERNET, E., GUTMAN, I. — In: Methods of enzymatic analysis. Vol. 3. Ed.: BERGMEYER, H. U. New York: Academic Press, Inc., 1985, p. 1499.
[3] MASON, M.: J. Am. Soc. Brew. Chem. 40 (1982) 2, 78.
[4] VERDUYN, C., Van DIJKEN, J. P., SCHEFFERS, W. A.: Biotechnol. Bioeng. 25 (1983) 4, 1049.
[5] NEISH, A. C. — In: Methods in Enzymology, Vol. 3. Eds.: COLOWICK, S. P., KAPLAN, N. O. New York: Academic Press, Inc., 1957, p. 255.
[6] CHEN, S. L., GUTAMANIS, F.: Biotechnol. Bioeng. 18 (1976), 1455.
[7] Sovjet Standard GOST 12787-81, from: Food Sci. Technol. Abstr. 16 (1984) 2, U76.
[8] РУХЛЯДЕВА А. П., ФИЛАТОВА Т. Г., ЧЕРЕДНИЧЕНА В. С.: Справочник для работников лаборатории спиртовых заводов. Москва: Пищ. Пром. 1979, 232.
[9] РУХЛЯДЕВА А. П.: Техно- химический контроль спиртового производства. Москва: Пищ. Пром. 1974, 355.
[10] VARMA, R., SAVANT, U. D., KARANTH, N. G.: Enzyme Microb. Technol. 6 (1984) 5, 233.
[11] ЙОФФЕ В. В.: Рефрактометрические методы химии. Ленинград: Химия, 1983, 349.

Acta Biotechnol. 7 (1987) 1, 100

Book Review

Umwelt-Mikrobiologie
Mikrobiologie des Umweltschutzes und der Umweltgestaltung

Berlin: Akademie-Verlag, 1985 (Reihe WTB, Band 290). 70 Abb., 4 Tabellen, 12,50 M

Es ist ein Anliegen dieses Buches, allen Vertretern wissenschaftlicher und technischer Disziplinen, die auf dem Gebiet des Umweltschutzes tätig sind, ein kurzgefaßtes Lehr- und auch Nachschlagewerk zu sein. Eine solche präzise Darstellung des Themengebietes ist zur Zeit nicht zu finden. Im ersten Teil wird die Komplexität der Umweltprobleme dargestellt und auf die Möglichkeiten der Mikrobiologie zur Umweltgestaltung eingegangen. Es werden einprägsame Übersichten und Schemata verwendet und wichtige Grundbegriffe erläutert. Dieser Teil könnte durchaus eigenständig als Kurzfassung stehen; er dient hier der Einordnung der nachfolgenden Detailausführungen.
Der zweite Abschnitt beschäftigt sich mit den Grundlagen der Mikrobenökologie, bringt prägnante Begriffserklärungen und bereitet zielgerichtet das Verständnis der weiteren Kapitel vor. Nach Abschnitten, die sich mit dem aeroben und anaeroben Abbau von Naturstoffen beschäftigen, folgt die ausführliche Behandlung des Fremdstoffmetabolismus. Erklärungen und Zahlenuntersetzungen der mikrobiellen Beteiligung an N-P-S-Kreisläufen unterstützen die schon vorher erfolgte Beschreibung des Kohlenstoffkreislaufes im globalen Maßstab.
Im dritten Komplex werden Anwendungsgebiete beschrieben. In kurzer Form sind die Prinzipien der areoben und der anaeroben Abwasserreinigung dargestellt, günstige Verfahrensvarianten und Kombinationen für Spezielfälle erwähnt worden. Die mikrobielle Stickstoff- und Phosphorelimination ist ebenso zu finden wie die neueren Verfahren oder Verfahrensvorschläge der mikrobiologischen Abluftreinigung oder auch die Kohleentschwefelung. Weitere Kapitel sind der mikrobiellen Abproduktnutzung (Eiweißsynthese, Biogasbildung, Metallanreicherung) gewidmet, den Abschluß bilden die Behandlung mikrobiell abbaubarer Pflanzenschutzmittel und die Darstellung der Möglichkeiten der mikrobiellen Bodenfruchtbarkeitssteigerung.
Nur wenige sich noch in der Entwicklung befindlichen Teile des Gesamtgebietes werden nicht behandelt oder erwähnt.
Es wäre günstig, wenn die noch wichtige mikrobielle Bautenkorrosion hätte zum Beispiel behandelt werden und auf Möglichkeiten zur Minderung der Bautenschäden eingegangen werden können.
Die ,,Umwelt-Mikrobiologie" wird einen wichtigen Platz nicht nur in der Lehre, sondern auch als orientierendes Nachschlagewerk einnehmen.

U. Stottmeister, P. Kuschk

Acta Biotechnologica

Volume 7 1987 Number 1

Contents

Short Communications

Acta Biotechnologica is indexed in Current Contents/ET & AT;
Chemical Abstracts; Biological Abstracts; Biotechnology Abstracts; Excerpta Medica

Bei Fragen zur Produktsicherheit wenden Sie sich bitte an:
If you have any questions regarding product safety,
please contact:

Walter de Gruyter GmbH
Genthiner Straße 13
10785 Berlin
productsafety@degruyterbrill.com

Sandhya K. Balaram • Levi Bassin

Cardiac Surgical Complications

Strategic Analysis and Clinical Review

Springer

Sandhya K. Balaram
Department of Cardiothoracic Surgery
New York-Presbyterian
Brooklyn Methodist Hospital
Weill Cornell Medicine
New York, NY, USA

Levi Bassin
Royal North Shore Hospital
St. Leonards, NSW, Australia

ISBN 978-3-030-71562-5 ISBN 978-3-030-71563-2 (eBook)
https://doi.org/10.1007/978-3-030-71563-2

This Springer imprint is published by the registered company Springer Nature Switzerland AG
The registered company address is: Gewerbestrasse 11, 6330 Cham, Switzerland

Dedicated to those who trust us with their lives -
our families
our patients

Preface

Countless amounts of vital information and advice are transmitted between experienced cardiac surgeons and their residents during the thousands of hours they spend together in the training environment. Given the importance of surgical outcomes, many of these conversations revolve around complications. Pointed discussions of both potential and actual complications are shared in the operating room, on the ward, and in the clinics. However, it is often the intensity of discussion, exchanged during scheduled Morbidity and Mortality conferences, that is ultimately remembered and internalized. This education, of course, does not end with the completion of residency and subsequent board certification.

It is well understood that decreased external distractions, balanced decision-making, and good communication help in the prevention of medical errors. Human factors play a key role in surgical outcomes, and a deeper understanding of prevention and treatment strategies helps avoid future catastrophes. It is frequently said that if a surgeon operates on a large number of patients throughout his or her career, nearly every complication will be experienced firsthand. This is an overwhelming thought for a young surgeon, particularly as decreased case volume and work hours act in tandem to diminish overall operative experience in training programs. In addition, not all cardiac surgery occurs in large academic centers, where options for 24-h a day intensive care specialists and immediate mechanical ventricular support are available. Prevention, particularly in these environments with few bailout options, is mandatory for good outcomes.

The purpose of this book is to provide structured knowledge regarding recognized dangers and pitfalls in the field of cardiac surgery. A sharp and prepared mind must accompany technical skill. The book takes a "premortem" approach: a focus on complications in cardiac surgery as a method of surgical preparation. Detailed, relevant, and memorable narratives provide scenarios that may perhaps replace experienced events, particularly for those just starting a career in cardiac surgery. Dramatic crises, survival checklists, technical pearls, basic science fundamentals, and strategic algorithms help firmly establish key surgical concepts. It is likely that young surgeons working with experienced colleagues have already heard some variation of these scenarios.

Cardiac surgery requires constant strategic thinking in a mentally and physically challenging environment. Reviewing complications to understand their prevention and treatment is a method of harnessing the experience of others. In addition to the technical and nontechnical elements of actual surgical procedures, human factors and time-sensitive decision-making play a significant role in outcomes. This is a specialty that strives for perfection with data-driven focus. However, there remain certain elements over which there is little control. Crisis scenarios in cardiac surgery have benefitted from the development of checklists and algorithms to create systematic treatment of several possible events. For a surgeon, crisis prevention with similar models is helpful, particularly in settings where limited backup assistance is available.

Published data tells us up to one-third of deaths after coronary surgery are potentially preventable. More importantly, the majority of events that occur during an operation may often be repaired without any noticeable effect on the patient. This critical information suggests that a resilient surgeon who is knowledgeable and flexible could carry on with a successful operation even after an initial unexpected event. A classic tenet of heart surgery, of which there are many, is that it is not the first unexpected event that harms the patient. Rather, harm often lies in the repair of injury, or the result of that repair, or even a further, more far-reaching result that occurs in the postoperative period. This cause-consequence is a key part of failure analysis. The rebalancing of the boat in a critical surgical situation is either lifesaving or devastating. Careful decision-making after early recognition of an error results in improved outcomes; every decision thereafter must be aimed at increasing this chance. Cardiac surgical decisions are made under strict time constraints and with serious consequences, both immediate and long term. It is essential to keep unexpected events in perspective.

High risk, coupled with life-protecting high stress, have led to comparisons between cardiac surgery and the airline industry. Both are fields in which unpredictable human error occur. Another similarity to high-consequence fields is that the familiarity of the same team working repetitively towards the same goal improves outcomes by both reducing teamwork failure and technical errors. Multiple complex factors contribute to the work itself, and analysis has led to the creation of models to improve patient safety. Furthermore, in the modern era, extensive data collection has been used to identify both surgeons and systems at risk.

Cardiac surgery has traditionally been compared to chess with its need for planned stepwise strategy, calm reactivity, constant reassessment, and time-sensitive decision-making. Similar to chess, all the information or rules are theoretically known prior to proceeding with an operation. Every move *should* yield a certain result. Sometimes subtle information about the patient, their hemodynamics, and the environment are being processed while the surgeon is deciding on the next move. These mental gymnastics occur whether a procedure is 30 min or 12 h long, and certainly account for the more mental than physical fortitude that is required. Given this variability, it is helpful to invoke the dictum of "playing a bad hand well." This concept exemplifies much of modern cardiac surgery, in which an intrinsically complex patient population with multiple comorbidities undergoes treatment in a complex environment. It is imperative to focus energy on careful preparation and smooth execution of operations, all the while preparing for unexpected issues that occur in real time.

Suboptimal outcomes affect more than just the patient and their families, as cardiac surgeons who have sat through decades of arduous morbidity and mortality conferences will attest. The truth is, many of these conferences are not about which needle to use, the size of stitch to place and where, or other mundane technical aspects of surgery. Rather, it is often judgment and decision-making that comprise the forefront of the discussion. Medical errors may be due to misinterpretation of information, poorly structured planning, or improper choices. Knowledge or system failures in the medical environment are not always individual issues, but team based. These challenging surgical situations, which are composed of constant anxiety and stress, are aided by an intuitive knowledge that promotes reasoning and minimal emotion with every decision made.

As a profession constantly striving towards perfection, cardiac surgery requires experience accompanied by maturity. To that end, crafting calm responses to events in the operating room is a critical skill. Given the high stakes involved, this can be very difficult at times. There is much to be learned; the simple act of thorough case review by an individual or group builds experience, knowledge, and intuition, even for completed successful cases. Formal thinking through each step of an operation leads to constant refinement and improvement.

Surgery is a focused exercise at its core, the goal of which is the right operation at the right time and in the right manner. Solid thought processes, repetitive safe actions, standardized reactions, and constant information gathering, all require attention to detail. The surgeon's life revolves around prevention, prompt detection, and management of recovery of unexpected crises. All of these may occur normally during a single operation. The creation of a habit of considering and preparing for complications will ultimately improve both decision-making and outcomes.

New York, NY, USA Sandhya K. Balaram
St. Leonards, NSW, Australia Levi Bassin

Acknowledgment

We would like to express our gratitude and acknowledgment to Dr. Paul Cullen, an advanced cardiac surgery fellow at Royal North Shore Hospital in Australia, who kindly assisted with the writing and preparation of the minimally invasive cardiac surgery chapters

About this Book

The ancient Stoic idea of *premeditatio malorum,* clearly visualizing what can go wrong in order to prevent that particular outcome, is an effective technique in preparing for cardiac surgery. There are an endless number of harmful events that have the potential occur in these settings, at virtually any large or small step of the operation. Even the initial skin incision can result in harm to a patient. Negative visualization of potential intraoperative problems prior to a specific procedure helps ensure that each step is safe and carefully considered. Sound decision-making and anticipation of outcomes are vital concepts in surgery.

Unexpected critical events disrupt the flow of an operation and cost time, sometimes vital time, during ischemic arrest. Psychological preparation for rare events such as air embolism or postoperative myocardial infarction is lifesaving in cardiac surgery. Education and experience allow for preparation and help avoid reactive decision-making in these situations.

This book is divided into chapters which emphasize the multifaceted nature of this specialty, as well as the impact of cardiac surgery on all organ systems. Some of these complications are extremely rare, and others are more common. The narrated sequence of case events presented are not actual but are realistic. The scenarios are written in a manner to help focus the surgery team of when to be aware, when to be cautious, and when to be substantially prepared for problems. The formal reviews of each complication also describe possible prevention and treatment. Included in each section are basic science, useful strategies, structured plans, and immediate actions. This book is not just a list of potential complications, it is an alternative way of thinking about cardiac surgical procedures that will help to improve outcomes.

This text is neither bedside manual nor formal textbook; it is based on the idea that the anticipation of a bad outcome can lead to a good outcome.

The scenarios and format are inspired by *Near Misses in Cardiac Surgery,* published in 1992 by Myles Edwin Lee (Butterworth-Heinemann, Newton MA). Although the narrative surgical scenarios are not similarly written in second person (nor with the same dramatic flair), it is our goal that they are current, relevant, meaningful, and memorable in expanding the reader's experience. This is a simple guide to prepare for war in times of peace, aim for excellence, and create connections and conclusions that are critical in cardiac surgery. We consider the book as a starting point in the development of technical and mental patterns that promote successful surgery in a complex environment.

Contents

About the Authors

Sandhya K. Balaram is the Chief of the Department of Cardiothoracic Surgery at New York-Presbyterian Brooklyn Methodist Hospital in New York City. She is an Associate Professor of Clinical Cardiothoracic Surgery at Weill Cornell Medicine. Dr. Balaram earned her PhD at Creighton University studying the molecular biology of atherosclerotic disease during her general surgical training. She completed her cardiothoracic surgery fellowship at the University of Wisconsin-Madison and became a diplomate of the American Board of Thoracic Surgery in 2004. Prior to her current position, she spent many years as the Director of Clinical Quality and Performance at Mount Sinai St. Luke's Hospital in NYC. A member of the AATS and STS, Dr. Balaram has dedicated the last 20 years to the cardiac surgical care of underserved patients, many of whom have had little previous access to consistent medical care. With a strong focus on surgical outcomes, her program has earned the distinguished three-star rating from the Society of Thoracic Surgery for excellence in patient care in coronary artery bypass grafting.

Levi Bassin is a cardiothoracic surgeon in Sydney Australia base at Royal North Shore Hospital. His practice focuses on robotic and minimally invasive valve and coronary surgery. He trained in Australia and followed with a fellowship at the Brigham and Women's Hospital in Boston, MA. He has a strong interest in improving safety in surgery and draws parallels with his passion of flying planes.

Abbreviations

ABG	Arterial blood gas
ACT	Activated clotting time
AMI	Acute myocardial infarction
APTT	Activated partial thromboplastin clotting time
ARDS	Acute respiratory distress syndrome
ASA	Aspirin
AT3	Antithrombin III
AVF	Arteriovenous fistula
BIMA	Bilateral internal mammary artery
BP	Blood pressure
CABG	Coronary artery bypass graft
CAD	Coronary artery disease
CI	Cardiac index
CO	Cardiac output
CO2	Carbon dioxide
CPB	Cardiopulmonary bypass
CTA	Computerized tomography angiogram
CVA	Cerebrovascular accident
CVP	Central venous pressure
Cx	Circumflex artery
CXR	Chest X-ray
DIC	Disseminated intravascular coagulopathy
DVT	Deep venous thrombosis
DSA	Digital subtraction angiography
EABO	Endo aortic balloon occlusion
EF	Ejection fraction
ETCO2	End tidal carbon dioxide
EVH	Endovascular vein harvesting
IABP	Intra-aortic balloon pump
ICU	Intensive care unit
IMA	Internal mammary artery
IMV	Internal mammary vein
LCA	Left circumflex artery
LIMA	Left internal mammary artery
LITA	Left internal thoracic artery
LM	Left main
LMWH	Low molecular weight heparin
LN	Lymph node
LPA	Left pulmonary artery
LSVC	Left superior vena cava
LUL	Left upper lobe
LV	Left ventricle

LVAD	Left ventricular assist device
MIDCAB	Mini thoracotomy coronary artery bypass
MR	Mitral regurgitation
MRI	Magnetic resonance imaging
MS	Mitral stenosis
MVR	Mitral valve replacement
NIRS	Near-infrared spectroscopy
NO	Nitric oxide
NSTEMI	Non-ST elevation myocardial infarction
OM	Obtuse marginal artery
OPCAB	Off pump coronary artery bypass
OR	Operating room
PA	Pulmonary artery
PAP	Pulmonary artery pressure
PAWP	Pulmonary artery wedge pressure
PCI	Percutaneous coronary intervention
PDA	Posterior descending artery
PE	Pulmonary embolism
PEEP	Peak end expiratory pressure
PH	Pulmonary hypertension
PPD	Packs per day
PVL	Paravalvular leak
PVR	Pulmonary vascular resistance
RA	Right atrium
RBC	Red blood cell
RCA	Right coronary artery
RCC	Right coronary cusp
RIMA	Right internal mammary artery
RITA	Right internal thoracic artery
RLN	Recurrent laryngeal nerve
RSVC	Right superior vena cava
RV	Right ventricle
RVAD	Right ventricular assist device
RVOT	Right ventricular outflow tract
SAM	Systolic anterior motion
SFA	Superficial femoral artery
SFJ	Sapheno-femoral junction
SGC	Swan-Ganz catheter
SIRS	Systemic inflammatory response syndrome
STEMI	ST elevation myocardial infarction
SVC	Superior vena cava
SVG	Saphenous vein graft
SVR	Systemic vascular resistance
TAVR/TAVI	Transcatheter aortic valve replacement/insertion
TEE	Transesophageal echocardiogram
TIA	Transient ischemic attack
TTE	Transthoracic echocardiogram
UFH	Unfractionated heparin
VAD	Ventricular assist device
VC	Vocal cord
VF	Ventricular fibrillation
VSD	Ventricular septal defect
ViV	Valve in valve

Cardiac Access

1

1.1 Introduction to Median Sternotomy

First proposed by Milton in 1897, the median sternotomy incision has become the standard exposure used in cardiac surgery since its reintroduction by Julian in 1957 [1]. The incision itself is simply the beginning of what will likely result in many hours of open surgery. Yet as such, even this most initial step has the potential to harm.

Prior to incision, the midline of the sternum is identified from the center of the sternal notch to the midline of the xiphoid process. The skin incision is made with the scalpel and then extended down to the fascia, usually with electrocautery. The muscle bellies of the pectoralis major may be identified in the midline. The next step is confirmation of the midline at the periosteal level, usually through palpation of symmetric intercostal and sternocostal spaces.

The sternoclavicular ligament is dissected from the undersurface of the sternum with electrocautery. Sufficient clearance is obtained to allow for the sternal saw. One must stay very close to the sternal notch when dissecting in this zone as the trachea and innominate vessels can be very close. Occasionally, the sternum and surrounding tissues are calcified. The pericardium is sometimes adherent to the underlying sternum. The entry site for the saw must be confirmed clear using finger palpation of the area underneath the sternal notch.

Next, the xiphoid is identified clearly, and the linea alba is marked with electrocautery. A commonly encountered bridging vein seen crossing the xiphoid is cauterized. Some surgeons resect the xiphoid process, while others simply divide it. To reduce the size of the skin incision, subcutaneous flaps are made superiorly and/or inferiorly and retracted for the period during which the saw is used. With ventilation held, an electric saw is used to divide the sternum starting from the top or the bottom, depending on surgeon preference.

1.2 Paramedian Sternotomy

After several weeks of observation and preparation, the new resident was ready to perform a median sternotomy. The patient was an elderly male with severe COPD, morbid obesity, and a barrel-shaped chest. The midline was difficult to ascertain. The sound of the saw motor is soon accompanied by a repetitive clicking sound, indicating division along the lateral ribs, rather than the sternal bone. A classic railroad sternotomy had been performed, far off the midline to the right, and uncontrolled bleeding ensued.

1.2.1 Discussion

Determining and then following the exact midline is the key to performing a good median sternotomy incision. Difficulty with this can herald problems later in an operation, in terms of both exposure and safe closure. The quality of bone and width of the sternum varies greatly in patients based on size, gender, genetics, and associated medical conditions. However, once problems occur, additional decisions will need to be made.

Thus, preventing a paramedian sternotomy is the first step. Opening off the midline affects both the strength of the closure and the blood supply necessary for healing at the periosteal level. The midline is usually identified and confirmed through palpation of the outer sternal bone edges at the second intercostal space over the pectoralis major muscles. Palpation can be difficult in both morbidly obese patients,with abundant subcutaneous tissue, and muscular patients, with muscle fibers crossing the midline. Electrocautery is used to dissect down to the bone itself, and the midline is marked prior to utilizing the saw. The linea alba in its inferior position is a clear marker and one that should be carefully identified. After sternotomy, the quality of the entry is checked by palpation of the underside of both sternal edges. Despite best efforts, some have reported an incidence of 9.3% of paramedian sternotomy, a number not

S. K. Balaram, L. Bassin, *Cardiac Surgical Complications*, https://doi.org/10.1007/978-3-030-71563-2_1

shown to be altered by surgeon experience [2]. Some authors have described employing a limited muscle flap developed at the second intercostal space in order to more clearly identify the exact midline [3].

The next step must be to cauterize any active sites of bleeding, but not without being circumspect, as bone can become devascularized at the periosteum with excessive cautery. Specific groups deserve special attention with regard to the healing of the sternal incision. These include those patients who are female, obese, or diabetic. Poor bone quality due to chronic steroid use, COPD, or osteoporosis also affects healing. Other patients who require more attention are those with bilateral internal mammary artery (IMA) harvest, those with chronic lung disease/cough, and those patients with sleep apnea who require positive pressure ventilation in the postoperative period.

High-risk patients are considered for alternative closing techniques rather than simple interrupted wire placement. If a sternal fracture is present, or if there is a paramedian incision, a reinforced closure is beneficial. This often involves the placement of multiple figure-of-8 wires or double wires. The Robicsek closure, first introduced in 1977, is commonly taught and used in practice for a paramedian entry (Fig. 1.1) [4]. A modified technique has been described for closure when multiple sternal fractures are present (Fig. 1.2) [5].

Hundreds of studies have been performed to try and determine the optimal sternal closure. Many of these studies are retrospective and include a degree of bias. A recent meta-analysis showed no significant difference in the use of standard wire closure for the majority of patients [6]. However, for those patients with greater than two risk factors (including multiple fractures), the technique of closure was found to be clinically relevant [6]. The use of plates and screws, polymer cables, and steel bands has been described and has been shown to be of benefit to specific patients [7].

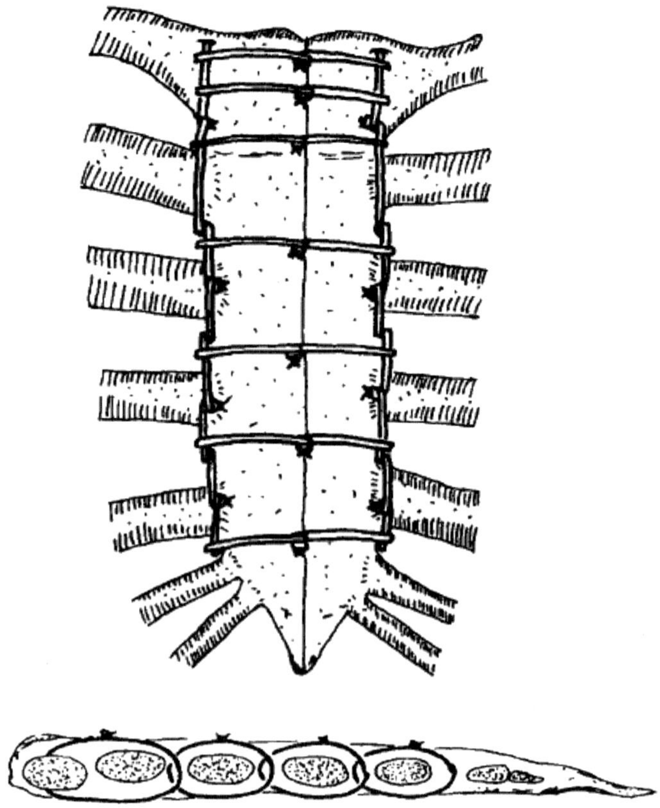

Fig. 1.2 Modified Robicsek closure. A parasternal chain is created with four interlocking wires on both sides of the sternum. This is followed by transverse interrupted peristernal wires to include the interlocking wires and provide lateral reinforcement. From "Modified Robicsek Technique for Complicated Sternal Closure," by Tavilla et al., 1991, *Ann Thorac Surg*, *Vol 52*, p. 1180. Copyright 1991by Elsevier Science & Technology Journals. Reprinted with permission

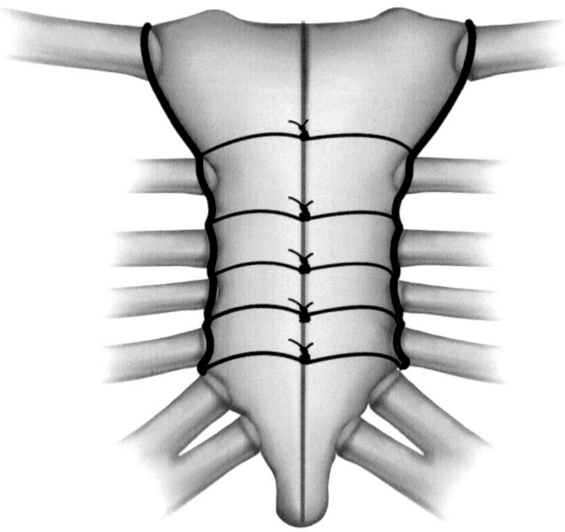

Fig. 1.1 Robicsek closure. Continuous parasternal wires placed bilaterally begin in a caudal direction and return in a cranial direction to encircle each rib. This is followed by standard simple interrupted parasternal wire closure

1.3 Tracheal Laceration

A simple aortic valve replacement with a pericardial prosthesis is planned for a 65-year-old emphysema patient with diabetes. During dissection around the sternal notch, the strap muscles are noted to be very thin. The electrocautery is reduced. While clearing the tissue beneath a very adherent and calcified sternal notch, the sound of a continuous cuff leak is suddenly audible. The anesthesiologist checks the tube as the surgeon observes air bubbles in the wound. A request for reinflation of the balloon is made just as the anesthesiologist reports that he is losing tidal volume.

1.3.1 Discussion

Intraoperative tracheal injury can occur during any thoracic surgery. It is more commonly seen with esophageal surgery, due to the anatomic location and proximity of these two structures. Any anterior chest exposures also place the trachea at risk. Careful attention to the device temperature and use of the electrocautery tip are paramount, particularly around the area of the protective strap muscles. The "spray mode" should not be used when dissecting near the manubrium, based on the vital structures present both superiorly and inferiorly. Prevention of this injury includes a preoperative non contrast chest CT that reveals the distance from the posterior surface of the sternum to the anterior mediastinal structures.

Most of the literature on direct tracheal injury during sternotomy are from isolated case reports [8]. Small, superficial injuries to the anterior trachea often heal without direct repair. A true full-thickness injury, however, must be identified and controlled.

The critical step in a large injury is airway control. If the cuff is injured at the time of tracheal injury, the tube must be changed and advanced past the site of injury to maintain adequate ventilation and allow time for repair [9].

For a small injury to the anterior trachea, early diagnosis is mandatory. Cuff leaks or air bubbles around the site of the trachea are clues that there is an injury. Actual volume loss during ventilation would signify a larger injury.

The principles of repair are similar to primary tracheal surgery. Superficial injuries are treated with conservative management, observation, and broad-spectrum antibiotics. If a gross defect is present, interrupted sutures of 4-0 Vicryl or 4-0 PDS are placed to repair the defect primarily, after the injury is carefully exposed. Only the defect should be exposed, if possible, to avoid interfering with the blood supply to the tracheal tissue laterally. Buttressing the repair with tissue such as strap muscle or mediastinal fat also helps cover over the defect and allow for improved healing. The endotracheal tube should be gently moved to make sure that it is not involved with the repair suture line prior to final closure. Completion bronchoscopy is helpful to examine the repair of large defects.

1.4 Carotid Artery Injury

The resident develops significant bleeding at the suprasternal notch during initial dissection, prior to sternotomy, and is holding finger pressure at the site. He calls the attending to the room. He has tried electrocautery and multiple clips to stop the bleeding, without success. He states he thinks the source is the innominate vein, but is unable to gain control without losing both a large amount of blood and visualization.

He removes his finger momentarily, and copious bleeding is confirmed. The bleeding is diffuse and difficult to control, with no sign of the injured vessel itself. The area is poorly exposed as the sternum remains closed at this point in the operation. The limited surgical site makes it difficult to determine if bleeding is arising from an artery or vein.

1.4.1 Discussion

Arterial or venous injury can occur during dissection around the sternal notch, even prior to sternotomy. The first step is to keep pressure on the site and determine whether the site of injury is amenable to repair with the available exposure. A full sternotomy is often required to improve exposure. Difficulty arises with finger pressure during use of the saw, and a sponge fashioned into a tight square may be held firmly on the site as the sternum is opened.

Once sternotomy is complete, digital pressure is resumed initially to control the bleeding. Arterial bleeding is easier to control with this method than venous bleeding. The carotid arteries, the aortic arch, and the innominate vein are all closely located here at the small area of the suprasternal notch. The innominate vein, if transected, sometimes retracts out of sight upon sternal spreading.

At this point in the operation, and throughout the duration of this process, the anesthesia team should be carefully resuscitating the patient. Good communication of ongoing blood loss and careful manual control are key. Patients who present for cardiac surgery, particularly for ischemic disease, are at significant risk from acute anemia and hypotension.

As with any uncontrolled bleeding, the safest next step is to obtain proximal and distal control of the injured vessel. In this case of carotid artery injury, the carotid sheath must be exposed and entered. To expose the upper portion of the aortic arch, innominate artery, and carotid artery, the pericardium should be opened up to the innominate vein. If needed for exposure for a high neck injury, this vein is typically encircled for retraction inferiorly. Vessel loops are then placed around the base of the innominate artery. If the injury is at the base of the innominate artery, cardiopulmonary bypass or even circulatory arrest is required for safe repair. Dissection further up the neck into the carotid sheath will help confirm an injury to the right carotid artery. Umbilical tapes are used to obtain proximal and distal control. Administration of heparin should be followed by clamping or snaring of the vessel to allow for direct repair.

An injury to the innominate vein may also occur at the time of sternotomy. Once controlled, if this injury cannot be fixed primarily, an autologous vein graft patch or small PTFE graft is used to reconstruct the vein.

Once safe repair is completed, a decision needs to be made regarding proceeding with the planned operation. Depending on blood loss, hemodynamic status, and reason for original surgery, this is a critical decision. Sometimes, it is prudent to wait for the patient to recover and show normal neurologic function (in the case of carotid injury), prior to proceeding. Hematoma and dissection of the carotid artery, which does not manifest clinically under general anesthesia, can occur at the time of injury, repair, anticoagulation, or reversal [10].

Key Points

1. Finger pressure is often the best way to maintain hemostasis and control after vessel injury, particularly if there is concern about placement of a side-biting clamp or if the vessel is not well exposed.
2. When struggling with bleeding, the first consideration—after immediate compression—should be given to increasing and improving exposure to the site while resuscitation is ongoing.
3. Proximal and distal control are critical to repair of any bleeding vessel.
4. Blood loss happens quickly, without team members aware of the actual volume; constant communication is mandatory during this time.
5. At times of an unexpected complication, it is wise to consider a delay or modification of the planned operation depending on the event (s), the patient, the operation, and the severity of underlying disease.

1.5 Introduction to Minimal Access

Although median sternotomy has been the workhorse for cardiac access for the last 60 years, this was not always the case. The first cardiac operations used an anterolateral thoracotomy. This was left-sided for mitral exposure and right-sided for atrial and ventricular septal defects. Generally feared by pioneering surgeons, the median sternotomy finally started to gain widespread acceptance in 1958 [1].

Minimally invasive cardiac surgery of the modern era has heralded a return to the thoracotomy for cardiac exposure. The key aspects of minimally invasive cardiac surgery are broadly broken up into (1) access, (2) peripheral cannulation, and (3) myocardial protection. The simple fact is that operating through a small hole increases the technical difficulty of any operation. Furthermore, minimal access limits the ability to deal with catastrophe such as ventricular distention or massive bleeding. Clearly, the anticipation of potential pitfalls in this setting is a critical skill, as restitution is far harder with limited exposure.

Common minimal access approaches include fourth/fifth interspace anterolateral thoracotomy (used for mitral, tricuspid, atrial septal defect, minimally invasive direct coronary artery bypass (MIDCAB) cases) and second/third interspace anterior thoracotomy and hemi-sternotomy (used for aortic and aortic valve cases).

The obvious increased cosmetic and decreased pain benefits come at a cost, as a smaller incision often necessitates the inclusion of peripheral cannulation. This carries its own set of complications. Small incision surgery can result in impaired visibility and difficult manipulation of surgical instruments. Myocardial protection is made more complex due to the inherent difficulties in placing a retrograde catheter and/or placing ostial cannulae for antegrade cardioplegia.

For these reasons, it is wise to advocate a stepwise progression of complexity in building a safe minimally invasive cardiac surgery program. For valvular work, the surgeon must be competent in the open setting, particular for specialized operations such as complex mitral valve repair. The next logical step would be to become comfortable with long-shafted instruments. This necessary instrumentation can reduce dexterity due to increased lever action, and technical expertise is best accomplished in a controlled open setting. Careful planning of peripheral cannulation and the use of long-duration cardioplegia solutions also increase familiarity prior to making a full transition to minimally invasive access.

Similarly, for mini-anterior thoracotomy bypass surgery (MIDCAB), the technique of off-pump coronary surgery through a sternotomy is key preparation. The margin for error is extremely small through these incisions. If a coronary shunt cannot be inserted quickly and the patient becomes unstable, there is no immediate recovery, unlike open procedures where internal cardiac massage or emergency cardiopulmonary bypass can be immediately initiated. Once these additional skill sets have been mastered, however, they are combined with minimal access incisions to facilitate safe surgery and decrease the learning curve. This challenging and rewarding part of cardiac surgery expands the toolbox of the modern surgeon to supply options in providing the best care to increasingly complex patients.

1.6 Mini-Sternotomy: Bleeding and Instability

An aortic valve replacement was performed via an upper hemi-sternotomy for a young patient who wished to have a small incision. Intraoperatively, there was injury to the right internal mammary artery on entry. This was ultimately controlled with careful placement of multiple clips. The surgeon did not place pericardial drains prior to the removal of the cross-clamp, and another sudden bleed occurred, this time from the anterior wall of the right ventricle during drain placement. This was repaired with a large suture. The patient did require transfusion, but was otherwise well postoperatively.

Six weeks postoperatively, the patient returns to the clinic complaining of a sternal click and significant pain. On examination, the incision appears to have healed with no sign of infection, but an audible click and instability are obvious on palpation. CT of the chest shows the right hemi-sternum to be displaced superiorly by 1 cm and a 5 mm gap along the entire length of the sternum.

1.6.1 Discussion

The upper hemi-sternotomy provides reasonable exposure for aortic and aortic valve surgery with a smaller incision than a traditional median sternotomy. A short skin incision is made, with flaps created if needed to expose the upper sternum. The upper part of the sternum is divided in the midline to the third or fourth interspace. Next, a perpendicular incision is made in the sternum to join the relevant interspace. This is best achieved with a narrow oscillating saw. A standard saw directed into this interspace with a gentle curve is another option. The hemi-sternotomy incision should either be "J'd" off to either side or "T'd" off into both sides.

Care must be taken to avoid the internal mammary artery, which is very difficult to control in the setting of poor exposure. The distance between the lateral sternal edge and the IMA is variable and is sometimes just a few millimeters away, particularly in small patients. Carefully dissecting the interspace and mobilizing the IMA laterally may be helpful if it appears close. Yet if an injury cannot be controlled quickly, and the patient is hemodynamically compromised, a completed full sternotomy and elevation of the sternal edges will provide excellent exposure to adequately ligate the IMA.

There have been many locations used for drains following an upper hemi-sternotomy. The usual sub-xiphoid location, common in full sternotomy cases, must be used with extreme care to avoid injury to the right ventricle, right atrium, and liver. To reduce the chance of injury, drains should be placed while still on bypass, sometimes with the cross-clamp in place. This allows increased space to maneuver between the posterior sternal table and the heart. Atrial and ventricular temporary wires, if needed, should be placed at this time; the heart is decompressed and still. After drain site skin incision, the abdominal fascia is pierced, and a long-shafted forceps is introduced. This needs to be held tightly, flush against the posterior table of the sternum, as it is advanced into the pericardium. Alternative sites for drain placement are supraclavicular (unsightly) or intercostal (sometimes more painful and result in bleeding).

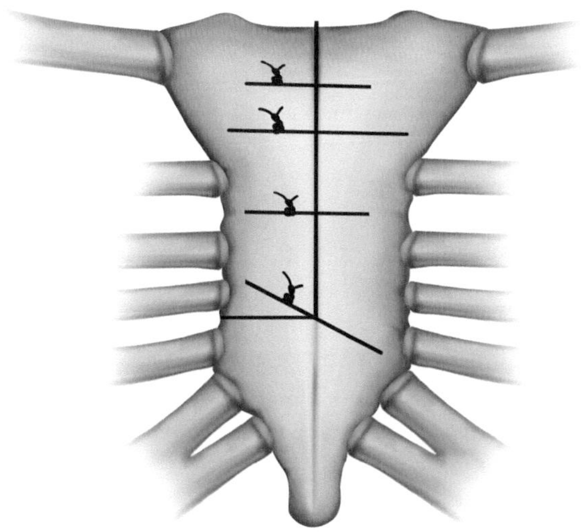

Fig. 1.3 Closure of upper hemi-sternotomy. Oblique wire at the inferior edge of the free bone increases stability

Establishing sternal stability is a key step for a hemi-sternotomy. It is essential that the free upper hemi-sternum is "locked" into the inferior corner that is created by the remaining bone. In the case of right hemi-sternotomy, a single standard wire is placed in an oblique fashion from the inferior edge of the free bone (Fig. 1.3). This achieves both lateral and longitudinal stability. Confirming sternal stability can be a more complex issue if the upper hemi-sternotomy is an inverted "T." This opens the interspaces bilaterally and leaves the sternum in three separate pieces. This method is less stable than a formal sternotomy. If conversion to full sternotomy is necessary for some reason in this setting, consideration should be given to plating the sternum at the completion of the surgery.

1.7 Right Anterior Mini-Thoracotomy

The team in the ICU notes that the mini-AVR patient was doing well and had been extubated in the OR. However, the patient suddenly requires high-dose vasopressors, and the hemoglobin is dropping, with 1 liter of dark blood in the pleural drain. The patient returns to the operating room, where double-lumen intubation is requested for complete right lung isolation. The patient is transfused packed red blood cells, platelets, and fresh frozen plasma. The right anterior thoracotomy is opened, and another liter of blood is seen in the right chest with an intact IMA and IMV and no evidence of chest wall bleeding. The pericardium is dry, as are all cannulation sites.

The patient remains unstable with ongoing bleeding. The drain site is suspected as the source, and a videoscope with a 30° angled camera is used to investigate. The drain site inferiorly is completely dry. The camera is directed at the diaphragm, and there is a small tear seen with dark blood coming through. The general surgery team is urgently called to perform an exploratory laparotomy that reveals a large tear in the liver. The bleeding was controlled and the patient has a slow, but steady, recovery.

1.7.1 Discussion

Whether it is performed for patient or surgeon preference, the anterior thoracotomy permits a smaller incision, enhanced cosmesis, and potentially less pain with earlier return to work for a patient undergoing aortic valve replacement. Providing safe minimal access surgery is a skill that can be of great benefit for certain patient populations, particularly in an era in which patient demand has grown. Generally speaking, however, the current evidence has not shown a clear benefit of one approach over another in terms of mortality and stroke risk [11, 12].

The right anterior mini-thoracotomy, or Chamberlain incision, was typically used for the evaluation of the mediastinum—often to sample level 5 or 6 lymph nodes. More recently, it has been adopted for minimally invasive aortic valve surgery. In addition to valvular disease, the anterolateral thoracotomy has been utilized for intracardiac pathology such as atrial septal defect repair or myxoma resection. This incision is placed in the right fourth interspace and does not require rib resection or division of the IMA or IMV. Aortic cannulation can be achieved by opening the interspace toward the sternum. While not necessary, double-lumen endotracheal intubation is helpful for surgeons and trainees to deal with unforeseen issues and to reduce the time pressure needed to get into the chest.

A medial anterior thoracotomy mandates division of the right internal mammary artery (IMA) and vein (IMV), which improves exposure particularly when the lower costal cartilage is divided. The IMA and IMV can be injured during the thoracotomy if not dissected out carefully. A good landmark is the location of the perforators into the pectoralis major. These vessels bleed significantly if a clip is incorrectly applied or if the soft tissue retractor pulls them off when it is removed. It is good practice to dissect out the IMA and IMV and clip each separately (twice on each side) and consider oversewing the IMA. If there is a large injury, the first step is to control with pressure. Next, ensure hemodynamic and respiratory stability, and finally, divide the upper or lower costal cartilage to improve exposure.

Minimally invasive cases via a right thoracotomy add the risk of bleeding from a few causes not found with traditional sternotomy. These include traction injury to the liver from diaphragm stay sutures, iliac vein injury from peripheral cannulation, and delayed chest wall bleeding. These cases require more vigilance postoperatively as the bleeding is often more occult than standard cardiac surgery. Diaphragm and liver injuries are serious complications of minimally invasive cardiac surgery due to the added effect of systemic heparinization. The possibility of these types of injuries is considered in those patients with a high right diaphragm. The preoperative CXR must be reviewed thoroughly, and consideration should be given to sternotomy.

Significant traction of the pericardium in a right thoracotomy can also result in injury to the phrenic nerve. This usually is discovered on the first post-extubation CXR. If a patient presents with a phrenic nerve palsy post-right thoracotomy cardiac surgery, and the surgeon is sure that the phrenic was not severed or had a thermal injury, the most likely cause is excessive traction on the pericardium. The nerve should recover over the next 6 months, and close observation with routine chest X-ray and fluoroscopy is required.

As with all postoperative bleeding, the surgeon should consider and exclude every surgical source. Bleeding is surgical until proven otherwise. As the surgeon approaches the operating theater in minimally invasive access, he or she should list in their mind every possible bleeding source from the skin down: pectoralis muscle, intercostal artery, IMA and IMV, innominate vein, aortic cannulation, aortotomy, pulmonary vein vent site, pacing wires, drain sites, and clamp site. If these are all dry, then the abdomen must be considered: direct diaphragm and liver injury, iliac vein injury from the femoral venous cannula, or even occult GI bleeding.

1.8 Lung Herniation

A 40-year-old female patient returns to the clinic, 6 weeks following a MIDCAB. She had undergone a robotic LIMA takedown followed by an off-pump left anterior descending bypass through a small left anterior thoracotomy incision. The patient now complains that her left chest "pops out" when she coughs and then settles over the next few minutes. On examination, there is a soft mass that protrudes with respiration, clearly increased with coughing. It is obvious that this is a lung hernia, and a CT chest is ordered to confirm.

1.8.1 Discussion

Lung herniation occurs in up to 2% of patients following minimally invasive cardiac surgery via limited access incisions [13]. These defects occur at the site of anterior mini-thoracotomy or at port access sites for robotic or thoracoscopic instrument insertion [14–16]. Generally speaking, these defects are benign. The chances of incarceration and strangulation are low, but have occurred, and are accompanied by pain and focal hemorrhage that can be life-threatening [17].

One risk factor is chronic obstructive pulmonary disease (COPD), a common condition in cardiac surgical patients [18]. The location of these defects is typically either anterior close to the sternum or posteriorly near the vertebra, where the intercostal muscles comprise a single layer [19]. Associated rib fracture or avulsion of costal cartilage at the site of mini-anterior thoracotomy for CABG may also contribute to lung herniation [13].

Prevention of this complication includes clear exposure of the rib space and an intact peri-costal closure, particularly for mini-thoracotomy. Smaller port incisions do not generally include division of the intercostal muscle, so these are considered to be at lower risk.

The diagnosis is made with physical exam: a distinct visible bulge underneath the incision that is increased with coughing or Valsalva maneuvers. Confirmation is typically obtained with two-view CXR and/or CT scan which help delineate the size of the defect for planned closure [14].

Conservative management is acceptable, but most patients elect to have this defect repaired for both cosmesis and discomfort [14]. Surgical correction involves lung isolation and redo thoracotomy. Management involves either (1) direct closure of involved tissues or (2) the use of absorbable and non-absorbable mesh. Symptoms and size dictate the choice of repair.

Adhesions from the thoracotomy itself need to be divided to obtain a clear space inside the chest wall for repair. The defect is then closed with mesh on the inner surface of the chest wall. This mesh material is typically used for abdominal wall hernia repair and includes polypropylene, PTFE, Gore-Tex, Marlex, or Dacron. There are differences described between the meshes. Gore-Tex is air- and water-tight and results in less adhesions and subcutaneous air; Marlex is coarser and allows for greater inflammation and scarring to help adhere the patch to surrounding tissue [20].

1.9 Reoperative Cardiac Surgery

One hour into a complicated reoperative CABG, the surgeon is carefully dissecting out the heart from multiple adhesions when the patient suddenly develops ventricular fibrillation. Attempted resuscitation with medications and cardiac massage fail to recover an effective rhythm or blood pressure. Emergency bypass through the groin is established. The patient undergoes a prolonged operation, which ends with the patient unable to wean from bypass due to biventricular failure.

1.9.1 Discussion

A number of problems could have resulted in cardiac failure for this patient: manipulation of the heart, distal embolization from vein grafts, ischemic arrest, a difficult dissection, or prolonged down time prior to bypass. Reoperative surgery is challenging at baseline. Surgical trends show these operations becoming less common in the current era of accessible complex percutaneous interventional techniques. Certainly, a reoperation adds an element of difficulty to cardiac exposure; it requires time and heightened technical skill. Previous surgery can make injury inherent due to extensive adhesions that alter tissue planes and preclude safe access to mediastinal structures.

The benefit of experience provides a more sophisticated understanding of what is needed to bring these patients through their operations safely. This procedure must begin with a *focused preoperative strategy*. This starts with imaging. Preoperative imaging and the early consideration of alternative techniques based on findings are essential for the prevention of major complications [21]. Rescue efforts in reoperative surgery are particularly difficult and arduous, sometimes ending unsuccessfully. An average incidence of sternal reentry injury is 2.7%, mostly due to adherent mediastinal structures, but this initial injury increases the mortality rate to 26% [22]. Although the right ventricle is most commonly injured (68%), the vein grafts, mammary artery grafts, and great vessels are also at risk [22]. As the operation progresses, there are many more possibilities for injury (Table 1.1). Manipulating patent vein grafts can result in distal embolization of laminar thrombus that leads to critical arrhythmia or myocardial infarction. The incidence of injury to a patent internal mammary artery is as low as 5.3% or as high as 50%, but the associated mortality is always significant [23, 24]. Immediate difficulty with myocardial perfusion/protection, often in the setting of hemodynamic instability, makes this particular injury one that should be carefully avoided.

One key to preparation is the review of the previous operative note(s). This is not always possible, given time, location, or historical constraints of medical records, but it is highly

Table 1.1 Common sites of injury in reoperative surgery

1. Right ventricle
2. Innominate vein
3. Left internal mammary artery
4. Aorta
5. Lung
6. Saphenous vein graft
7. Pulmonary artery
8. Left ventricle

advisable. The sequence, technique, and findings at the first operation are critical to all subsequent interventions. The timing from the last operation is often a marker as to the density of adhesions expected on reentry. Reoperations that occur within 1 year of the previous surgery, specifically within a few months, are expected to have dense adhesions and inflammatory tissue present. This tissue obscures normal planes and anatomy, resulting in a difficult and sometimes dangerous dissection. Significant postoperative bleeding or coagulopathy at the first surgery may also cause mediastinal inflammatory reactions that further complicate the sternal reentry.

The second key to reoperative surgery is the careful review of preoperative imaging, starting with a CT chest scan with contrast. This provides a road map that needs to be carefully reviewed well before the operation. The gold standard of preop CTA can lower the complication/injury rate to 0.6% from 3.6% without CT [22]. For a coronary bypass operation, CTA supplies critical information that adds to the catheterization films: where the LIMA lies, patency of a right internal mammary artery, the presence of ascending aortic calcifications, and additional data (Table 1.2) [21].

Some authors have suggested that reoperative surgery be addressed with a unique protocol [25]. Specific steps of this preparation would include use of preoperative CT scans, open exposure or wiring of the femoral vessels, a specific plan and backup plan for exposure, temporarily clipping or avoiding a patent LIMA, and the possible use of cardiopulmonary bypass or circulatory arrest *prior* to sternal reopening. Protocols are a mechanism by which forward stepwise progression occurs, even in complex dynamic situations. Care and attention to preoperative planning are critical for these reoperations, and the execution of a specific plan makes a successful outcome more likely.

One crucial early decision to be made is whether the patient requires peripheral cannulation. The preemptive placement of percutaneous lines in the femoral vessels for rapid cannulation using Seldinger technique is one method to prepare for a difficult reentry. Common reasons for planned full peripheral cannulation include known increased risk of sternal reentry based on preoperative imaging, multiple previous surgeries, or hemodynamic instability [22].

Table 1.2 CT scan review in reoperative surgery

1.	Retrosternal adhesions
2.	Distance from the sternum to ascending aorta (3–5 mm cutoff)
3.	Distance from the sternum to RV <5 mm (critical <3 mm)
4.	Distance from the innominate vein to sternum
5.	Ascending aortic calcifications, diameter, lumen
6.	Location of RV in relation to the xiphoid
7.	Location of previous bypass grafts and patency
8.	Diaphragm insertion and height
9.	Location of the pericardial edges

Specific findings such as the presence of a Type A dissection or a severely calcified aorta make this access mandatory. Peripheral cannulation also has its share of complications including lymphatic leak, infection, neuropathy, bleeding, hematoma, or limb ischemia [22].

Myocardial protection is of paramount importance in the reoperative setting. Although the incidence of LIMA injury in the recent literature has been decreased to approximately 5%, managing this situation requires a distinct plan [23]. If the IMA graft injury occurs prior to the initiation of cardiopulmonary bypass, prompt peripheral cannulation and immediate CPB will help maintain hemodynamic stability. As there is often much dissection remaining for adequate exposure of the mediastinal structures, it is critical that reperfusion of the LIMA occurs promptly in this setting, usually once antegrade and retrograde blood cardioplegia have been delivered [23]. This is the safest course to prevent myocardial injury, although the expected infarction rate in this situation can be high.

The technical skill of what is termed the "no-touch" technique around old saphenous vein grafts is a preventive measure to prevent distal atheroembolization from aged vein grafts [26]. The close dissection of these vein grafts should be reserved for after the patient is on cardiopulmonary bypass, and potentially after aortic cross-clamping, so that emboli will not adversely affect postoperative function.

Alternative incisions should be considered for easier access and to prevent damage of mediastinal structures. This is particularly relevant in patients who have had previous valve surgeries. For instance, the presence and location of a patent IMA graft support use of a right thoracotomy in the setting of a required reoperative mitral valve surgery [27]. An AVR may be done through a mini-sternotomy, and left-sided grafts may be created through a left thoracotomy [27].

Minimally invasive procedures, including TAVR and valve-in-valve techniques, could help avoid many of these difficult situations. For every high-risk operation, one must consider alternative techniques, incision sites, and perfusion access sites.

Although focused preventive strategies have been helpful in decreasing the high morbidity and mortality of reoperative surgery, intraoperative events continue to occur. A study by Roselli et al. found that the majority of intraoperative adverse events (IAE) occurred during sternal reopening (23%) and pre-bypass dissection (39%) [28]. Yet those patients with a preoperative strategy in place (such as peripheral bypass) had a higher rate of rescue than those without, leading to a 12% mortality in the IAE group and 4% mortality in the non-IAE group [28]. What was considered a lapse in the preventative strategy occurred in 55% of patients with an IAE [28]. Although most lapses were due to inadequate preoperative imaging, some were also due to technical aspects, such as use of the electrocautery, removal of wires, and lack of the no-touch technique [28]. In this particular study, higher mor-

Table 1.3 Preparation for future reoperative surgery

1.	Use few or no pledgets
2.	Cover the innominate vein with surrounding tissue/fat
3.	Keep the LIMA away from the midline and place it into the left chest by incising or making a defect in the left lateral pericardium above the phrenic nerve
4.	Consider loosely closing at least the upper pericardium/pericardial fat
5.	Position right-sided grafts away from the underside of the sternum and next to the right lateral side of the heart
6.	If reoperation is planned and expected, consider using a synthetic patch to cover the anterior mediastinum

tality was seen in those patients with multiple reoperative surgeries and previous chest wall radiation [28].

Although reentry complications and bleeding are common issues surrounding reoperations, these patients must also be monitored closely in the postoperative period. Sudden hypotension and changes in cardiac function, even after a successful weaning of CPB, may be due to subtle direct injuries to the grafts and heart or suboptimal myocardial protection. The possibility of a protamine reaction, due to previous exposure, requires close monitoring of the heart and hemodynamics with mandatory slow administration in the operating room and afterward. Sudden spasm of the IMA can result in acute ischemia, particularly if the graft exposure required close dissection in order to apply a temporary occlusive clip.

Lastly, the possibility of future surgery should always be considered at the time of initial surgery. In general, a clear and detailed operative note along with certain technical additions can make reoperative, especially planned reoperative surgery, much easier (Table 1.3).

References

1. Gerbode F, Braimbride MV, Melrose DG. Median sternotomy for open cardiac surgery during total heart-lung by-pass. AMA Arch Surg. 1958;76(5):821–3. discussion 823–4.
2. Zeitani J, Penta de Peppo A, Moscarelli M, Guerrieri Wolf L, Scafuri A, Nardi P, Nanni F, Di Marzio E, De Vico P, Chiariello L. Influence of sternal size and inadvertent paramedian sternotomy on stability of the closure site: a clinical and mechanical study. J Thorac Cardiovasc Surg. 2006;132(1):38–42.
3. Cannata A, Russo CF, Vitali E, Bruschi G. Technique to prevent inadvertent paramedian sternotomy. J Card Surg. 2009;24(3):290–1.
4. Robicsek F, Daugherty HK, Cook JW. The prevention and treatment of sternum separation following open-heart surgery. J Thorac Cardiovasc Surg. 1977;73(2):267–8.
5. Tavilla G, van Son JA, Verhagen AF, Lacquet LK. Modified Robicsek technique for complicated sternal closure. Ann Thorac Surg. 1991;52(5):1179–80.
6. Pinotti KF, Cataneo DC, Rodrigues OR, Cataneo AJM. Closure of the sternum with anchoring of the steel wires: systematic review and meta-analysis. J Thorac Cardiovasc Surg. 2018;156(1):178–86.
7. Nenna A, Nappi F, Dougal J, Satriano U, Chello C, Mastroianni C, Lusini M, Chello M, Spadaccio C. Sternal wound closure in the current era: the need of a tailored approach. Gen Thorac Cardiovasc Surg. 2019;67(11):907–16.
8. Takeshita J, Nishiyama K, Fukumoto A, Ohira S, Beppu S, Sasahashi N, Shime N. Tracheal injury detected immediately after median sternotomy by inexperienced surgeons: two case reports. J Med Case Rep. 2018;12(1):49.
9. Lui N, Wright C. Intraoperative tracheal injury. Thorac Surg Clin. 2015;25(3):249–54.
10. Marvasti MA, Parker FB Jr, Bredenberg CE. Injuries to arterial branches of the aortic arch. Thorac Cardiovasc Surg. 1984;5(32):293–8.
11. Balmforth D, Harky A, Lall K, Uppal R. Is ministernotomy superior to right anterior minithoracotomy in minimally invasive aortic valve replacement? Interact Cardiovasc Thorac Surg. 2017;25(5):818–21.
12. Semsroth S, Matteucci Gothe R, Raith YR, de Brabandere K, Hanspeter E, Kilo J, Kofler M, Müller L, Ruttman-Ulmer E, Grimm M. Comparison of two minimally invasive techniques and median sternotomy in aortic valve replacement. Ann Thorac Surg. 2017;104(3):877–83.
13. Ng PC, Chua AN, Swanson MS, Koutlas TC, Chitwood RW, Elbeery JR. Anterior thoracotomy wound complications in minimally invasive direct coronary artery bypass. Ann Thorac Surg. 2000;69:1338–41.
14. Athanassiadi K, Bagaev E, Simon A, Haverich A. Lung herniation: a rare complication in minimally invasive cardiothoracic surgery. Eur J Cardiothorac Surg. 2008;33(5):774–6.
15. Bhamidipati CM, Iyalla KI, Seymour KA, Lutz CJ. Lung hernia following robotic-assisted mitral valve repair. J Card Surg. 2012;27(4):460–3.
16. Gouda H, Multz AS, Khan A, Rossoff LJ, Green LM, Graver D. Lung hernia as a sequela to limited-access mitral valve surgery. Tex Heart Inst J. 2002;29(3):203–5.
17. Schroeter T, Bittner H, Subramanian S, Hansig M, Mohr F, Borger M. Life-threatening hemothorax resulting from lung hernia after minimally invasive mitral valve surgery. Thorac Cardiovasc Surg. 2011;59:252–4.
18. Ross RT, Burnett CM. A traumatic lung hernia. Ann Thorac Surg. 1999;67:1496–7.
19. Glenn C, Bonekat W, Cua A, Chapman D, McFall R. Lung hernia. Am J Emerg Med. 1997;15:260–2.
20. DeBord JR. The historical development of prosthetics in hernia surgery. Surg Clin North Am. 1998;78:973–1006.
21. Valente T, Bocchini G, Rossi G, Sica G, Davison H, Scaglione M. MDCT prior to median re-sternotomy in cardiovascular surgery: our experiences, infrequent findings and the crucial role of radiological report. Br J Radiol. 2019;92(1101):20170980.
22. Imran Hamid U, Digney R, Soo L, Leung S, Graham AN. Incidence and outcome of re-entry injury in redo cardiac surgery: benefits of preoperative planning. Eur J Cardiothorac Surg. 2015;47(5):819–23.
23. Gillinov AM, Casselman FP, Lytle BW, Blackstone EH, Parsons EM, Loop FD, Cosgrove DM 3rd. Injury to a patent left internal thoracic artery graft at coronary reoperation. Ann Thorac Surg. 1999;67(2):382–6.
24. Elami A, Laks H, Merin G. Technique for reoperative median sternotomy in the presence of a patent left internal mammary artery graft. J Card Surg. 1994;9:123–7.
25. LaPar DJ, Ailawadi G, Harris DA, et al. A protocol-driven approach to cardiac reoperation reduces mortality and cardiac injury at the time of resternotomy. Ann Thorac Surg. 2013;96(3):865–70.
26. Savage EB, Cohn LH. No touch dissection, antegrade retrograde blood cardioplegia, and single aortic cross clamp significantly reduce operative mortality of reoperative CABG. Circulation. 1994;90:II140–3.
27. Cohn LH. Evolution of redo cardiac surgery: review of personal experience. J Card Surg. 2004;19(4):320–4.
28. Roselli EE, Pettersson GB, Blackstone EH, Brizzio ME, Houghtaling PL, Hauck R, Burke JM, Lytle BW. Adverse events during reoperative cardiac surgery: frequency, characterization, and rescue. J Thorac Cardiovasc Surg. 2008;135(2):316–23.

2 Cardiopulmonary Bypass

2.1 Introduction to Cardiopulmonary Bypass

It was 1953 when John Gibbon performed the world's first successful open heart surgery using cardiopulmonary bypass (CPB) [1]. Yet the story began in 1931, when Gibbon was motivated to create a machine to bypass the heart and lungs after having seen a young woman die with a massive pulmonary embolism [2]. This gap reveals the many years of small steps that eventually culminated in a successful operation:18-year-old Cecelia Bavolek had her atrial septal defect repaired during 26 minutes of support by the cardiopulmonary bypass machine [3]. Two other children were operated on with CPB that same year, but their deaths led Dr. Gibbon to never again perform cardiac surgery [3]. In subsequent years, many giants of the field contributed to cardiopulmonary bypass; the circuit as we now know it has benefitted greatly by modifications made by Dr. John W. Kirklin at Mayo Clinic in the 1950s and by Dr. C. Walton Lillehei's use of cross-circulation [4–7].

Cardiopulmonary bypass is clearly one of the principal medical advances of the twentieth century. The concomitant inventions of heparin, a closed circuit, roller pumps, safe oxygenators, and the management and understanding of the blood-surface interface were all necessary. Yet each of these technical aspects required years of study and refinement. Knowledge can grow through trial, error, and failure. Constant surveillance and minute attention to detail were necessary to bring the patient through safely, both then and now. Even prior to the addition of cardioplegia and ischemic arrest, the bypass run itself took the patient on a journey of systemic inflammation, coagulopathy, and hemodynamic changes, all in an attempt to protect the brain and body during the time of cardiac repair.

CPB remains remarkable to those of us who use this technology daily, without the ability to truly improve upon it. CPB involves both risk and the suppression of risk. Confidence in the machine alone is not enough, but is absolutely necessary to maintain. A deep understanding and the ability to solve immediate problems are required. Standardization after analysis has been key to this technique, and specific recovery processes have been described that can be lifesaving. It is imperative to separate process from outcomes. The dramatic complications of CPB require understanding of the technical aspects by a surgical team that has knowledge, experience, a sense of prevention, and reliable recall.

2.2 Heparin Resistance

An AVR/CABG is scheduled for a 60-year-old woman with aortic stenosis who presented with a NSTEMI and was transferred in from another hospital where she had received long-acting platelet inhibitors. She was maintained on intravenous heparin for 5 days prior to surgery and had no chest pain. She was brought to the operating room with a platelet inhibitor assay of >200.

After conduits were harvested, 40 units/kg of heparin was given with an ACT result of 350. Heparin is re-dosed and rechecked. The perfusionist reports a repeat ACT of 360. At this moment, the patient becomes hemodynamically unstable, with rising PA pressures, ST segment elevations in leads V1–V3, and clear anterior wall motion abnormalities. The blood pressure drops.

2.2.1 Discussion

Heparin resistance is a unique aspect of cardiac surgery in which the target ACT required for cardiopulmonary bypass is not achieved despite the administration of high-dose heparin. This complication occurs with an expected incidence of 10–20% [8, 9]. An early study found three main predictors of resistance: (1) antithrombin III (AT-III) activity, (2) heparin pre-treatment, and (3) platelet count >300,000 [9]. 85% of the activity of heparin rests on its binding to AT-III, thereby inhibiting the binding of AT-III to thrombin [9]. Only 15% of its action comes from *direct* anti-Xa inhibition [9].

S. K. Balaram, L. Bassin, *Cardiac Surgical Complications*, https://doi.org/10.1007/978-3-030-71563-2_2

The problem with ultra-fractionated heparin, a heterogeneous molecule, is its variable anticoagulation. The actual mechanism of heparin is well known: to potentiate the action of AT-III. After binding to AT-III to form an AT-III/heparin complex, this complex binds to thrombin, and then heparin breaks away. This leaves the thrombin/AT complex that is then cleared by the reticuloendothelial system. There is marked variation in the functionality of heparin based on chemical properties:

1. The specific sequence that controls binding is found in only 1/3 of heparin molecules.
2. There are differences between bovine and porcine heparin sources.
3. The pharmacokinetics of heparin in the heparin/AT-III complex depends on cardiopulmonary bypass, previous exposure, organ function, clearance, hemodynamic instability, and hemodilution [10].

Exposure of patients to intravenous heparin prior to surgery can decrease AT-III levels by 5–7% per day during treatment [11]. This disappearance of AT-III is considered to be the basis of heparin resistance. Preoperative enoxaparin (low molecular weight heparin LMWH) treatment has also been shown to increase the incidence of heparin resistance [12]. But in fact, there are both AT-III-mediated and non-AT-III-mediated etiologies to heparin resistance (Table 2.1) [9, 10].

In addition, the wide variation in the evaluation of heparin action can be partially attributed to the machines by which we measure its effect [10]. The activated clotting time (ACT) is a test that works by mixing fresh, whole blood with a contact activator (e.g., kaolin) which changes Factor XII (prekallikrein) to Factor XIIa, (kallikrein). When a subtherapeutic ACT is reached, one must remember that other factors affect this measurement. These include temperature (hypothermia), hemodilution, medications, platelet inhibitors, gross platelet count, anti-fibrinolytics, clotting factor deficiencies, fibrinogen, and specific antiphospholipid or anticardiolipin antibodies [10]. The preoperative use of warfarin and direct thrombin inhibitors also plays a role.

Table 2.1 Common mechanisms of heparin resistance

AT-III-mediated	Non-AT-III-mediated
Decreased production of AT-III	**Increased heparin binding**
Increased clearance	Platelets (thrombophilia, activation, release)
Preoperative use of heparin or LMWH	Chemokines
Increased consumption	Enzymes
Mechanical: IABP, assist device, ECMO	Extracellular matrix proteins
Disseminated intravascular coagulopathy	Medications
Endocarditis	Growth factors
DVT/PE	Extracellular matrix proteins

Heparin is eliminated by two mechanisms. The first is the reticuloendothelial system, and the second is renal elimination [10]. Cardiopulmonary bypass itself interferes with drug levels due to anticoagulation, hypothermia, hemodilution, and changes in organ function. Furthermore, the half-life of heparin increases from short, after a small bolus, to long, after a large bolus. The initial doses given for cardiac surgery sometimes do not require re-dosing during a CPB run.

In the setting of heparin resistance, there are four therapeutic options:

1. Give additional heparin, and consider further additional doses. Usage up to 1200 U/kg has been reported.
2. Consider giving FFP, which contains 1 IU of AT per 1 cc of FFP [13] (2 units = approx. 500 IU of AT).
3. Give AT-III concentrate.
4. Accept a slightly lower ACT, remembering that both the measurement and loading dose based on weight are standardized, but the measurement of ACT is not always perfectly accurate. This option would not be preferable for the majority of surgeons.

The problem of heparin resistance, in the current era, most commonly results in a delay. Many surgeons will initially treat the patient with AT-III without knowing the mechanism of resistance, and this is effective in those patients with AT-III deficiency due to preoperative heparin use. Obtaining an AT-III level is not always feasible, given time and cost constraints.

Giving additional heparin to the patient is usually attempted first. Large doses of AT-III can increase anticoagulation activity in non-heparin-resistant patients, making it difficult to know if the dose had a direct effect [14]. Immediate AT-III administration, however, is not best in all situations. Price is a consideration in its use as 500–1000 IU can cost up to several thousands of dollars [10]. In addition, heparin rebound and postoperative bleeding can occur. AT-III administration is not always successful in raising the ACT, particularly in those patients who do *not* have heparin resistance. Increased doses of heparin initially may be all that is required for those patients with an AT-III-independent mechanism [10].

Preoperative measurement of AT-III levels is an option for those patients with prior treatment with heparin or LMWH >48 h, liver disease, malnutrition, mechanical support such as ECMO/IABP/VAD, increased consumption such as DVT/PE and DIC/sepsis, or suspected loss of AT-III as seen in nephrotic syndrome [10]. The level of AT- III remains unknown in most situations. No studies have been performed to truly understand the utility of preoperative levels, but knowledge of an abnormality increases the preparation for heparin resistance and the initial treatments necessary. Normal or supratherapeutic AT-III levels can be expected in 7.2% of patients with heparin resistance, with the severity of the resistance being less [15].

The inability to move forward because of a subtherapeutic ACT at the exact moment the surgeon is ready to go on bypass is a test of patience as it does require a mandatory break in the operating progress and can be a precarious situation for an unstable patient. Predicting the problem of heparin resistance may help prepare the patient for timely treatment, so that moving forward onto CPB proceeds smoothly.

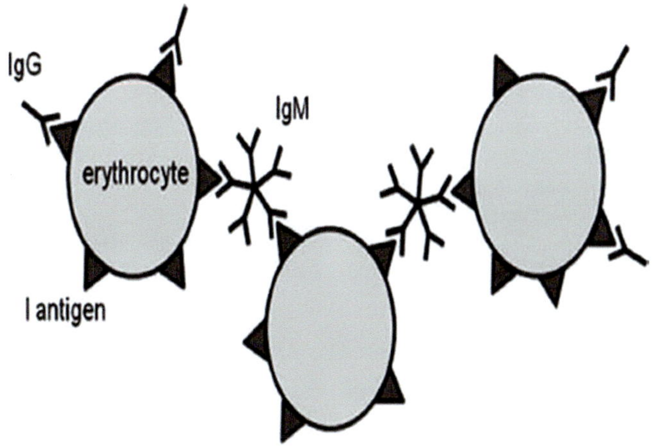

Fig. 2.1 Mechanism of cold agglutination. IgM autoantibodies bind with I antigens on the surface of circulating erythrocytes to result in agglutination. *IgM immunoglobulin M, IgG immunoglobulin G.* From "Incidental Cold Agglutinins in Cardiac Surgery: Intraoperative Surprises and Team-Based Problem-Solving Strategies During Cardiopulmonary Bypass," by Patel et al., 2017, *J Cardiothorac Vasc Anesth, Vol 31,* p. 1110. Copyright 2017 by Elsevier Science & Technology Journals. Reprinted with permission

2.3 Cold Agglutinin Syndrome

A 62-year-old patient has severe lung disease and a history of lung resection for stage I adenocarcinoma in the remote past. He was recently hospitalized for mycoplasma pneumonia and chest pain. An elevated troponin level led to cardiac catheterization, which demonstrated triple vessel disease. The patient presents now for elective CABG. Cardiopulmonary bypass is initiated with an adequate ACT, and the patient is cooled to 32 °C. While examining the targets, the surgeon asks the perfusionist to flush through the cardioplegia lines. The perfusionist immediately reports high line pressures. The tubing is examined and thrombus is seen inside the cardioplegia lines.

Table 2.2 Secondary causes of cold agglutinin syndrome

Etiology	Examples
Autoimmune mediated	Warm/cold antibodies
Trauma	Mechanical valve
Bacterial toxins	Chlamydia, mycoplasma, clostridium
Neoplasms	Adenocarcinoma
Infection	Malaria
Drug-induced	Cephalosporins, primaquine

2.3.1 Discussion

Cold agglutinin syndrome (CAS) is a type of autoimmune hemolytic anemia. Activated IgM antibodies from the body recognize their own red blood cells as foreign, but only at cold temperatures. The presence of these cold antibodies is rare and comprises a quarter of all autoimmune hemolytic anemias [16]. An antigen on the surface of the RBC binds with IgM antibodies that are activated at cold temperatures and result in clumping of RBCs (Fig. 2.1) [17]. This reaction can occur at temperatures below 37 °C or at any temperature equal to or below room temperature [16]. Patients present with cool distal extremities or digits; the syndrome results in both agglutination and hemolysis.

The occurrence of this syndrome in cardiac surgery, which often involves systemic cooling, is 0.08% [17]. Clotting can sometimes be seen first in the cold cardioplegia flush. For a standard on-pump case with even mild hypothermia, an inability to recognize this problem could mean disaster. The blood aggregates due to the antigen-antibody reaction, and subsequent complement system activation soon results in hemolysis. Clotting of the microvasculature results in end-organ damage. Clinically, this manifests as cerebrovascular accident, hepatic or renal infarcts, acute myocardial infarction, or severe hemolytic anemia.

Primary and secondary forms of CAS exist. Secondary CAS occurs when the body's immune response prior to exposures causes it to form antibodies to I antigens on the red blood cell. Some of these exposures include infections (both viral and bacterial) and tumors such as adenocarcinoma (Table 2.2) [17]. These antibodies can last up to 4–6 weeks in the body, but often this manifestation of CAS is less pronounced than in those that have a primary etiology.

For *known* CAS, prevention includes the use of preoperative IV IgG administration to coat the I antigens on the surface of the RBC and prevent the interaction with IgM. A monoclonal antibody can be given to block the immune response [18]. If recognized prior to surgery, a specific intraoperative strategy should be carefully discussed with perfusion, anesthesia, and the entire team (Table 2.3) [19, 20]. The thermal amplitude, the highest temperature above which antigen binds to antibody, is considered significant if it is below 30 degrees [17].

Table 2.3 Strategies for treatment: known cold agglutinin syndrome

Preop IV IgG
Off-pump surgery (warm)
Consider PCI
Systemic normothermia of mild hypothermia with warm blood cardioplegia
Continuous retrograde warm blood cardioplegia
Intermittent antegrade warm blood cardioplegia
Warm crystalloid followed by intermittent cold crystalloid cardioplegia

Table 2.4 Strategies for treatment: intraoperative discovery of CAS

1. Ischemic arrest on bypass with continuous warm blood cardioplegia
2. Intermittent warm blood cardioplegia given every 20 minutes
3. Cold crystalloid antegrade or retrograde cardioplegia with a warm heart
4. Off-pump surgery
5. Mild systemic hypothermia
6. Removal of coronary microemboli with retrograde warm crystalloid

In most cases, CAS is first recognized in the operating room. When the patient is initially cooled on CPB, CAS may be diagnosed by the presence of clot in the CPB circuit or in the cardioplegia line in the setting of a therapeutic ACT. Increased pressure measured during the CPB run or during flushing of the cardioplegia line is often the first clue, rather than direct visualization of clot [18].

Upon recognition of the abnormality, the entire team should be immediately notified. It is useful to know at what exact temperature the thrombus appeared, if possible. First steps include modifying the goal temperature back to normothermia, resuming the air-warming device, warming the room, and carefully considering surgical options. The anesthesiologists should warm the intravenous fluids and anesthetic gases. Monitoring of both the heart and core temperatures is critical at this time. If CPB has already been initiated, the perfusionist should raise the patient temperature to 36 degrees. If ischemic arrest is in place, cardioplegia also needs to be warmed. This choice does incur a risk of suboptimal myocardial protection and thus decreased function in the postoperative period (Table 2.4) [18, 19].

Thus once cold agglutinins are encountered, the current recommendations are to keep the patient at normothermia and adjust the operation as needed [21]. Bicaval cannulation and LV venting are methods to prevent blood from cooling within the heart [21].

Although these choices work for a CABG procedure, an aortic arch replacement with deep hypothermic circulatory arrest (DHCA) presents a more difficult problem. The only real option would be moderate cooling with antegrade cerebral flow. Staying above the thermal amplitude with anterograde cerebral perfusion and above the temperature where the clot was noted are key actions. Obviously, this is less than ideal and only would be considered in an emergency scenario such as a Type A dissection [18]. If circulatory arrest is required in known primary CAS, consideration should be given to preoperative plasmapheresis, perioperative plasmapheresis, or IV IgG [22, 23]. Transcranial Doppler and emergency selective cannulation of the left common carotid during antegrade cerebral perfusion at mild hypothermia have also been described [23].

2.4 Coronary Sinus Rupture

A frail 82-year-old female with a body surface area (BSA) of 1.3 presents for valve replacement with rheumatic mitral valve stenosis. A calcified posterior mitral annulus was observed on angiography, echo, and CT scan, and preoperative planning was extensive.

The first step in replacement was establishing CPB. The surgeon attempted coronary sinus cannulation, using echo guidance. There was some concern that the catheter had not slipped in correctly, as the echo views were not optimal. From a surgical standpoint, the mild aortic insufficiency made retrograde cardioplegia absolutely critical, particularly for the time the mitral retractor is in place. The coronary sinus orifice is eventually visualized on echo, and the catheter is seen. The anesthesiologist remarks that the sinus is small.

After cross-clamp and delivery of the initial antegrade cardioplegia dose, the switch is made to retrograde. Unfortunately, the perfusionist does not initially adjust the flow rate while switching to retrograde. The cardioplegia blasts in, with coronary sinus pressures flashing up to 80 mmHg. Bright red blood is seen coming from behind the heart.

2.4.1 Discussion

Located on the posterior portion of the heart between the left atrium and ventricle, the coronary sinus is the terminal portion of the great cardiac vein that drains the great, middle, and small cardiac veins into the right atrium. The importance of the coronary sinus in cardiac surgery is as the delivery site for retrograde cardioplegia, providing protection for the heart during ischemic arrest.

There are multiple advantages of coronary sinus retrograde cardioplegia [24]. This method of delivery provides better distribution of protection in the setting of diffuse and distal coronary artery lesions. Retrograde cardioplegia provides good protection for the subendocardium in particular.

Furthermore, in valve surgery, the flow of the operation for both aortic and mitral surgery is not halted for a need to deliver cardioplegia. A specific advantage when aortic insufficiency is present is the ability to achieve arrest after cross-clamping. There are also disadvantages: delayed arrest in comparison to coronary ostial antegrade, suboptimal protection for the right ventricle, the need for bicaval cannulation if direct visual insertion is necessary, and direct injury [25].

Placement of this catheter is challenging in some cases. A known site of variant anatomy, the coronary sinus is sometimes partially obstructed by a Thebesian valve or a Eustachian valve close to the orifice [26]. The incidence of coronary sinus injury is approximately 0.6% [27]. The nature of the injuries range from a simple hematoma to laceration to perforation to frank rupture of the coronary sinus or right ventricle. These injuries most commonly occur during placement of the coronary sinus cannula due to the stylet or catheter itself. The coronary sinus can also be injured indirectly by high pressures and flows, resulting in tissue edema, rupture, and hemorrhage. Injury may also occur if the heart is retracted forcefully during delivery of cardioplegia with the balloon inflated. Specific predictors of this complication include advanced age, female gender, weight, and small BMI [25].

The key to coronary sinus injuries, like almost all complications in heart surgery, is preparation and avoidance. Patients who are at risk for this injury should be recognized prior to proceeding with insertion (Table 2.5) [28]. Prevention of the injury is assisted by both careful maneuvering and careful monitoring (Table 2.6) [28].

Movements to place the cardioplegia catheter within the coronary sinus catheter must be gentle, careful, and subtle. If there is a stylet of a self-inflating balloon in place, this must be carefully adjusted. Moving the catheter once it is in place, particularly on bypass when the heart is empty and collapsed, is more challenging. There are two additional options available if the catheter is dislodged and clamp is already on: switching to antegrade cardioplegia or opening the right atrium and inserting it under direct vision.

Table 2.5 The "at-risk" coronary sinus

Balloon overinflation
Size and direction of the venous system
Blind cannulation
Self-inflating balloon
Small women with low BMI
Small coronary sinus
Inexperience/forceful insertion in the presence of a web

Table 2.6 Prevention of injury: insertion of the coronary sinus catheter

1. Use gentle motion and anatomic knowledge to maneuver the cannula into place
2. Use TEE guidance
3. Consider a smaller cannula
4. Inflate the balloon slowly while observing the waveform
5. Maintain coronary sinus pressure under 40 mm hg
6. Open the right atrium and place the cannula directly

The intraoperative TEE is a helpful tool. The course of the sinus is oblique; thus, guidance toward the left shoulder to align the cannula with the TEE is useful. In minimally invasive surgery, with no tactile feedback, it is mandatory to use the TEE to guide the cannula toward the sinus opening (Fig. 2.2) [29].

Confirmation of placement is key. This can be done in various ways:

1. Assistance from the anesthesiologist/TEE.
2. Monitoring of waveform and perfusion pressures.
3. If the aorta is open, check for blood returning from the coronary ostia.
4. LV filling from the aorta during retrograde delivery for mitral surgery.
5. Temperature probe to confirm cold cardioplegia myocardial delivery.

When an injury occurs, the first step is to prevent further injury. An immediate intraoperative plan regarding the best method requires consideration of heart function, necessary treatment, and current stage of the operation. The first step is always to stop the retrograde cardioplegia and remove the cannula. The heart must be protected at this point if the cross-clamp is already on, and antegrade cardioplegia and/or direct coronary ostial perfusion catheters are used to administer cardioplegia. Once the heart is protected, the next step is to determine the extent of the injury.

When the heart is lifted, what is seen is either a small defect with a contained hematoma or a large rupture. With a small hematoma, often nothing has to be done in terms of manipulation and repair. Removal of the coronary sinus cannula, use of an alternative method of cardioplegia delivery, and monitoring are all that is required [30].

If the injury is more extensive, the first decision to be made is whether to attempt to fix it directly or to go on bypass to decompress the ventricle. Sometimes, placement or replacement of the cross-clamp is required. Hemodynamic shifts occur with exposure of this inferior surface of the heart. Much of the decision regarding repair depends on the patient and the stage of the operation at which the injury is discovered. A failed initial repair is a more difficult problem. With a large injury, a decision must be made if the patient is still able to tolerate the initially planned operation. This is based on the severity of injury and extent of repair, as well as underlying cardiac function. Off-pump techniques and instruments may be helpful.

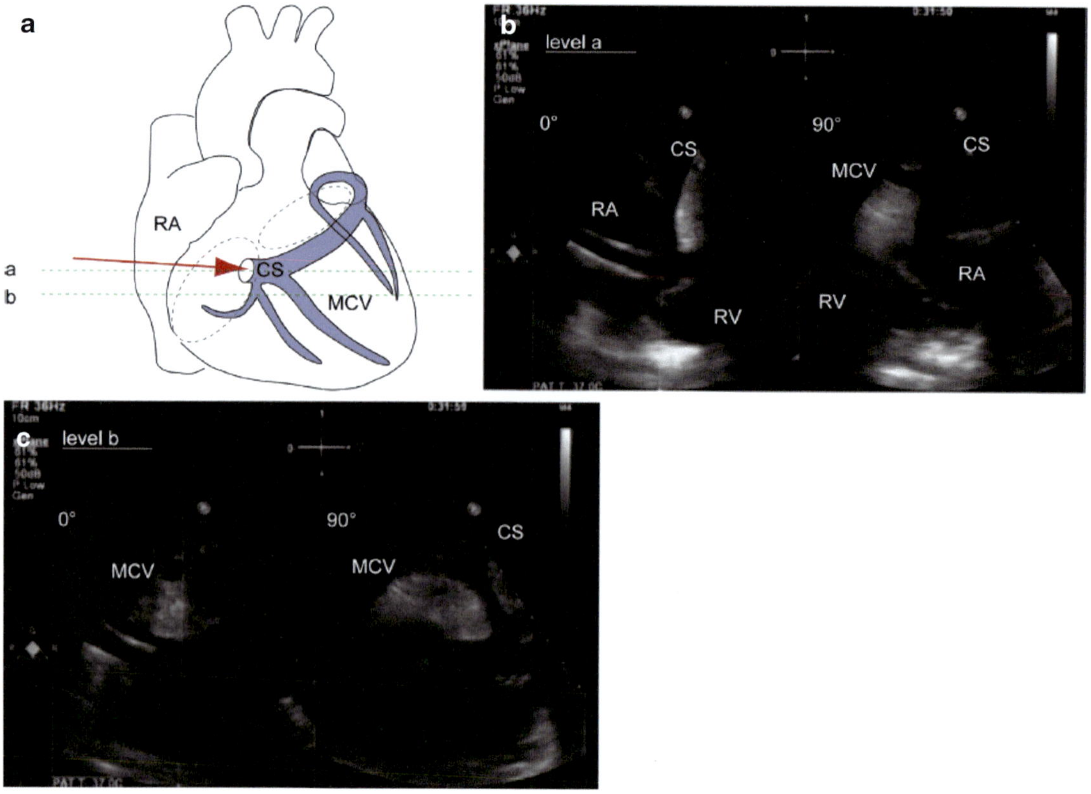

Fig. 2.2 Diagram of coronary sinus anatomy and TEE visualization of coronary sinus. *CS, coronary sinus; MCV, middle cardiac vein.* (**a**) Two scanning levels: CS level (level a) and MCV level (level b). (**b**) TEE view at level a. (**c**) TEE view at level b. (*RA, right atrium; RV, right ventricle*). From "Avoidance of Coronary Sinus Injury During Retrograde Cardioplegia," by Orihashi K et al., 2016, *Ann Thorac Surg, Vol 102*, p. E584. Copyright 2016 by Elsevier Co. Reprinted with permission

Questions to consider after coronary sinus injury:

Has it been a long operation?
Is there coagulopathy?
Is the ejection fraction or RV function poor?
Is the patient unlikely to tolerate the lift based on mitral prosthesis or hemodynamic instability?
Has a fix been attempted once already with poor results?
Is there a mitral prosthesis in place?

A variety of techniques have been described to repair the coronary sinus. One option is to use 7–0 Prolene suture on the undersurface of the heart, with or without a pericardial patch [31, 32]. Obviously, this is quite difficult exposure, particularly if a mitral prosthesis is already in place. AV groove rupture, often a fatal event, has occurred with repeated lifting of the heart in this setting. Repairing a large injury with the heart beating can be difficult and lead to poor outcomes, and return to cardiopulmonary bypass or ischemic arrest is often necessary.

The patency of the coronary sinus is important [33]. Repairs with partially occluding stricture or complete thrombosis remove the natural source of drainage of the cardiac venous system. Complete closure may result in a loss of left ventricular function, and a balloon pump or other mechanical assist device may be required [33]. Various topical hemostatic patches have also been used and reported in the literature. This averts the problem of suturing in this difficult location [34, 35].

2.5 Persistent Left Superior Vena Cava

A young attending has scheduled an elective CABG in a morbidly obese man. After sternotomy, the innominate vein is not well visualized on initial dissection due to significant adipose tissue in the mediastinum. The cannulation proceeds routinely using a 23F aortic cannula and a dual-stage large venous cannula. The coronary sinus cannula slips in easily and is seen to be large on echocardiography. The cross-clamp is placed and cardioplegia is delivered, both antegrade and then retrograde. Good coronary sinus perfusion pressures are not achieved despite reinflating the catheter balloon; the back pressure is low. Arrest is eventually achieved with a high volume of cardioplegia. The heart is positioned and the distal anastomoses are begun. The perfusionist states repeatedly that the drainage is poor. In addition, the surgeon notes that the heart is distending and begins beating just 3 minutes after the cardioplegia dose. An LV vent is placed after multiple readjustments of the venous cannula prove unsuccessful to improve drainage. There is real concern about myocardial protection.

Upon being called to review the CXR and after appraising the situation, the senior surgeon on the team prepares for surgery. Readying a 4–0 suture, he lifts the patient's heart, creates a purse string, and inserts a 20F venous drainage cannula back to the pump. The heart is now empty, silent, and still.

2.5.1 Discussion

The rare anatomical abnormality of a left superior vena cava is not often seen, but it suddenly, and most inconveniently, appears in cardiac surgery. Persistent left superior vena cava (LSVC) affects 0.5–2% of the general population and is in fact the most common venous anomaly [36]. This anomalous vein is usually connected to the coronary sinus and is present because the left anterior cardinal vein did not correctly regress in utero [37]. The LSVC crosses anterior to the aortic arch, the left PA, and pulmonary veins, to enter the pericardium in the posterior AV groove where it drains into the coronary sinus [38]. A right SVC is usually also present with a LSVC, but it is not connected to the left system.

Other genetic anomalies are sometimes seen within the mediastinum at the time of cardiac surgery with a LSVC. In rare situations (0.09–0.13%), patients have an absent right SVC [39]. There are known anatomic abnormalities associated with a LSVC. These include atrial septal defect, ventricular septal defect, tetralogy of Fallot, mitral atresia, and dextrocardia [40].

Table 2.7 Clinical suspicion of left SVC

1. Observation of a vertical shadow on the left upper border of the cardiac silhouette
2. EKG shows a leftward deviation of P axis
3. Screening preop CT scan
4. Direct angiography
5. Large coronary sinus (normal size is usually <1 cm in the parasternal TEE view)
6. Lack of visualization of the innominate vein upon initial dissection
7. Very small innominate vein
8. Poor drainage on CPB with any of the above findings

Although there are some factors that lead to suspicion of a left SVC, the classic diagnosis is made with contrast echo with right vs left arm injection (Table 2.7) [38, 41]. If the coronary sinus is unroofed, the left atrium will be seen first. Angiogram, venogram, TEE, or CT of the chest also definitively demonstrates this anomaly [42].

It is helpful to rule out associated conditions. ASD is common and must be recognized preoperatively, given the possibility of paradoxical air emboli. Cardiac arrhythmia and conduction system defects have been shown to occur due to the associated large coronary sinus; the opening stretches the AV node and bundle of His [43, 44].

Problems associated with LSVC during surgery include inadequate cardioplegia delivery and large volumes of venous return. There are also occasions when this anomaly is not recognized. A CABG or valve surgery with a large single venous cannula may not demonstrate a significant intraoperative difference with LSVC except for increased filling on rotation of the heart for exposure. Bicaval cannulation with a right atrial opening may require a third cannula to drain the right atrium and prevent large blood return into the field [37, 42]. The placement of additional drainage cannula prevents massive venous return (Fig. 2.3) [38]. This improves myocardial protection by preventing filling, warming, and distention of the heart during cold ischemic arrest. A number of other treatment options have been described (Table 2.8) [38].

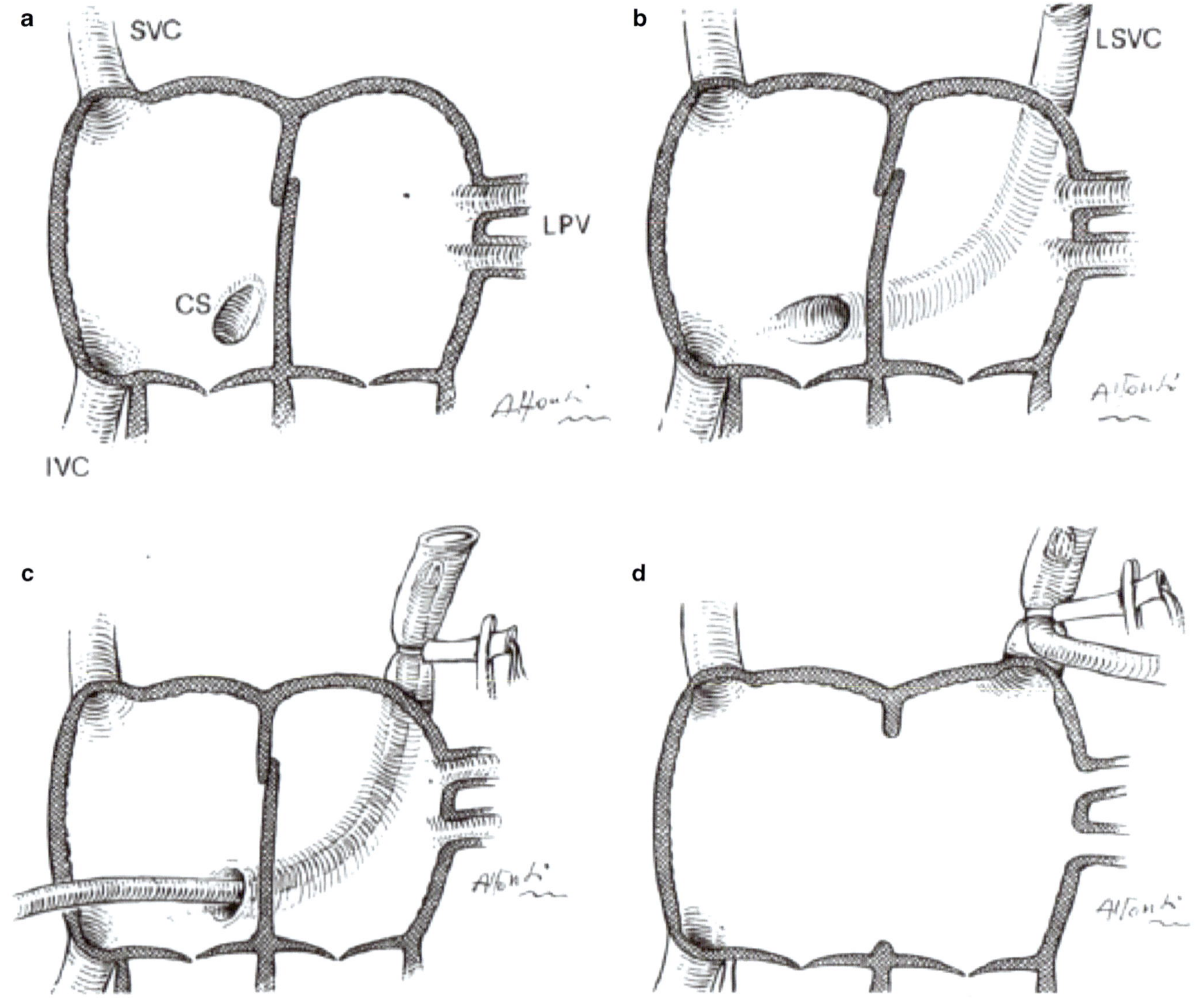

Fig. 2.3 Methods of left superior vena cava cannulation. (**a**) Normal systemic venous connection (*SVC, superior vena cava; CS, coronary sinus; IVC, inferior vena cava; LPV, left pulmonary vein*). (**b**) Left superior vena cava (LSVC) draining into coronary sinus (CS). (**c**) Cannulation of LSVC through the right atrium. (**d**) Direct extra-cardiac cannulation of LSVC with a right angle cannula. From "Anomalies of the systemic venous return: a review," by Mazzucco A et al., 1990, *J Card Surg, Vol 5,* p. 124. Copyright 1990 by John Wiley & Sons. Reprinted with permission

Table 2.8 Options for the treatment of LSVC during CPB

1. Expose the LSVC behind the appendage and put in a small cannula after snaring
2. Drain the coronary sinus to the LSVC and place a snare
3. Occlude the vein outside the heart with a snare
4. Place a catheter into the right atrium through the coronary sinus and into the left SVC
5. Directly cannulate the left SVC and drain into the CPB circuit
6. Deep hypothermia with total circulatory arrest
7. LSVC ligation

2.6 Air Embolism

During a routine AVR/CABG, the LV vent is placed after the patient is on bypass. The surgeon temporarily looks away from the field to examine the conduits as the patient goes on bypass. Suddenly, a large column of air enters the vent line and is injected directly into the LV. The cross-clamp is not yet on and the heart is still ejecting.

2.6.1 Discussion

Reversal of the vent line is a classic catastrophic emergency. There is no time to ponder this situation that may have never been seen in thousands of previous cases. The basic treatment of air in the line must be immediate: clamp inflow of air, put the patient in deep Trendelenburg position, and determine the next steps. Instantaneous action and good communication are essential.

The proposed solutions for massive air embolism are a classic case of preparing for war in times of peace. It is a difficult complication in cardiac surgery and carries a high mortality. Air embolism occurs in 0.1–0.2% of cardiac cases [45]. The treatment plan needs to be reviewed by the entire team for the immediate response to this rare problem to be rapid and systematic. Practice runs of management can be lifesaving.

The pathology is well described; the introduction of air into the arterial system results in immediate vasoconstriction. Resulting stasis and thrombosis in addition to cellular congestion and endothelial damage can quickly lead to stroke, shock, and death [46]. Intra-arterial air can cause severe cerebral edema based on the *rate* of delivery as opposed to the volume [46]. Air within the arterial tree, similar to the cardiopulmonary bypass circuit itself, contains the basic problem of foreign body to blood surface reaction. As it travels through the vessels, the air breaks into smaller sizes in the arterioles and ends with perivascular hemorrhage and cellular congestion. Ultimately, tissue hypoxia ensues. Even after the air dissipates, there remains cellular congestion and endothelial disruption [46]. This cellular injury is most poorly tolerated by the brain.

Sources of massive air embolism are well described (Table 2.9) [47]. Air entry can occur in the setting of deflating an IABP during aortotomy for cannulation, excessive gradients on CPB while cooling or warming, use of roller pumps, or through a side port of a venous line.

Every aspect of air removal has been extensively studied, starting from the 1950s to the 1970s [47]. Safety checklists have been created for the sole purpose of avoiding this complication. It is key that prevention is performed in a systematic manner for every CPB case (Table 2.10) [47, 48]. Air can enter with clamping of the aorta, the use of LV and aortic vents, and return from within the pulmonary veins. Immediate positioning of the patient with head down, rotating the table left to right, use of an apical needle to directly withdraw air, and the use of CO_2 in the operative field, are regular methods of de-airing. These routine cardiac de-airing techniques vary in time and intensity from surgeon to surgeon but also include ballottement, ventilation of the lungs, and filling of the heart. In the current era, TEE has been helpful in finding residual, pooled, or unexpected air (Table 2.11).

Once an air embolism occurs, there is a clear time-sensitive algorithm that should be followed. In addition, there are general principles that make sure that all the steps are addressed (Table 2.12) [49].

The keys to controlled treatment of air embolism include checklists, simulation, and good communication [21]. If the three main steps are retained in memory, the finer details of the necessary steps can be more easily recalled (Table 2.13) [50, 51]. For instance, stopping air entry includes clamping

Table 2.9 Sources of massive air embolism

Reversal of vent lines/tubing
Defective oxygenator or arterial tubing
Pressurized cardiotomy reservoir
Inattention to venous reservoir level
Introduction during aortic/venous cannulation or cardioplegia delivery
Intracardiac shunts
Inadequate de-airing maneuvers

Table 2.10 Prevention of air embolism: equipment and human error

1. Checklist of the CPB circuit by two people
2. Use of one-way valves
3. Mark tubing direction with arrows
4. Keep left ventricular vent clamped or do not place until on bypass or after cross-clamp
5. Consider one standardized setup for perfusion among all the surgeons
6. Perform simulation exercises for cardiac catastrophes
7. Check for tubing punctures, cuts, poor connections, and open stop cocks
8. Avoid excessive gradients of cooling or rewarming as >10–12 degrees can cause gas to come out of solution
9. Use CO_2 to displace air in the field
10. Ensure the needle vent in aorta is in correct position

Table 2.11 Prevention of air embolism: routine de-airing maneuvers

1. Clamping of aortic or antegrade cardioplegia line
2. Elective controlled ventricular fibrillation with cooling
3. Use of LV vent or PA vent
4. Needle vent in the aorta
5. Head down position
6. Flooding field with crystalloid or CO_2
7. CO_2 aspirated by pump suction
8. Pulmonary vein clearance
9. Ballottement of the heart
10. Use of TEE to check for residual air prior to cross-clamp removal

Table 2.12 Immediate steps in the management of air embolism

1. Stop air entry and identify the problem
2. Protect the brain by changing metabolic demand
3. Consider mechanical removal of air through retrograde or antegrade cerebral perfusion
4. Control inflammatory progression

Table 2.13 Treatment of massive air embolism

Stop air entry	Protect the brain	Control inflammation
Clamp the line	Deep Trendelenburg position	Cool the patient to 25 degrees
Turn off CPB circuit	Mannitol 1 g/kg × 1	Complete operation?
Identify the problem	Methylprednisolone 30 mg/kg IV × 1	Rewarm slowly to 35 degrees
Aspirate air from aortic vent (1 min)	Dexamethasone 20 mg IV x 1	Hyperbaric 02
Reprime lines to resume bypass	Ventilate 100% Fio2	Vasopressors to keep MAP 60–70 mmHg
	RCP at 20 degrees 1–2 min	
	Pack head in ice	

Table 2.14 Retrograde cerebral perfusion: technical steps

1. Clamp the aortic cannula and the aortic line tubing and separate the lines
2. Take the aortic line tubing and connect it to the venous line or place cannula directly into SVC
3. Begin retrograde perfusion at 20 °C hypothermia at 1–2 L/min for 1–3 min
4. Tell the perfusionist to flow 200–500 cc/min and keep the CVP <25
5. Put the aortic vent on high suction
6. Fill the aortic cannula and restart standard arterial perfusion
7. Manual compression of carotids

the line and turning off the circuit. Protecting the brain includes ice, increasing oxygenation, increasing perfusion pressure, medications, and retrograde cerebral perfusion (RCP). The key steps of RCP should be memorized (Table 2.14) [48, 50]. If possible, establishment of RCP with concurrent organization of an available hyperbaric oxygen chamber is beneficial [52].

2.7 Surgical Fire

There was an innocuous small cauterization of the lung during opening of the left pleura during a routine CABG surgery. The surgeon had noticed large blebs in the apex when dividing the upper pleura. While in the mid-portion of the mammary takedown, struggling with a large mammary vein, the surgeon hears the anesthesia resident describe some volume loss from the ventilator. The resident increases the gas flow and FiO_2 to compensate. After encountering some bleeding, the surgeon asks for a dry gauze and continues working with the cautery. Suddenly, there appeared a spark, smoke, flash, and a feeling of warmth.

The chest was on fire.

2.7.1 Discussion

Fires occur at a rate of approximately 100 cases per year in the operating rooms of the United States, with other reports showing 20–650 per year [53, 54]. The first thing to remember in this situation is how a fire begins. The classic "fire triangle/triad" includes fuel, oxygen, and an ignition source (Table 2.15) [54]. The concept of the fire triangle deals with all aspects of why the fire has occurred and thus how it is best controlled.

Once a fire occurs, specific actions are mandatory. The surgeon should determine the size of the fire. Sometimes, all that occurs initially is a smell or a feeling of warmth. In view of the fire triangle, these subtle hints help to immediately contain the fire before it escalates. By the time initial flames are spotted, a small fire can be controlled with a glove or a wet towel [55].

A large fire is a different story. A useful approach in these times of extreme crisis is to memorize a mnemonic so that no steps are forgotten. The surgeon must think fast and act faster using the mnemonic **CPR**: #1 **C**ontrol fire, #2 **P**atient, and #3 **R**ecovery (Table 2.16) [55]. Communication is key in this complication, and all team members must be taught their responsibilities during an event of this kind, *prior* to an actual fire occurring.

Like nearly every complication, but perhaps more dramatically, a surgical fire is an example of an event for which

Table 2.15 The classic fire triangle

Fuel	Prepping agent (contains alcohol), swabs, drapes, endotracheal tube, gloves, sheets, gowns, towels, dressings
Oxygen	Anesthetic gases, O2 (high concentrations above 70%), nitrous oxide
Ignition source	Lasers, electrocautery (85%), light source, defibrillators, camera

Table 2.16 Immediate treatment of a surgical fire

1. Stop the flow of oxidizers and gases by disconnecting the anesthesia circuit. Cut the flow of oxygen
2. Remove any burning materials: Swabs, drapes, etc. (such as a sponge in the chest). If possible, use an instrument to do so. (Do steps 1 and 2 together)
3. Extinguish the burning materials to the side or within the patient. This can be done with water or saline. A fire extinguisher can be used if large, but a wet towel may be enough
4. Provide care for the patient. Restore ventilation by re-establishing the airway, controlling bleeding, and checking for injuries
5. Evacuate the room if needed, depending on the amount of smoke or the size of the fire
6. Call the fire department if the entire room is involved

Table 2.17 Keys to prevention

Be aware where all heat sources are at all times on the operative field
Listen to alarms while operating. Ask for the source if it something new
Make sure all prepping agents are dry and prevent pooling on the operative field
Remember that oxygen is heavier than air and can collect in low areas, including the open chest cavity
Adjust O_2 to a minimum required on the ventilator for adequate PO_2
Control all heat sources (know where foot switches and cautery are at all times)
Consider fire drills for the OR team members. Focus on communication and practice
Have a mental and acting checklist that is systematic

the best treatment is prevention. Preparation and rehearsal are keys to this complication (Table 2.17) [56]. When it does occur, the unexpected and alarming nature of an operating room fire needs to result in an automatic and structured response by the cardiac surgical team.

2.8 Inferior Vena Cava Injury

A reoperative mitral replacement at 3 months post-initial surgery presents with difficult severe adhesions. The adhesions were particularly dense inferiorly, and the inferior vena cava (IVC) had been partially denuded of adventitia during the dissection. The IVC was thin-walled and nearly transparent. The surgeon lifted the heart slightly and placed a wet umbilical tape used as a snare under the short IVC. Suddenly, dark blood rushed from behind the IVC. Anesthesia is notified immediately as a cell saver suction is used to clear the blood. The heart is lifted to better appreciate the defect. A large posterior hole in the IVC is bleeding continuously.

2.8.1 Discussion

IVC injuries during cardiac surgery can be associated with a high mortality because of the difficulty of exposure for repair and the delicate tissue involved. This tear can easily extend into inaccessible regions below the diaphragm. Posterior, juxtadiaphragmatic, and retrohepatic tears are extremely difficult to fix surgically.

Prevention of what could be a potentially fatal injury includes an understanding of those patients that are at risk. These injuries are more frequently reported in the reoperative setting when anatomy and tissue planes are altered. Some common characteristics seen in these patients include women, reoperations, small left atrium, rheumatic heart disease, and prolonged ischemic times [57]. A short, adherent IVC can cause great difficulty [58]. Congenital surgery includes several procedures that are prone to IVC injury based on not only a reoperative setting, but also the right heart physiology/abnormal hemodynamics that result in changes in anatomy.

Direct injuries to the IVC can happen in any surgery, but can be more common in valve surgery when snares are placed around the IVC to control drainage and prevent air entry. Injuries can occur during dissection, particularly in reoperations, but can also happen during lifting of the heart, passing of a clamp or snare, and placement or repair of a purse string for IVC cannulation [57]. This injury can also happen during placement of the retrograde catheter, with aggressive suctioning in proximity to fragile tissue, or with placement of IVC cannula itself.

Depending on the extent of injury, there are a number of considerations, calculations, and decisions that need to be made quickly. Because the volume of blood lost from the venous system can be extensive and difficult to control, the principal tenet is how to best provide a quick and effective repair. This helps prevent further tearing in which the repair is ultimately successful, but can easily contribute to an overall poor outcome in terms of multiple transfusions and hypotension.

If an injury does occur, the treatment algorithm must include choices for closure based on findings. First, a decision must be made on how to *expose* the defect. This often includes careful lifting. Use of a side-biter clamp or vascular clamp around the defect to minimize immediate blood loss is one option, but metal clamps on thin IVC tissue have the potential to cause more injury. The possibility of posterior, retrohepatic, or infradiaphragmatic extension is always present. A takedown of the diaphragm and subcostal incision may be required depending on the injury. The site should be packed temporarily, if possible, and options for repair should be reviewed while calling for assistance.

Based on location of the injury, a repair technique must be chosen. If the injury is small, anterior, and easily accessible, one can consider direct repair. A purse-string suture is often the best option, sewn around a cannula or Foley placed directly into the defect. This provides the most controlled setting, preventing ongoing blood loss and creating only a short period of hypotension that the patient can tolerate (Fig. 2.4) [58]. Several other techniques have been described (Table 2.18) [57]. The common surgical principles of these repairs include careful handling of the tissue, minimal hypotension, and a repaired IVC that is not narrowed.

Unlike a trauma setting where cavoatrial shunt or total vascular isolation is required, the setting of median sternotomy and cardiac surgery provides some benefits in exposure and options. It is vital to be aware of the phase of the operation that the injury occurs and its relationship to CPB, which can provide some hemodynamic stability and drainage during repair. For example, a reoperation in which the major mediastinal structures have not yet been dis-

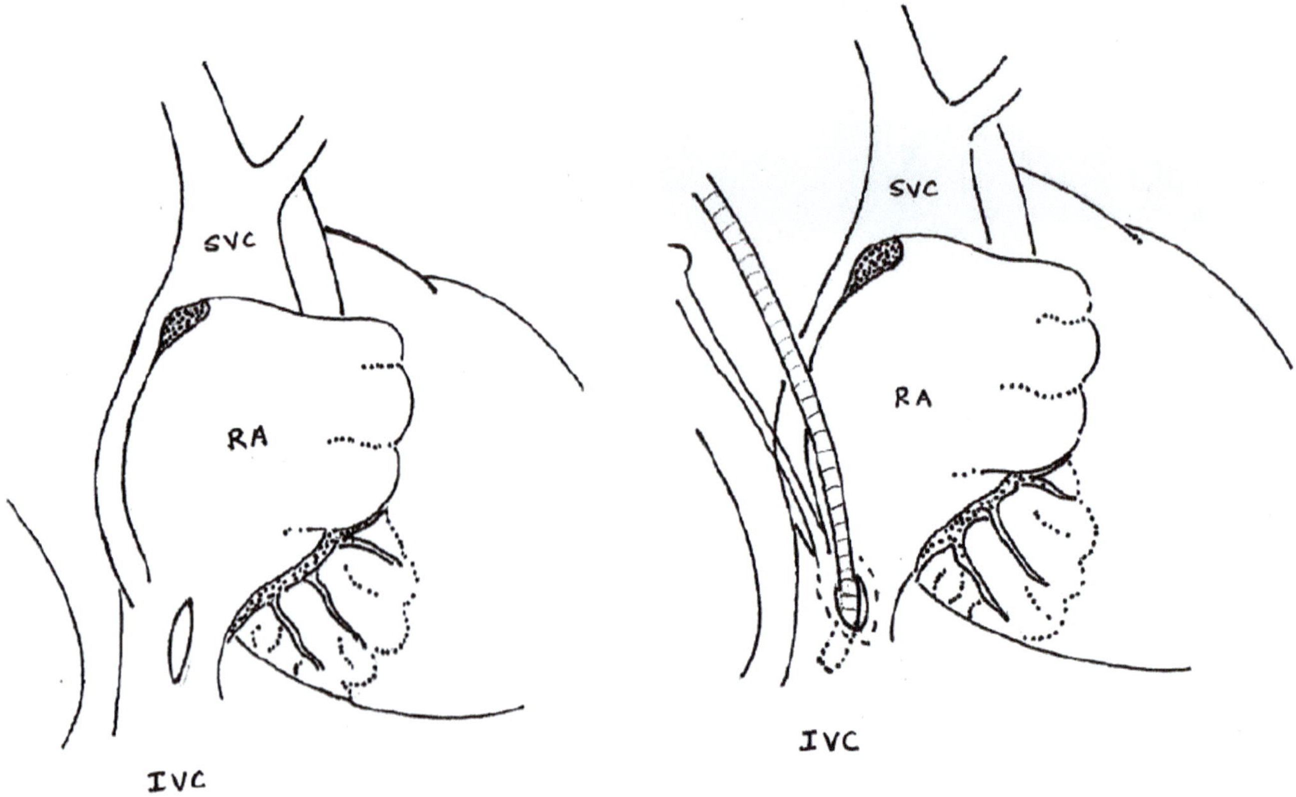

Fig. 2.4 Direct repair of IVC injury. *SVC, superior vena cava; RA, right atrium; IVC, inferior vena cava.* A curved IVC cannula is inserted in the tear to control bleeding and a purse-string suture placed around the tear. From Narayan P et al., Successful repair of iatrogenic inferior vena cava injury during cardiac surgery. Perfusion. 2015 Mar;30(2):166–8. Copyright 2015 by Sage Publications Ltd. Journals. Reprinted with permission

Table 2.18 Techniques of IVC repair

1. Direct repair
2. Atriocaval shunt
3. Lateral phleborrhaphy
4. Vein patch angioplasty
5. Tube graft
6. Vascular occlusion
7. Saphenous vein interposition graft
8. Use diaphragm or pericardial patch to repair
9. Consider endovascular covered stent graft for retrohepatic or inaccessible venous bleeding

sected supports immediate peripheral access for bypass. Bypass can commence through the SVC to give time to perform a repair, depending on the stage of the operation. A completed reoperative mitral replacement patient who has already been decannulated sometimes requires return to bypass for an inaccessible IVC injury. Clearly, risks and benefits have to weighed more heavily at this point. Return to bypass is not optimal, but good hemodynamic management is ultimately more critical for the patient in this period.

Finally, a patch repair is considered to prevent stenosis of a large defect (Fig. 2.5) [59]. Technically, the surgeon should leave 25% of the supradiaphragmatic IVC diameter to prevent stenosis and potential thromboembolism [60]. Endovascular repair can also be considered, depending on the location of the injury [61]. Whatever technique is chosen, the surgeon must constantly think about preventing *extension* of the tear. The IVC tissue, particularly in a reoperative setting, can be very delicate and does not tolerate suture placement well. Extension of a small defect can result in retrohepatic caval injury and subsequent catastrophe.

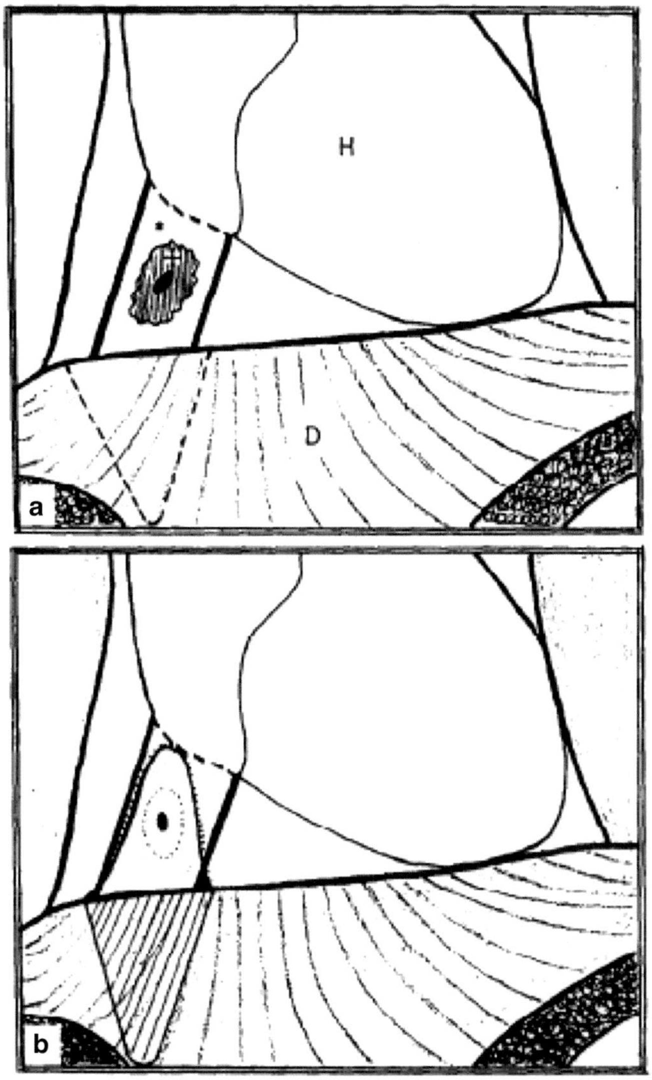

Fig. 2.5 Diaphragmatic patch repair of IVC injury. (**a**) Tailoring of a diaphragmatic patch to repair a low IVC injury. (**b**) Turning down a pericardial patch for repair. Both repairs use 5–0 Prolene suture. *D, diaphragm; H, heart.* Demirtas M, Aykut S, DaĞgsali S. Repair of the supra- diaphragmatic inferior vena cava lacerated during redo open-heart surgery. Cardiovasc Surg 1995;3:345–347. Copyright 1995 by Sage Publications Ltd. Journals. Reprinted with permission

2.9 Protamine Reaction

A routine AVR/CABG x 2 was planned for an elderly woman with mild pulmonary hypertension. The case proceeds smoothly, and the patient weans off CPB without difficulty, demonstrating normal cardiac function and rhythm. Protamine administration is requested.

All appears to be in order as the surgeon prepares to place drainage tubes. Suddenly, the right heart starts distending. The monitor shows rapidly rising pulmonary artery pressures from 40 to 60 with a concomitant drop on systemic blood pressure. A glance directly at the heart shows an RV that is massively dilated, slow, and struggling. As the resident announces the near completion of the protamine dose, everything escalates. The patient has profound hypotension. She requires large boluses of epinephrine and vasopressin followed by open cardiac massage.

2.9.1 Discussion

The use of protamine to reverse the effects of heparin at the conclusion of cardiopulmonary bypass operations is mandatory and occurs at a rate of millions of operations per year worldwide. Protamine molecules, required for spermatogenesis, play an essential role in cardiac surgery. The biochemical properties of this highly *positively* charged polypeptide allow it to bind to negative phosphate groups. It is exceedingly effective for neutralizing heparin, a highly *negatively* charged glycosaminoglycan [62]. This bond of highly charged chemicals, containing a combination of positive and negative ion groups, results in multiple effects.

What is observed clinically in the operating room varies dramatically. A simple drop in blood pressure to some degree is a common occurrence after protamine infusion. The addition of protamine alone or to counter systemic heparin can decrease systolic/diastolic pressures, cardiac output, heart rate, and systemic vascular resistance (SVR); it is this drop in afterload that causes the main drop in BP [63–65]. A release of histamine is thought to be responsible for the decrease in SVR, as protamine activates mast cells by multiple mechanisms [62]. An increase in thromboxane, complement, and antibodies may also occur [62]. In addition, protamine has also been shown to have a direct effect on the myocardium by binding to the sarcolemma and decreasing myocardial contractility, particularly if given alone, without circulating heparin to bind to preferentially [66].

Protamine has other effects after CPB. It can increase pulmonary artery pressures, documented in the setting of hemodynamic monitoring. The "classic" pulmonary hypertension that is observed after protamine administration is due to the release of thromboxane that occurs with cardiopulmonary bypass. The neutralization of heparin causes a fall in phospholipase and prostaglandins (which also increase with CPB) but does not decrease thromboxane. This can lead to severe pulmonary vasoconstriction, with high thromboxane levels causing endothelial constriction [67] . Protamine has also been shown to increase nitric oxide production [68]. This exact mechanism is unknown but can also contribute to decreasing SVR [68]. Furthermore, complement pathways are activated during every CPB operation. As the CPB circuit itself activates the *alternative* complement pathway, heparin-protamine complexes activate the *classic* complement pathway, all adding to the hemodynamic effects [64].

In summary, the phrase "protamine reaction" indicates a range of effects, from zero effect to complete cardiovascular collapse. Catastrophic events include profound hypotension due to rapid infusion, a true anaphylaxis with complement production, an allergic reaction due to the presence of IgG or IgM antibodies, or a severe pulmonary vasoconstriction with resultant right heart failure [65]. A sudden hemodynamic change should be addressed during every protamine administration.

Specific groups have been shown to be at risk for significant adverse effects from protamine. Classic teaching is that patients with a fish allergy, prior vasectomy, or exposure to NPH insulin are at increased risk. This is based on the natural source of protamine, salmon semen. Reoperative patients who have been exposed to protamine previously should also be given the drug cautiously. Not all studies, however, support these traditional risk factors. Use of NPH, fish allergy, and *non*-protamine drug allergy were found to be significant on multivariate analysis in one study [69]. If a patient has a known history of protamine reaction, a measurement of protamine antibodies is helpful. The use of corticosteroids and histamine blockers prior to administration may prevent a second reaction.

A *true* protamine reaction usually occurs within 20 minutes of administration, without any other cause, and is described as a drop of BP of at least 25% that requires the addition of inotropes [69]. Severe and moderately severe adverse events from protamine are all associated with some degree of hemodynamic changes. The first step in support is to start inotropes and pressors to increase the blood pressure. Calcium is often administered. If there is no marked improvement in hypotension, re-institution of CPB is required. Depending on the infused dosage of protamine, heparin needs re-dosing prior to CPB. Waiting for the reaction to clear while on pump is the safest next step.

At this point, one must make sure there are no concurrent or confounding surgical issues such as new ischemia and graft or valve dysfunction. An IABP is sometimes necessary for hemodynamic support. Readministering of protamine is not recommended in the setting of a true protamine reaction. The alternative, however, is waiting for the heparin to metabolize and clear on its own. This necessitates a long period of waiting and potential bleeding in the operating room, and the risks and benefits of this option must be carefully weighed.

2.10 Right Ventricular Failure

The case was a routine MVR/CABG x 1 (SVG to RCA) with a straightforward bypass run of 90 minutes. The patient is weaned off bypass easily. Protamine is administered, and the patient requires only low-dose inotropes at this point. The heparin has been reversed, and the patient was off-pump for at least 45 minutes with what appeared to be normal biventricular function. After obtaining hemostasis, the cell saver is returned to the patient, and sternal wire placement is begun. Suddenly, the right heart becomes massively dilated and hypokinetic. The monitor shows PA pressures of mid-60s and a systolic pressure drop to mid-80s. Calcium and pressors are administered without significant effect. The arterial pressure continues to drop, and the anesthesiologist reports massive RV dilation, moderate to severe tricuspid regurgitation, and marked RV hypokinesis on the TEE with new progressive LV dysfunction.

2.10.1 Discussion

The differences between the right ventricle (RV) and left ventricle (LV) are varied and complex in terms of anatomy, physiology, and pathology.

Essential RV facts to remember include:

1. The RV coronary circulation fills during both systole and diastole. In fact, 2/3 of filling occurs during systole so measured aortic pressure is vital for adequate RV perfusion.
2. As a result of this coronary flow, the RV is more sensitive to changes in afterload and wall tension.
3. The LV dysfunction that can occur after RV dysfunction can be due to a change in the septal and geometric shape of the LV, in addition to a decrease in forward flow.
4. Alpha agonists used to increase blood pressure can decrease the contractility of the RV.
5. The RV can adapt against a pressure gradient of pulmonary hypertension to a much greater degree than the LV can adapt to high LV afterload. This adaptation occurs through hypertrophy, but may result in relative ischemia of the myocardial tissue [70].

It is well understood that CPB can cause damaging effects to the right heart. The onset or increase of pulmonary hypertension is common while weaning off cardiopulmonary bypass due to pulmonary endothelial cell dysfunction, hypothermia, inflammatory mediators, nonpulsatile flow, activated free radicals, and complement [71]. Pressure and volume overload, ischemia, and poor protection can all cause maladaptive work of the RV. Contraction itself occurs in a delay between the apex and the conus sections of the right ventricle. This can appear at times as a peristalsis effect to the eye of the surgeon who has the RV well exposed in the operative field. The contraction of the RV can reflect subtle changes in PA pressures during medication administration and a change in loading conditions. The ultimate outcome of this can be difficulty in weaning of CPB [71, 72] (Fig. 2.6).

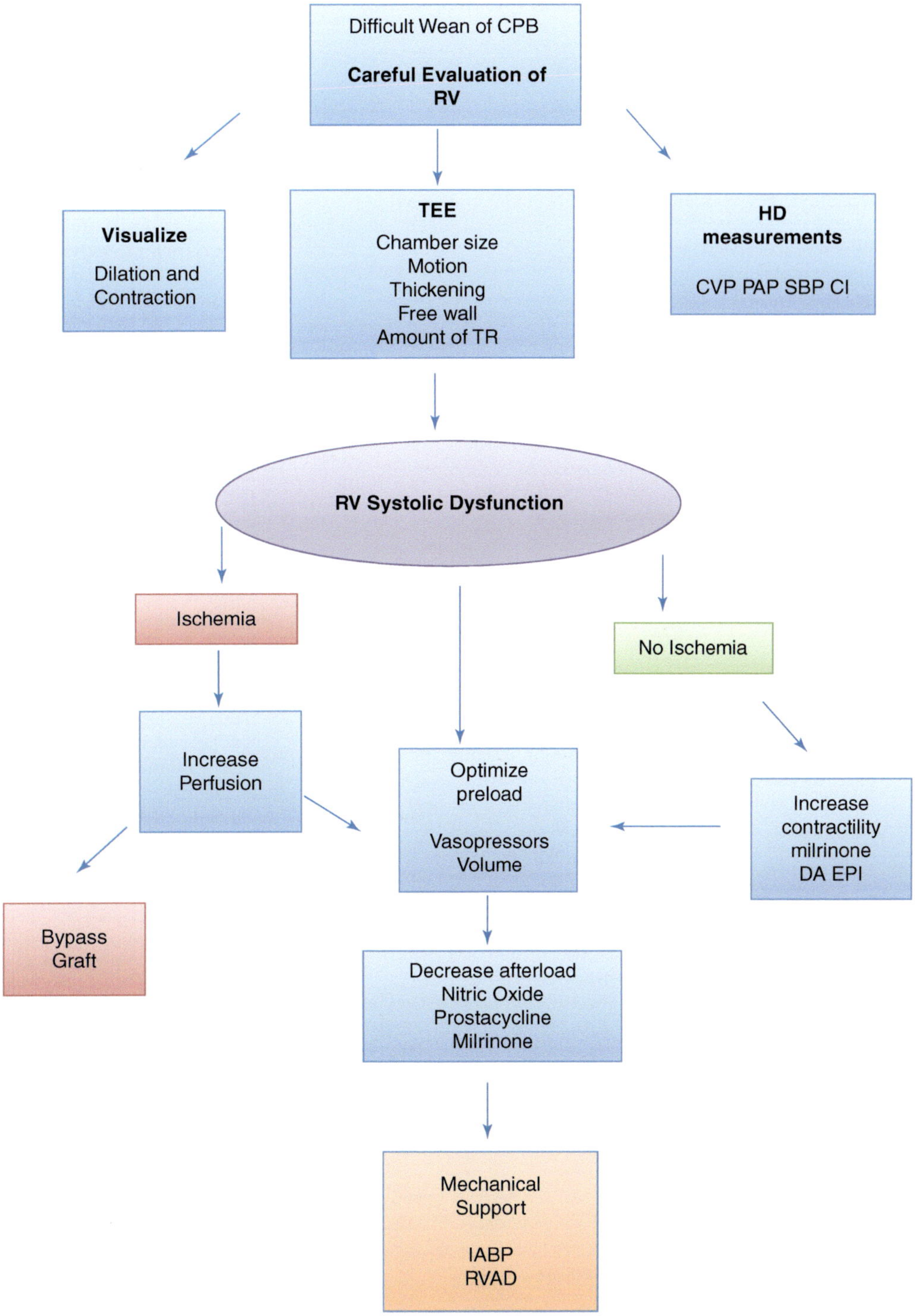

Fig. 2.6 Management of intraoperative RV systolic dysfunction. *RV, right ventricle, TEE, transesophageal echo, CVP, central venous pressure, PAP, pulmonary artery pressure, SBP, systolic blood pressure, CI, cardiac index, IABP, intra-aortic balloon pump, RVAD, right ventricular assist device, TR, tricuspid regurgitation, DA, dopamine, EPI, epinephrine*

Right ventricular dysfunction can be unforgiving at an early stage, but has been shown to improve with careful management (Table 2.18) [71–73]. The options for treatment must be reviewed deliberately and sequentially. The immediate onset of dysfunction is a critical time, as a spiral into severe RV and then LV dysfunction may occur. Problems with RV function can occur during weaning of CPB, after separation from bypass, or at any time throughout the immediate postoperative period. Actions such as packing the chest with a lap sponge while checking the sternal edges for bleeding, return of blood from the pump into the aortic cannula prior to decannulation, or rapid protamine administration, can push the RV over its optimal filling into a picture of dilation and severe hypokinesis. An IABP, conventionally used for decreasing LV afterload, is also used to augment pressures and increase the pulmonary-LA pressure gradient. Considering the potential causes while providing support is necessary in the management of these cases (Table 2.19).

Table 2.19 Management of RV dysfunction

1. **REVERSE**
Fix reversible factors for pulmonary HTN: Hypoxia, hypercarbia, acidosis, hypokalemia
Adjust ventilator settings to decrease peak inspiratory pressures
2. **OPTIMIZE**
Improve RV hemodynamics with preload without overdistension
Reduce afterload with addition of vasodilators (milrinone, nitric oxide, epoprostenol)
Add inotropes
Keep RV perfusion pressures high
Normalize and optimize rhythm and rate
Increase RCA perfusion
Use CVP measurements to evaluate preload
Afterload reduction
3. **SUPPORT**
Consider coronary bypass for signs of RCA ischemia
Add alpha agonist to raise perfusion pressures
Insert IABP
Consider RVAD

References

1. Gibbon JH Jr. The application of a mechanical heart and lung apparatus to cardiac surgery. Minn Med. 1954;37:171.
2. Passaroni AC, Silva MA, Yoshida WB. Cardiopulmonary bypass: development of John Gibbon's heart-lung machine. Rev Bras Cir Cardiovasc. 2015;30(2):235–45.
3. Romaine-Davis A. John Gibbon and his heart-lung machine. Philadelphia, Pa: University of Pennsylvania Press; 1991.
4. Stoney WS. Evolution of cardiopulmonary bypass. Circulation. 2009;119(21):2844–53.
5. Cohn LH. Fifty years of open-heart surgery. Circulation. 2003;107(17):2168–70.
6. Hessel EA 2nd. A brief history of cardiopulmonary bypass. Semin Cardiothorac Vasc Anesth. 2014;18(2):87–100.
7. Edmunds LH Jr. The evolution of cardiopulmonary bypass: lessons to be learned. Perfusion. 2002;17(4):243–51.
8. Staples MH, Dunton RF, Karlson KJ, Leonardi HK, Berger RL. Heparin resistance after preoperative heparin therapy or intraaortic balloon pumping. Ann Thorac Surg. 1994;57:1211–6.
9. Ranucci M, IsgroÁ G, Cazzaniga A, et al. Predictors for heparin resistance in patients undergoing coronary artery bypass grafting. Perfusion. 1999;14:437–42.
10. Finley A, Greenberg C. Review article: heparin sensitivity and resistance: management during cardiopulmonary bypass. Anesth Analg. 2013;116(6):1210–22.
11. Matthai WH Jr, Kurnik PB, Groh WC, Untereker WJ, Siegel JE. Antithrombin activity during the period of percutaneous coronary revascularization: relation to heparin use, thrombotic complications and restenosis. J Am Coll Cardiol. 1999;33:1248–56.
12. Pleym H, Videm V, Wahba A, Asberg A, Amundsen T, Bjella L, Dale O, Stenseth R. Heparin resistance and increased platelet activation in coronary surgery patients treated with enoxaparin preoperatively. Eur J Cardiothorac Surg. 2006;29:933–40.
13. Mintz PD, Blatt PM, Kuhns WJ, Roberts HR. Antithrombin III in fresh frozen plasma, cryoprecipitate, and cryoprecipitate- depleted plasma. Transfusion. 1979;19:597–8.
14. Levy JH, Despotis GJ, Szlam F, Olson P, Meeker D, Weisinger A. Recombinant human transgenic antithrombin in cardiac surgery: a dose-finding study. Anesthesiology. 2002;96:1095–102.
15. Ranucci M, Isgrò G, Cazzaniga A, Ditta A, Boncilli A, Cotza M, Carboni G, Brozzi S. Different patterns of heparin resistance: therapeutic implications. Perfusion. 2002;17:199–204.
16. Talstad I, Scheie P, Dalen H, Roli J. Influence of plasma proteins on erythrocyte morphology and sedimentation. Scand J Haematol. 1983;31:478–84.
17. Agarwal SK, Ghosh PK, Gupta D. Cardiac surgery and cold-reactive proteins. Ann Thorac Surg. 1995;60:1143–50.
18. Patel PA, Ghadimi K, Coetzee E, Myburgh A, Swanevelder J, Gutsche JT, Augoustides JGT. Incidental cold agglutinins in cardiac surgery: intraoperative surprises and team-based problem-solving strategies during cardiopulmonary bypass. J Cardiothorac Vasc Anesth. 2017;31(3):1109–18.
19. Barbara DW, Mauermann WJ, Neal JR, et al. Cold agglutinins in patients undergoing cardiac surgery requiring cardiopulmonary bypass. J Thorac Cardiovasc Surg. 2013;146:668–80.
20. Atkinson VP, Soeding P, Horne G, Tatoulis J. Cold agglutinins in cardiac surgery: management of myocardial protection and cardiopulmonary bypass. Ann Thorac Surg. 2000;85:310–1.
21. Mukherji J, Hood RR, Edelstein SB. Overcoming challenges in the management of critical events during cardiopulmonary bypass. Semin Cardiothorac Vasc Anesth. 2014;18(2):190–207.
22. Youngblood SC, Deng Y, Chen A, et al. Perioperative therapeutic plasmapheresis. Anesthesiology. 2013;118:722–8.
23. Panos A, Murith N, Myers PO, et al. Aortic arch repair and cold-reactive agglutinins: what to do? Ann Thorac Surg. 2007;84:1403–4.
24. Menasche P, Piwnica A. Cardioplegia by way of the coronary sinus for valvular and coronary surgery. J Am Coll Cardiol. 1991;18(2):628–36.
25. Sabzi F, Zokaei A. Factors predicting coronary sinus rupture following cannula insertion for retrograde cardioplegia. Clin Med Insights Cardiol. 2012;6(1–6):954.
26. Netter FH. Atlas of human anatomy. 2nd ed. East Hanover, NJ: Novartis; 1997. plates 202, 208
27. Menasché P, Subayi JB, Piwnica A. Retrograde coronary sinus cardioplegia for aortic valve operations: a clinical report on 500 patients. Ann Thorac Surg. 1990;49(4):556–63. discussion 563–4.
28. Kurusz M, Girouard MK, Brown PS Jr. Coronary sinus rupture with retrograde cardioplegia. Perfusion. 2002;17(1):77–80.
29. Orihashi K, Miyashita K, Tashiro M, Kihara K, Kondo N, Yamamoto M, Hirose N, Fukutomi T, Nishimori H. Avoidance of coronary sinus injury during retrograde cardioplegia. Ann Thorac Surg. 2016;102(6):e583–6.
30. Panos AL, Ali IS, Birnbaum PL, Barrozo CA, al-Nowaiser O, Salerno TA. Coronary sinus injuries during retrograde continuous normothermic blood cardioplegia. Ann Thorac Surg. 1992;54(6):1137–8.
31. Aigner C, Wolner E, Mohl W. Management of central coronary sinus ruptures using the pericardial patch repair technique. Ann Thorac Surg. 2006;81:1275–8.
32. Economopoulos GC, Michalis A, Palatianos GM, Sarris GE. Management of catheter-related injuries to the coronary sinus. Ann Thorac Surg. 2003;76:112–6.
33. Kazui T, Lin TM, Lick SD. Coronary sinus rupture repair: patency is important. Ann Thorac Surg. 2018;106(1):e25–6.
34. Weltert L, Wolf LG, Turani F, De Paulis R. Repair of ruptured coronary sinus by sole apposition of self-adhesive sealing hemostatic patch. J Thorac Cardiovasc Surg. 2019;157(4):e189–91.
35. Panagiotopoulos N, Patrini D, Adams B, Lawrence D, Roberts N. A novel sutureless technique for the repair of coronary sinus injuries. J Thorac Dis. 2015;7(12):2359–61.
36. Edwards J, DuShane J. Thoracic venous anomalies. Arch Pathol. 1950;49:514–37.
37. Tan JH, Ng ZQ, Vendargon S. Persistent left superior vena cava and absence of innominate vein during coronary artery bypass surgery. BMJ Case Rep. 2018;2018:bcr2018225271.

38. Mazzucco A, Bortolotti U, Stellin G, Gallucci V. Anomalies of the systemic venous return: a review. J Card Surg. 1990;5(2):122–33.
39. Soward A, ten Cate F, Fioretti P, et al. An elusive persistent left superior vena cava draining into left atrium. Cardiology. 1986;73:368–71.
40. Nsah EN, Moore GW, Hutchins GM. Pathogenesis of persistent left superior vena cava with a coronary sinus connection. Pediatr Pathol. 1991;11:261–9.
41. Sheikh AS, Mazhar S. Persistent left superior vena cava with absent right superior vena cava: review of the literature and clinical implications. Echocardiography. 2014;31(5):674–9.
42. Lentini S, Recupero A. Recognition of persistent left superior vena cava in non-congenital patients undergoing cardiac surgery. Perfusion. 2011;26(4):347–50.
43. Hwang C, Wu TJ, Doshi RN, et al. Vein of Marshall cannulation for the analysis of electrical activity in patients with focal atrial fibrillation. Circulation. 2000;101:1503–5.
44. Hanson EW, Hannan RL, Baum VC. Pulmonary artery catheter in the coronary sinus: implications of a persistent left superior vena cava for retrograde cardioplegia. J Cardiothorac Vasc Anesth. 1998;12:448–9.
45. Spampinato N, Stassano P, Gagliardi C, Tufano R, Iorio D. Massive air embolism during cardiopulmonary bypass: successful treatment with immediate hypothermia and circulatory support. Ann Thorac Surg. 1981;32(6):602–3.
46. Chase WH. Anatomical and experimental observations on air embolism. SGO. 1934;59:569–77.
47. Kurusz M, Butler B, Katz J, Conti VR. Air embolism during cardiopulmonary bypass. Perfusion. 1995;10(6):361–91.
48. van der Zee MP, Koene BM, Mariani MA. Fatal air embolism during cardiopulmonary bypass: analysis of an incident and prevention measures. Interact Cardiovasc Thorac Surg. 2014; 19(5):875–7.
49. Quintero OL, Giraldo JC, Sandoval NF. Successful management of massive air embolism during cardiopulmonary bypass using multimodal neuroprotection strategies. Semin Cardiothorac Vasc Anesth. 2019;23(3):324–32.
50. Tarola CL, Hirji S, Yule SJ, Gabany JM, Zenati A, Dias RD, Zenati MA. Cognitive support to promote shared mental models during safety-critical situations in cardiac surgery (Late Breaking Report). 2018 IEEE Conf Cogn Comput Asp Situat Manag CogSIMA (2018). 2018;2018:165–167.
51. Mills NL, Ochsner JL. Massive air embolism during cardiopulmonary bypass. Causes, prevention, and management. J Thorac Cardiovasc Surg. 1980;80(5):708–17.
52. Gibson AJ, Davis FM. Hyperbaric oxygen therapy in the treatment of post cardiac surgical strokes–a case series and review of the literature. Anaesth Intensive Care. 2010;38(1):175–84.
53. Rinder CS. Fire safety in the operating room. Curr Opin Anaesthesiol. 2008;21(6):790–5.
54. Yardley IE, Donaldson LJ. Surgical fires, a clear and present danger. Surgeon. 2010;8(2):87–92. https://doi.org/10.1016/j.surge.2010.01.005. Epub 2010 Feb 18
55. When a fire occurs... Knowing how to respond in the event of a surgical fire. Health Devices. 2006;35(2):58–63.
56. Kezze I, Zoremba N, Rossaint R, Rieg A, Coburn M, Schälte G. Risks and prevention of surgical fires : a systematic review. Anaesthesist. 2018;67(6):426–47.
57. Güney MR, Aksoy N, Sezerman O. Repair of lacerated inferior vena cava with the intravascular tube graft in redo cardiac surgery. J Thorac Cardiovasc Surg. 2001;121:187–8.
58. Narayan P, Das D, Saha A, Das M. Successful repair of iatrogenic inferior vena cava injury during cardiac surgery. Perfusion. 2015;30(2):166–8.
59. Demirtas M, Aykut S, DaĞgsali S. Repair of the supra- diaphragmatic inferior vena cava lacerated during redo open-heart surgery. Cardiovasc Surg. 1995;3:345–7.
60. Buckman RF, Pathak AS, Badellino MM, Bradley KM. Injuries of the inferior vena cava. Surg Clin North Am. 2001;81:1431–47.
61. Briggs CS, Morcos OC, Moriera CC, Gupta N. Endovascular treatment of iatrogenic injury to the retrohepatic inferior vena cava. Ann Vasc Surg. 2014;28(7):1794.e13–5.
62. Carr JA, Silverman N. The heparin-protamine interaction. A review. J Cardiovasc Surg. 1999;40(5):659–66.
63. Shapira N, Schaff HV, Piehler JM, White RD, Still JC, Pluth JR. Cardiovascular effects of protamine sulfate in man. J Thorac Cardiovasc Surg. 1982;84(4):505–14.
64. Kirklin JK, Chenoweth DE, Naftel DC, Blackstone EH, Kirklin JW, Bitran DD, Curd JG, Reves JG, Samuelson PN. Effects of protamine administration after cardiopulmonary bypass on complement, blood elements, and the hemodynamic state. Ann Thorac Surg. 1986;41(2):193–9.
65. Horrow JC. Protamine: a review of its toxicity. Anesth Analg. 1985;64:348–61.
66. Hird RB, Wakefield TW, Mukherjee R, Jones BU, Crawford FA, Andrews PC, Stanley JC, Spinale FG. Direct effects of protamine sulfate on myocyte contractile processes. Cellular and molecular mechanisms. Circulation. 1995;92(9 Suppl):II433–46.
67. Nakamura H, Kim DK, Philbin DM, Peterson MB, Debros F, Koski G, Bonventre JV. Heparin-enhanced plasma phospholipase A2 activity and prostacyclin synthesis in patients undergoing cardiac surgery. J Clin Invest. 1995;95(3):1062–70.
68. Li JM, Hajarizadeh H, La Rosa CA, Rohrer MJ, Vander Salm TJ, Cutler BS. Heparin and protamine stimulate the production of nitric oxide. J Cardiovasc Surg. 1996;37(5):445–52.
69. Kimmel SE, Sekeres MA, Berlin JA, Ellison N, DiSesa VJ, Strom BL. Risk factors for clinically important adverse events after protamine administration following cardiopulmonary bypass. J Am Coll Cardiol. 1998;32(7):1916–22.
70. Sanz J, Sánchez-Quintana D, Bossone E, Bogaard HJ, Naeije R. Anatomy, function, and dysfunction of the right ventricle: JACC state-of-the-art review. J Am Coll Cardiol. 2019;73(12):1463–82.
71. Riedel B. The pathophysiology and management of perioperative pulmonary hypertension with specific emphasis on the period following cardiac surgery. Int Anesthesiol Clin. 1999. Spring;37(2):55–79.
72. Wanner PM, Filipovic M. The right ventricle-you may forget it, but it will not forget you. J Clin Med. 2020;9(2):432.
73. Haddad F, Couture P, Tousignant C, Denault AY. The right ventricle in cardiac surgery, a perioperative perspective: I. anatomy, physiology, and assessment. Anesth Analg. 2009;108(2):407–21.

3 Coronary Artery Bypass Surgery

3.1 Carbon Dioxide Embolism

The alarm starts slowly, and the surgeon looks up from the operative field. At the same time, there are multiple bleeding mammary vein branches that require ligation and the surgeon's immediate attention. Unknown to him, the physician assistant had difficulty with CO_2 flow on the endoscopic vein harvest machine and had increased the insufflation pressure to 15 mmHg.

Meanwhile, the patient undergoes subtle changes. Acute hypercarbia and a bulging left diaphragm are noted. The end-tidal monitor shows a rising CO_2 to 45 mmHg, and the alarm continues. The pulmonary artery pressures rise slowly but steadily, and the patient develops a respiratory acidosis. Suddenly, there are markedly unstable hemodynamics. A quick glance at the TEE shows a large bubble of air in the right ventricular outflow tract.

3.1.1 Discussion

There are events that can be difficult to manage at the initial stages of a coronary bypass operation, a time when the surgeon and assistants are busy procuring the conduit required. The most serious of these involve hemodynamic instability. Acute cardiac ischemia is often and correctly the first suspicion, but there are other possibilities. CO_2 embolism is rare, but well described in the literature [1–4]. It can be accompanied by hemodynamic compromise. Many surgeons have never experienced it, even in the modern era of universal endoscopic vein harvesting (EVH).

The true incidence of CO_2 embolism is likely underestimated; the presentation is quite variable. One observational study showed that massive CO_2 embolus with hemodynamic shifts occurs in 0.5% of patients undergoing EVH, but up to 3.5% of patients were found to have moderate bubbles in the IVC [1]. On TEE, this same study showed the presence of CO_2 bubbles can occur in up to 17% of EVH patients [1].

The mechanisms by which CO_2 enters the circulation during EVH are either (1) direct absorption into tissue and subsequently the circulation or (2) direct entry into a vessel. With moderately large or massive embolism, this direct entry occurs most commonly at the time of ligation or manipulation of branches of the greater saphenous vein [1]. A tear may occur just at the saphenofemoral junction, leaving a clear trajectory to the IVC while the harvester is working at the upper thigh [1].

In order to avoid direct entry by this mechanism, CO_2 pressure is carefully regulated. Many programs use 12 mmHg as their limit. When using a strict pressure limit of 15 mmHg in the EVH system, severe embolism occurred in 4% of patients as reported in one study [1]. The emergent nature of CO_2 embolism is based on its rapid, unexpected onset and immediate need for diagnosis. The presentation varies from no clinical effect to full cardiogenic shock (Table 3.1) [1, 2]. The anesthesiologist observes increased $ETCO_2$, while the surgeon sees rising CVP, elevated PA pressures, and RV strain. One must be aware that $ETCO_2$ does not increase in all cases of embolism. In fact, as seen in cardiac arrest situations, $ETCO_2$ can actually decrease with cardiogenic shock [1].

It has been suggested that CO_2 absorption plays very little role, compared to CO_2 embolism entering directly into the peripheral vein and traveling to the IVC. A very large bolus of air can migrate to the heart and cause an airlock in the venous cannula if present or in the actual right ventricular outflow tract.

Table 3.1 CO_2 embolism: presentations

Sudden hypotension
Possible rise of $ETCO_2$
Rising pulmonary artery pressures
Rising CVP
Recent increase in insufflation for EVH
Acute RV failure
Cardiac arrest/shock
Mill wheel murmur
Increased peak airway pressures

S. K. Balaram, L. Bassin, *Cardiac Surgical Complications*, https://doi.org/10.1007/978-3-030-71563-2_3

Table 3.2 CO_2 embolism: first steps in management

1.	Turn off CO_2 insufflation
2.	Compress the saphenofemoral junction
3.	Reposition the patient: Trendelenburg head down and airplane to the left
4.	Review the TEE with mid-esophageal bicaval view and look for air embolism, including at the RV apex
5.	Remember that CO_2 can be quickly absorbed due to its solubility, so rapid assessment will help to determine the next steps

Table 3.3 CO2 embolism: treatment for shock

1.	Increase preload with volume; fill the right heart to displace the air
2.	Increase blood pressure with vasopressors
3.	Use inotropes to increase cardiac output
4.	Needle extraction can be used for a large air embolus visualized in the RV apex
5.	Consider IV heparin and initiation of CPB
6.	Initiate emergent CPB for large air embolism in the setting of a PFO/ASD

Prevention of this potentially catastrophic event includes focused education of the cardiac team regarding safe EVH insufflation pressures and methods of branch ligation, particularly above the knee. An insufflation pressure of 12 mmHg reduces the chance of embolism during EVH [1]. Early consideration of an unexpected CO2 embolism during *any* endoscopic radial or vein harvest must occur if there is an unexpected hemodynamic change during the time of conduit harvest. The operative technique itself must be careful and direct, particularly at the saphenofemoral junction. Every time a branch is ligated, good hemostasis with proximal and distal control should be insured. Some have suggested empirically dividing the saphenous vein at the proximal and distal ends during the initial phase of the dissection to avoid any entry of CO2 during manipulation of the vein and its branches [4].

The treatment of CO2 embolization is based on the severity of presentation, which in turn is based on the volume and rate of embolus. Initial management is similar to any air embolism (Table 3.2) [1]. If the embolism is large, the patient will need treatment for shock (Table 3.3) [5].

Other described complications of the EVH in coronary revascularization include intraluminal clot stranding, compartment syndrome, systemic CO2 absorption with hypercarbia, pneumoperitoneum, and scrotal distention [6].

3.2 Coronary Steal

A nephrologist in the dialysis unit calls the surgeon's office to discuss a 60-year-old patient that underwent CABG 1 year ago. The patient is now being dialyzed through a left upper arm arteriovenous fistula that he received 2 months prior. Each time the patient is placed on dialysis, he has acute substernal chest pain and is mildly diaphoretic. He has some hypotension, but more significantly, he has experienced episodes of acute pulmonary edema after dialysis, despite volume removal. The chest pain he is having is impacting all his dialysis sessions.

3.2.1 Discussion

Coronary steal has been reported in some cases when the left internal mammary artery (LIMA) is anastomosed to the left anterior descending (LAD) artery, a technique repeated in nearly all coronary revascularization procedures. Various forms of reversal of flow or "steal" have been reported in up to 4.5% of patients [7]. A simple example of this is when patent anterior intercostal branches remain after in situ LIMA harvest for CABG. The first intercostal branch can be a culprit, redirecting blood away from the sutured anastomosis present on the epicardial surface. Steal has also been shown to occur in the presence of proximal subclavian stenosis in which blood flows backward from the heart toward the vertebral system [8, 9].

This steal phenomena can also occur in the setting of an upper extremity arteriovenous fistula (AVF) with an ipsilateral internal mammary artery graft in place. The diagnosis can be made when chest pain, anterior wall hypokinesis, or EKG changes occur during dialysis. Some patients with steal do not have angina, and other patients with angina do not have steal.

An interesting phenomena exists when considering a high-flow left upper arm arteriovenous fistula (AVF) for dialysis access and its connection to the delivery of oxygenated blood to the anterior myocardial wall. Physiologically, after ligation of the distal LIMA and the anastomosis, the flow in the artery changes from occurring during both systole and diastole to occurring mainly during diastole. The LIMA becomes part of the coronary circulation, dependent on the drop in peripheral resistance for adequate flow during diastole. This physiologic principle generally protects the myocardium from steal [10], but the addition of an ipsilateral high-flow AVF sometimes actually drops resistance in the other direction, causing a reversal of flow from the IMA to the subclavian [5].

In 2003, this phenomena was demonstrated by Gaudino and colleagues in five patients with a patent LIMA to LAD anastomosis and ipsilateral AVF [11]. The study utilized TTE and Doppler studies during dialysis treatment to show increased flow through the fistula with reduced peak systolic flow, end- diastolic velocity, and increased pulsatility index in the ipsilateral ITA but not in the contralateral artery [11]. The measurement of time average mean velocity (TAMV) drop was associated with a hypokinetic anterior wall in all five patients and chest pain in three patients, all of which returned to normal after dialysis was complete [11]. The contralateral flow in the mammary artery was not changed, making systemic hypotension or hemodynamic shifts of dialysis not the likely etiology [11]. A separate study showed retrograde flow during HD through a left arm AVF even with an unharvested LIMA, prior to CABG [12].

However, subsequent studies, although including few patients, found no increased mortality or cardiac risk based on ipsilateral LIMA and high AVF flow [10, 13, 14]. Rahbar and colleagues looked at 15 patients with ultrasound Doppler flow in the LIMA and an unused RIMA and found no difference between them during HD [13]. Coskun et al. studied 22 patients and did not find a significant change in LIMA flow although three patients had ventricular wall hypokinesia, EKG changes, and angina [10]. Furthermore, the largest published study in 2013 of 155 patients with ipsilateral IMA and AVF found no clinical significance in early or late outcomes over 5 years [14].

Although mortality has not been shown to be affected, these patients may be more at risk for chest pain with dialysis related to coronary steal, and other authors recommend free LIMA grafts for this reason [15]. AV fistulas can increase their flow as they mature over time, thus causing more of a potential steal on proximal vessels. In 2018, a study of 85 HD patients found that those with in situ ipsilateral grafts had more chest pain with dialysis with an odds ratio of 5.528 [15]. In fact, the incidence of chest pain with dialysis was 45.5% for noncomposite in situ grafts, 19.4% for in situ composite grafts, and only 5.7% for free grafts ($p < 0.009$) [15].

Ahn et al. looked at the choice and placement of ipsilateral AVF in the left arm created *after* LIMA to LAD CABG surgery [16]. Twenty-two patients received ipsilateral AVF, and although none of these had documented chest pain, three (12%) of these patients had clinically significant steal syndrome from the AVF and were treated with closure of the fistula [16]. This group thus does *not* place fistulas on the same side as LIMA. They suggest that low EF is a risk factor and overall short- and long-term mortality is higher than in other groups [16]. Other specific risk factors that predispose to steal are subclavian stenosis, location and maturity of the AV fistula, graft conduit choice, and a change in IMA anatomy [15].

Decades after this effect was first described, the issue remains controversial. In practice, both in situ and free graft LIMAs are used in these dialysis patients with left arm AVF. Recently presented as a best evidence topic in *Interactive Cardiovascular and Thoracic Surgery*, a total of 478 parents with 219 in situ IMA grafts and ipsilateral upper arm AV fistula were studied [17] . While mortality did not appear to be affected, there are sufficient increases in reports of chest pain during dialysis to conclude this is due to in situ ipsilateral grafts [17]. The overall clinical evidence finds that steal did occur and that patients did have evidence of malperfusion during dialysis. Clearly, dialysis patients are high risk at baseline. However, experimental and anecdotal evidence suggest ipsilateral LIMA grafts can affect outcomes for dialysis patients with left upper extremity AVF. It is advisable to consider contralateral mammary or free grafts in this patient population [17].

3.3 Radial Artery Spasm

The first case of the new specialist is an all-arterial CABG x 4 with bilateral IMA and left radial artery grafts. The surgery goes well, and the patient is stable postoperatively. On postoperative day #1, the patient is transferred out of the ICU to the step-down unit. At 5 p.m. that night, a cardiac arrest code is called on the patient. The patient is resuscitated with less than 5 min of chest compressions and low-dose epinephrine, but remains hypotensive on multiple drips. Echo shows new hypokinetic inferior wall changes, and the EKG suggests RCA ischemia. The cath lab is immediately notified.

The angiogram shows severe spasm with complete mid-occlusion of the right radial artery graft (aorta to the distal RCA target). The patient is improved hemodynamically with direct intravenous nitroglycerine infusion and immediate heparinization. He proceeds to a full recovery, but returns to the office 1 month later, complaining of persistent pain in his left hand.

3.3.1 Discussion

In 1973, Carpentier introduced the radial artery as a conduit; the innovation was subsequently discarded for concerns of early graft closure and vasospasm [18]. The radial graft was revived by Acar and colleagues in 1993 [19]. Over the last 30 years, radial artery use, and the data regarding its use, has multiplied several-fold. Cardiac surgeons continue to search for the most optimal conduit, aside from the LIMA, to provide patients with the best long-term survival. The original study of excellent long-term outcomes of bilateral internal mammary artery grafts [20] is planned to be replicated with the radial artery. Concurrently, however, the distal radial artery has also become the favored site of entry for percutaneous coronary angiograms and interventional procedures.

There are three key issues regarding the radial graft as a conduit for surgeons: (1) safety and side effects of harvest, (2) arterial spasm, and (3) long-term patency. The early data of the 1990s was not convincing. The potential complications of radial harvest and the presence of arterial spasm were documented [21]. The use of postoperative vasodilators to prevent spasm were recommended to accompany radial

Table 3.4 Potential risks of radial harvest

Peripheral neuropathy; numbness and paresthesias, dysesthesia
Ischemia needing revascularization
Regional pain
Exercise-induced ischemia

grafting [22]. Other studies reported neurologic complications of the hand. In fact, early data suggested that neurologic complications from radial artery harvest occurred in 30.1% of patients with a sensory abnormality in 18.1% and motor dysfunction with decreased thumb strength in 5.5% [23] (Table 3.4).

Although observational evidence would suggest that bilateral mammary artery grafting results in better long-term outcomes than vein grafts, the initial randomized control ART trial did not fully support this finding [24]. The search continues for an alternative arterial graft that has a simple harvest and provides excellent long term flow. In the modern era, the advent of *endoscopic* radial harvesting has decreased neurologic and bleeding complications significantly, and radial artery use has continued to grow [25, 26].

In fact, data over the recent several years has supported the radial artery as an excellent and lasting arterial conduit, preferable over venous conduit, for its durability and improved long-term survival [27–29]. In comparison to bilateral mammary use, the radial graft avoids two of the known potential complicating issues of BIMA harvest: wound infection and increased bleeding. Bleeding at the harvest site with radial artery harvest is rare and infection even more so. Technical aspects of safe harvest include dissection at the medial border of the brachioradialis and the use of clips or electrocautery to control branches. After completion, the arm is typically wrapped and tucked in the functional anatomic position for the remainder of the operation.

Local ischemia of the hand is uncommon due to an increase in ulnar flow that occurs after harvest [30]. Although forearm blood supply is technically cut in half with this procedure, the ulnar artery has been shown to increase in diameter and blood flow velocity 8 weeks after surgery [31]. This central aspect of safety, the concept of ulnar compensation, is a documented effect of radial harvest [30, 31]. Classic anatomy is present in only 10% of patients [32], and an incomplete superficial palmar arch with no direct supply to the thumb and index finger occurs in 34% of patients [33]. Yet overall, harvest of the radial artery has been shown to be safe in large studies [34].

There have been some concerns about neurologic effects, particularly in patients who use their hands frequently. One study showed that 5 years after surgery, there were no significant increases in neurologic symptoms; however, testing was limited to 5 min of exercise, and abnormal values of pO_2 and CO_2 in the operated arms were seen when compared to controls [35]. Exercise did not lead to complaints of forearm symptoms of hand ischemia, and increased relative flow in the ulnar artery was seen with exercise that was not as significant in the non-operated hand [35].

The most common neurologic effect is radial sensory neuropathy, which can occur even with ABG and radial arterial line placement. Allen et al. found in multivariate analysis that it is PVD and not the harvest of the radial artery that is key in the development of this complication after CABG [32]. In their study, no patients had hand claudication or ischemia, but a radial sensory neuropathy was present in 9.9% of patients; this did not reach statistical significance with controls [32]. Risk factors included obesity, age, diabetes, and peripheral vascular disease [32]. Neuropathy may not be based on ischemia or vascular deficit, but long-term sensitivity disorders can remain over many years at radial harvest sites [36]. Sensory loss in the lateral cutaneous nerve occurs at a rate of 2.1%, and 15.5% of patients reported sensation alteration [37]. These sensory losses can occur in not only the lateral antebrachial cutaneous nerve but also the superficial branch of the radial nerve [31]. Care must be taken with electrocautery and mechanical traction on the radial sensory nerve.

Vasospasm and vasoreactivity of the radial graft have been of concern since the initial use of the artery. 5–10% of grafts can be expected to have some element of spasm [28]. Early fears of vasospasm are not frequently reported in the current era, although the reasons for this are not known. There may be a difference in the no-touch technique and/or the use of endoscopic harvesting. There is now routine use of vasodilation solutions intraoperatively to directly flush the graft and prevent vasospasm. These agents have included calcium channel blockers, nicardipine, papaverine, milrinone, and nitroglycerine, often continued in the postoperative period [38–42]. If postoperative spasm occurs, as noted in the above scenario, use of local vasodilation agents and anticoagulation can be helpful in relieving spasm.

As in all heart surgery, patient selection is key. Obviously, some contraindications are clear: patients with a positive Allen's test, previous injury, severe PVD, calcification, and renal failure are not always good candidates. Peripheral abnormalities, scleroderma, Raynaud's, poor glucose control, and lupus may add more risk than benefit. Most surgeons also consider recent angiography access to be a contraindication [43].

The radial artery has emerged as a key player in the ultimate all-arterial revascularization, a modern strategy to provide maximal patient benefit. This is in part due to endoscopic harvesting, a low rate of complications, and ease of use. The initial findings of neurologic changes, patient dissatisfaction, and arterial spasm are not as commonly reported as in the past, and the use of the radial artery as a second arterial conduit has graduated to a class IIa recommendation in coronary revascularization [44].

3.4 Leriche Syndrome

A 75-year-old three-pack-a-day smoker presents with new chest pain, elevated troponin, and chronic lower extremity claudication. His history is significant for Leriche syndrome, with leg and buttock claudication at rest. An examination of the patient's legs reveals thin muscle wasting with no palpable pulses. Coronary angiogram performed via right radial access shows multi-vessel coronary disease and a low ejection fraction. A CTA confirms complete cutoff of the distal aorta with collateral perfusion to the lower extremities. The planned surgery is CABG x 4 using LIMA, left radial, and two veins. The LIMA was unusually large in caliber, with excellent flow. The patient weans off-pump without difficulty and is sent to the ICU.

Postoperatively, the patient is hypotensive with an elevated lactate. Capillary and venous filling of the left leg are delayed, and the leg is cool to touch from the thigh down. IV heparin is started, but within 24 hours, the patient has a cold, immobile, and cyanotic left leg. He requires massive doses of vasopressors and inotropes, despite normal left ventricular function. A vascular surgery consult states the patient's only option is a left above-the-knee amputation. His large left IMA, which now provided excellent flow to the anterior wall of his heart, had previously held the critical job of supplying collaterals to the left leg.

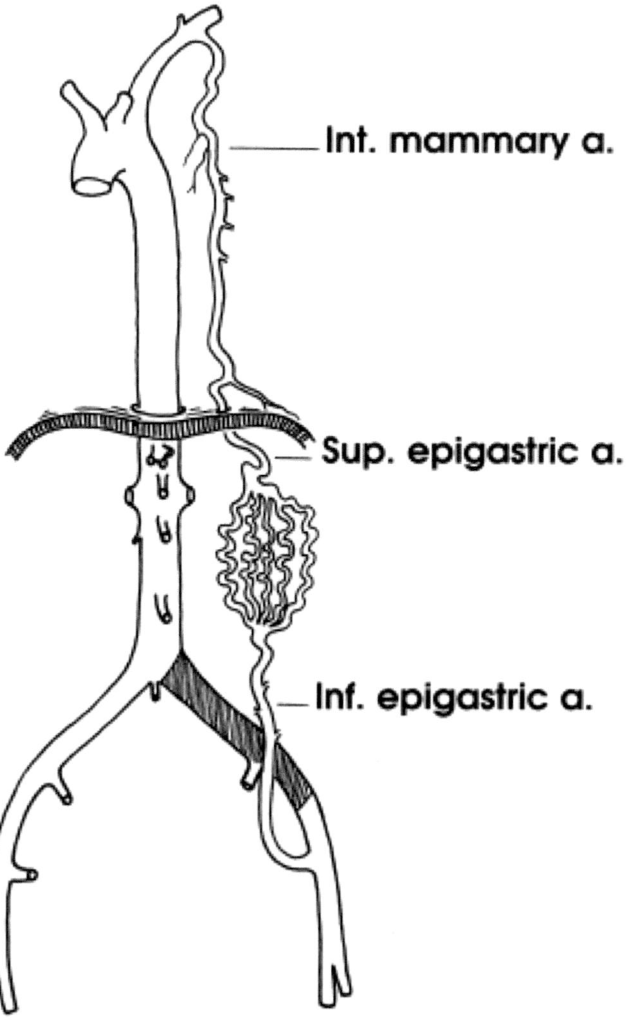

Fig. 3.1 Collateral pathway for occluded left common and internal iliac arteries. *Int. Mammary a, internal mammary artery; Sup. Epigastric a., superior epigastric artery; Inf. Epigastric a, inferior epigastric artery*. From "Internal mammary artery perfusing the Leriche's syndrome," by Arnold JR et al., 2000, *Ann Thor Surg, Vol. 69,* p. 1245. Copyright 2000 by Elsevier Inc. Reprinted with permission

3.4.1 Discussion

Leriche syndrome, the triad of buttock claudication, impotence, and lower extremity claudication, is well described and is not uncommon in a population of patients with severe cardiovascular disease [45]. In the setting of known Leriche syndrome or significant distal aortic or aortoiliac occlusion, the possibility of collateral circulation from the internal mammary artery must be considered [46].

When *complete* aortic occlusion is present, there are multiple collateral pathways to the lower extremities that may develop [46]. These have been termed the *systemic-systemic* pathway and the *visceral-systemic* pathway [47]. The systemic-systemic pathway is a collateral route between the subclavian artery, the internal mammary artery, the superior epigastric artery, and the inferior epigastric artery [47, 48] (Fig. 3.1). This drains directly into the ipsilateral collateral external iliac branch into the common femoral and is termed the Winslow pathway [49]. This anatomical solution can provide more than 40% of necessary flow to the lower extremities [50]. The intercostal arteries from the IMA can also collateralize to the superior and inferior epigastric arteries and then to the external iliac through the lumbar and iliolumbar arteries [51]. This can present as "rib notching" on chest X-ray.

Alternatively, the visceral-systemic collaterals can provide blood flow through the mesenteric arteries, to the artery of Moskowitz and marginal artery of Drummond, inferior mesenteric arteries to the superior rectal and middle rectal artery branches, and finally into the internal iliac artery. In this way, the visceral circulation can collateralize with the systemic circulation of the external iliac and femoral arteries [51]. An additional known visceral collateral pathway is from the renal arteries or the aorta through the gonadal arteries [51]. It is difficult to ascertain the degree to which each of these collateral pathways contributes to lower limb perfusion in the setting of aortic occlusion, emphasizing the importance of selective IMA angiography in these situations.

The danger of this collateral circulation in coronary revascularization is that the IMA is routinely ligated distally prior to anastomosis to the LAD. This ligation destroys all collaterals from the superior and inferior epigastric arteries. Ligation of one or both of the IMAs can lead to acute limb ischemia and limb loss if this particular anatomic anomaly is not recognized preoperatively [48, 52, 53]. In patients with signs of severe PVD or non-palpable lower extremity pulses, a thorough history to determine symptoms of iliac artery claudication (pain, impotence, and pulseless lower extremities) should be obtained. One study of 15 patients with one or both mammary arteries supplying collaterals to the lower extremities showed that 80% of this group presented with symptoms of intermittent claudication [52]. Of these patients, 60% had occlusion of the distal aorta, and 33.3% had occlusion of left or right iliac arteries [52].

Injection of both IMA is typically not recommended for routine coronary angiograms. Complications can occur with wires manipulated in a diseased aorta that can compromise the IMA itself. However, an angiogram can supply vital information about lumen size and the presence or absence of collaterals, as well as the presence of subclavian artery stenosis. Given the high incidence of collaterals in the setting of aortoiliac occlusion, this selective injection should be done preoperatively to prepare for CABG (Table 3.5). Noninvasive CTA may also demonstrate these collaterals. Noninvasive Doppler ultrasound of the inferior epigastric artery (IEA) will show reversal of flow. Yurkadul et al. found 95% of patients with aortoiliac occlusion had an ITA-IEA pathway by prospectively studying 64 patients and doing Doppler exams that showed reversal of IEA flow in a caudal direction in 95% of patients with aortoiliac occlusion [50]. This collateral pathway appears less affected by atherosclerosis and becomes critical in a patient with aortoiliac occlusion [48] (Fig. 3.2).

In the setting of emergency surgery, it is imperative to document the status of peripheral pulses. If severe aortoiliac disease is suspected based on history and physical exam, the finding of an unusually large IMA adds suspicion for aortoiliac occlusion. A recent retrospective study demonstrated lower limb ischemia in a large group of patients who underwent skeletonized in situ BITAS. In nearly 2000 patients with PAD who underwent CABG, the incidence of acute lower limb ischemia in patients with peripheral vascular disease was higher by 14% in BITA use than in those who received LITA alone, although mortality was not significantly different [53]. It is important to remember PAD in itself is a predictor of reduced long-term survival. It is possible that sudden interruption of both ITA in the face of other poor collaterals could cause alterations in distal blood flow.

In terms of treatment, preoperative recognition is key. Several options exist in order to secure adequate LE flow in the setting of surgical coronary revascularization (Table 3.6).

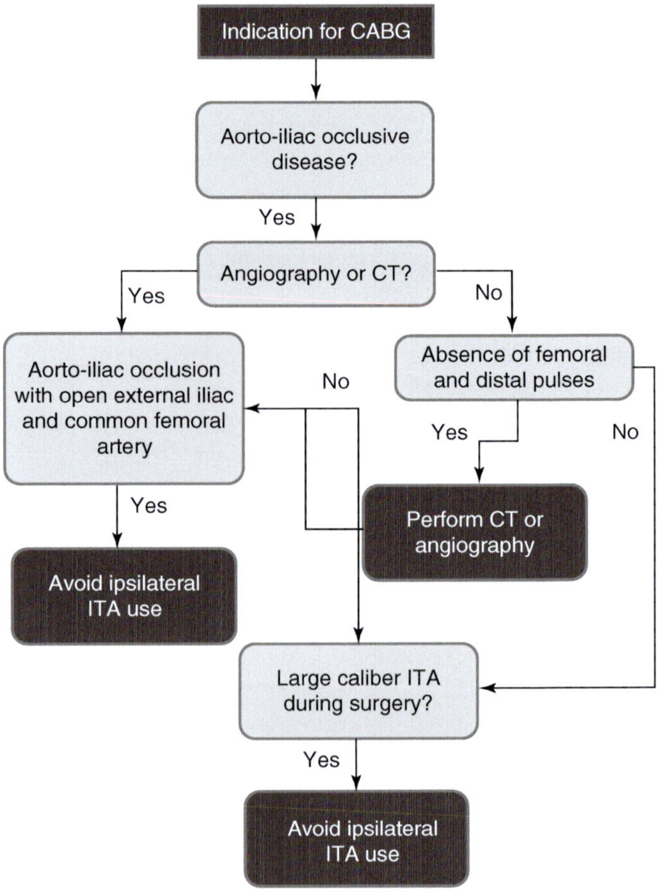

Fig. 3.2 Management of aortoiliac occlusion and CABG. *CABG, coronary artery bypass grafting, CT, computerized tomography, ITA, internal thoracic artery.* From "Acute limb ischemia after internal thoracic artery harvesting: a case report," by de Heer LM et al., 2011, *J Thorac Cardiovasc Surg, Vol 141,* p. e6. Copyright 2000 by Elsevier Inc. Reprinted with permission

Table 3.5 Vascular indications for selective IMA angiography prior to CABG

1.	Suspected subclavian artery stenosis
2.	Aortoiliac occlusive disease
3.	Severe lower extremity peripheral arterial disease with non-palpable femoral and/or distal pulses
4.	History of Leriche syndrome

Table 3.6 Options for the treatment of Leriche syndrome with IMA collaterals

1.	Use contralateral mammary alone in combination with radial/vein
2.	Use SVG and radial arteries alone; use EVH to allow for LE healing in the setting of generally compromised blood supply
3.	If cardiac disease is stable, fix LE ischemia first with extra-anatomic (axillary-femoral) bypass or aorto-bifemoral bypass
4.	PCI of the iliac artery
5.	No CABG; plan for multi-vessel PCI
6.	Concomitant CABG and extra-anatomic bypass

3.5 Postoperative Myocardial Infarction

The anatomic reality of a tiny left internal mammary artery in a petite 46-year-old diabetic woman is daunting for any surgeon, but the patency and long-term survival benefits of the LIMA to LAD anastomosis are critical in coronary revascularization. In this case, the surgeon harvests a small-diameter graft that appears to have acceptable flow. The artery is prepared and anastomosed onto the LAD. According to the transit time flow numbers, the anastomosis is open, but the flow is comparatively low with a slightly high resistance. The patient remains hemodynamically stable after weaning from bypass. The anterior wall is somewhat sluggish initially, but improves with higher mean perfusion pressures. There are some slight ST segment elevations on the monitor, but without any corresponding hypokinesis on echo. The patient requires minimal inotropes and vasopressors.

After arrival in the ICU, the patient immediately has some brief runs of ventricular tachycardia. Then she goes into rapid SVT, at times up to 160, with associated hypotension. She is given amiodarone boluses, and a continuous drip is started. EKG demonstrates new anterior ST segment elevations. The patient is hemodynamically stable now, but remains markedly tachycardic with multiple ectopic beats.

3.5.1 Discussion

The extreme anxiety surrounding postoperative myocardial infarction (POMI) revolves around the fact that the mortality for this complication is extraordinarily high. The incidence of infarction due to early graft failure after CABG has been reported to occur in 1–3% of patients [54]. Furthermore, the numbers of immediate graft occlusion rates reported are not inconsequential. Studies of vein grafts have demonstrated failure rates up to 12% [55]. Other studies of occlusion rates prior to discharge show vein grafts 3–12%, radial arteries 3–4%, and IMA grafts 1–2.5% [55–59]. POMI can also be caused by incomplete revascularization and poor myocardial protection. In addition to the inability to wean from bypass and profound postoperative cardiogenic shock, postoperative MI should be given recognition as a potential disaster situation. Although the accepted mortality for elective CABG is approximately 1%, one can expect, in the face of postop MI, that the mortality rate may increase up to 10% [54].

Thus, concerns surrounding postop EKG changes, particularly when associated with hemodynamic instability and/or ventricular arrhythmias, are serious in the immediate time period after CABG. There are a number of key questions: What are the roles of troponin measurements, echocardiograms, and immediate angiograms? What is the best treatment and in what time frame? The main tenet in these situations is to rapidly save muscle, which will lead to saving cardiac function and help in saving the patient.

The key to saving heart muscle, as in a ST-elevation MI, is time. The current European Society of Cardiology/European Association for Cardio-Thoracic Surgery (ESC/EACTS) guidelines recommend immediate coronary angiography in the setting of perioperative MI in order to detect the underlying cause, plan for intervention, and limit myocardial damage [58].

Postoperative ischemic events that require intervention are caused by the LIMA graft in 1/3 to 1/2 of cases [54]. The LIMA supplies a large area of muscle, and thus its occlusion often results in immediate hemodynamic symptoms. Based on clinical suspicion and concern for instability, these patients are likely to return immediately to the OR for revascularization. Of those patients who go back immediately to the OR, one study showed the LIMA graft as the culprit lesion in 43.8% of cases [57]. Of those periop MI patients who underwent angiography and PCI, SVG occlusion was shown to be the most common finding (68.8%) [57].

A meta-analysis of 1104 patients who underwent angiography for ischemic symptoms after CABG reports the incidence of postop MI as 3.5% [60]. Of these, 46.3% of the patients had repeat revascularization; 54.2% had surgical reoperation, and 45.8% underwent PCI [60]. The findings of this meta-analysis showed that patients who did not undergo angiography but returned directly to the OR had a mortality of 43.6% [60]. Those that did have immediate angiography had a positive finding of graft closure in 62.5%, but the associated mortality was only 8.9% [60]. Of course, there is a selection bias in that patients who returned directly to the OR were more unstable, without time for angiography and definitive diagnosis. Interestingly, this analysis also found that 1/3 of patients who underwent angiography did *not* have structural graft problems, but more likely had poor runoff, poor protection, distal embolization, or a subendocardial MI termed type V MI [60].

Thus, angiography can be expected to show a problem in the grafts in nearly 2/3 of the patients and is used as a guide for immediate plans for revascularization [60]. The use of angiography itself drops the early mortality rate regardless of treatment. The benefits are obvious: immediate diagnosis of the problem, potential treatments of graft spasm or graft occlusion, placement of IABP or percutaneous assist devices, and surgical planning if required [60]. However, potential risks must be weighed: moving the patient from the ICU to a less controlled environment and the use of dye and instrumentation in a fresh postoperative patient. A variety of technical problems with the grafts have also been found by angiography (Table 3.7) [54, 61].

Table 3.7 Potential graft abnormalities

Acute spasm
Kinking
Graft thrombosis
Anastomotic occlusion or stenosis
Dissected graft
Stretch (graft too short)

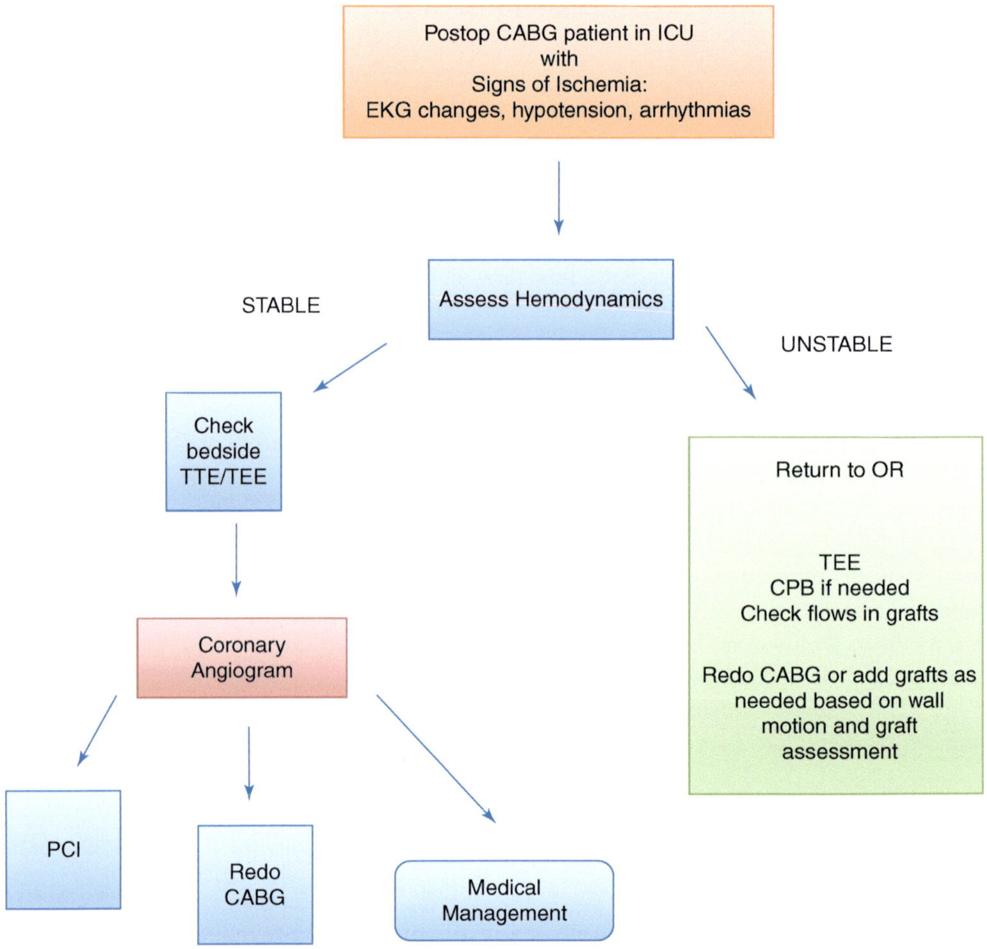

Fig. 3.3 Treatment of postoperative CABG myocardial infarction. *ICU, intensive care unit, EKG, electrocardiogram, TTE, transthoracic echo, TEE, transesophageal echo, CPB, cardiopulmonary bypass, PCI, percutaneous coronary intervention, CABG, coronary artery bypass grafting, OR, operating room*

The algorithm to follow in this situation is not exact, but guidelines for therapy are available (Fig. 3.3) [62]. If the patient is hemodynamically stable and postoperative ischemia is suspected, evaluation with EKG, echo (usually TEE), and troponin is warranted. These markers can indicate, but not prove, a problem. The signs of MI are sometimes as subtle as tachycardia alone. There is also no uniform cutoff number recommended for troponin or CKMB. If the patient is stable and not immediately post-surgery, one can consider CTA as a reliable noninvasive technology that carries less complications [63]. However, artifact from clips on the mammary or vein may make it difficult to view the grafts and distal anastomoses.

Patients with arrhythmias and cardiogenic shock usually return immediately to the OR for hemodynamic support as well as evaluation of grafts. In the operating room, Doppler analysis of arterial and venous grafts should be performed, particularly if there is any question about anastomotic patency. This information should be used with regional wall motion abnormalities (RWMA) and EKG changes to help determine the culprit graft, if present.

Regional wall motion abnormalities have a low positive predictive value (33%) but high negative predictive value (100%) [64]. Wall motion can also be affected by loading conditions, particularly in the postoperative period. However, marked abnormalities, if present, should be carefully examined immediately after initial revascularization. One study found that using RWMA could have led to repair of 20% of graft failures *prior* to leaving the OR [64]. Regrafting in the OR based on clinical suspicion and flow monitoring is the surest method to protect the myocardium.

3.6 Chylothorax

A morbidly obese patient 4 days after CABG was doing well. While listening to the resident present the case in the hallway, the surgeon makes the mental decision to move the patient forward by removing the chest tubes. He is surprised to hear the drainage was persistently high overnight. As the team enters the patient's room, all of them notice the drainage collection device at the same time. In the second column, layered on top of the expected serosanguinous drainage, there is a long column of new, milky white fluid.

3.6.1 Discussion

The incidence of chylothorax after cardiovascular surgery ranges from 0.25% to 0.5% [65]. Chylopericardium, even more rare than chylothorax, can also occur [66]. Based on the anatomy and the course of the thoracic duct, chylothorax in thoracic surgery is a complication that more commonly occurs during esophagectomy or descending aortic procedures. It has also been described in congenital cardiac procedures.

The anatomy of the thoracic duct as it traverses the chest cavity is well known. The duct originates with the cisterna chyli which lies on the anterior surface of the first and second lumbar vertebrae in the retroperitoneum. This collects lymph from the lower extremities and peritoneal cavity. The thoracic duct begins at the cisterna chyli and traverses through the aortic hiatus on the right, crossing over at T4-5 to travel on the left side, to its insertion at the junction of the left IJ and left subclavian vein. This exact course is only found in 50–60% of the population, and anomalies are present in 12–15% of cases [67].

A number of variations in anatomy have been described [68]. There exists a mediastinal lymph node chain near the LIMA anteriorly, at the origin, that can directly join the thoracic duct or can go separately into the junction of the left IJ and subclavian [69]. This surrounding area is rich in what are termed anomalous perithymic lymphatics. This is critical to know in LIMA harvest, particularly when the mammary artery is to be placed as a free graft and a high ligation is required. These mediastinal and thymic lymphatics, after dissection with electrocautery, may become exposed to increased venous pressure after surgery and result in lymph drainage into the superior mediastinum.

Chylopericardium is even more rare. But lymphatics also are present in tissues at the posterior mediastinum, mediastinal pleura, pericardial fat, and parietal pericardium [70]. A study of the tributaries of the thoracic duct in 530 cadavers showed that 8.9% of lymphatic vessels come from the heart to connect with the thoracic duct, likely accounting for lymph leaks within the pericardium [70]. Injury can occur during the time of placement of the aortic cross-clamp or when dissecting between the pulmonary artery and the aorta.

Suspicion for chylothorax occurs in the setting of high, prolonged chest tube output that is either serous, turbid, or milk colored. Only a certain percentage of chylothoraces have the characteristic appearance of milky, white fluid. In fact, studies have described that less than half have this milky appearance [71]. Simply an abnormally high serous output from chest tubes of >400 cc/day may indicate a lymph leak [72].

The first step is diagnosis and involves sending the pleural drainage to the lab for triglycerides and cholesterol levels. The key value to demonstrate chylothorax is a triglyceride level of >110 mg/dl. For those patients who are intermediate range (50–100 mg/dl), it is appropriate to order a lipoprotein electrophoresis [73].

The mainstay of therapy is restricting oral intake. However, all patients with this pathology are post-surgery, and early postoperative recovery can be greatly influenced by malnutrition and immunocompromise. Restriction of oral intake to decrease lymph drainage can lead to significant hypovolemia and malnutrition. Parenteral nutrition will be required if the lymph leak persists. The fluid load of surgery can decrease sodium and calcium with a protein loss that also affects the metabolic state. Furthermore, chyle contains T lymphocytes; patients who are already immunosuppressed from surgery will be further at risk as these are lost in drainage. In a patient already in a catabolic state, dehydration, malnutrition, pneumonia, or hypoxia can be observed. In children, loss of antithrombin and episodes of thrombosis have been described [74].

There are two main choices of management for lymphatic leak: conservative management vs invasive therapies (Fig. 3.4) [73]. Historically, the initial main treatment has been conservative and nonoperative, based on (1) strict NPO, (2) alternative sources of nutrition such as TPN, and (3) use of octreotide. Octreotide is a synthetic somatostatin analogue that inhibits growth hormone, glucagon, and insulin [75]. It usually works within a couple of days and is used in conjunction with TPN and bowel rest. Drain output is expected to decrease with its use. Drainage >500 cc/day is not likely to heal spontaneously [73]. When oral nutrition is resumed, only medium-chain triglycerides are given. Long-chain triglycerides increase chylomicron transport to the intestinal lymphatics.

In the past, failure of prolonged medical therapy led to operative intervention. This treatment requires a right thoracotomy/thoracoscopy and mass ligation of the lymphatic duct just proximal to the aortic hiatus. This can be a highly morbid procedure, however, and does not guarantee improvement. For this reason, many surgeons wait at least 7–14 days after diagnosis planning for conservative therapy. Another option for treatment is reoperation using a biologic sealant or suturing of the leaking area, usually through a left VATS or left thoracotomy. Pleurodesis with tetracycline, bleomycin, or talc has also been used, but can cause a severe, painful inflammatory reaction that is poorly tolerated by patients.

In the current era, particularly with the advent of improved imaging and minimally invasive therapies, the next step after diagnosis should be the localization of the leak. This is done with lymphangiography with a high-fat content solution [76]. Lymphoscintigraphy, which has a lower side effect profile, can also be used. Chylothorax as a complication has markedly benefited from technologic progress in imaging and percutaneous therapy. In the same setting of diagnosis of the exact site of leakage, a biologic sealant, coils, or glue can

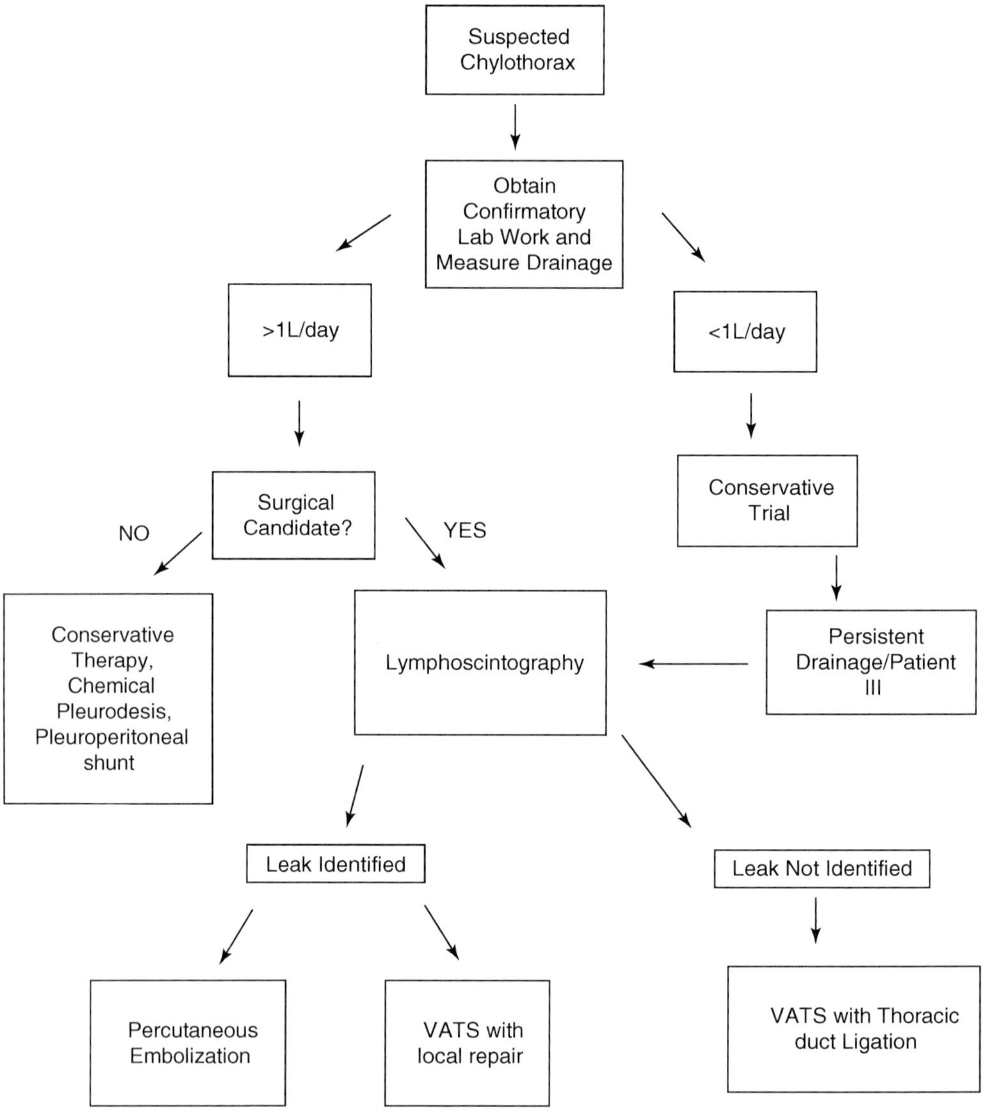

Fig. 3.4 Management of chylothorax

be used to embolize and stop the leak [76]. This minimally invasive option for treatment will decrease the number of days of waiting for medical management to suffice and will lead to a quicker surgical recovery for the patient.

3.7 Post-Pericardiotomy Syndrome

A 46-year-old man post-CABG with bilateral IMA grafts is frustrated postoperatively by the continued presence of his draining chest tubes. Prior to surgery, he was anticoagulated with platelet inhibitors. He had significant postoperative bleeding requiring two blood transfusions. Serous drainage from his mediastinal tubes was persistently high starting on postoperative day 3. He was now on postoperative day 6, and despite aggressive diuresis, the tubes continue to drain. He also developed a mild leukocytosis, complaints of chest pain, and a clear pericardial rub.

3.7.1 Discussion

Post-pericardiotomy syndrome (PPS) is an inflammatory process, most often presenting as fever, pleural or pericardial effusion, and localized pain. It can affect 10–40% of patients after cardiac surgery [77]. Recent studies have suggested this number is less than 10% in the current era [78]. It has been described more commonly after valve or aortic surgery, perhaps due to the amount of dissection required, or the operative time leading to a higher inflammatory state [78]. It is seen more commonly in younger patients and generally occurs within the first month after cardiac surgery. This generally benign and self-limited process has been shown in one study to increase mortality by 1.7-fold [78]. This is perhaps due to its association with acute cardiac tamponade postoperatively or with the chronic pericardial constriction that can occur over time. A risk factor for PPS is blood transfusion, which likely increases the overall inflammatory state after heart surgery [79].

The European Society of Cardiology refers to this as post-cardiac injury syndrome and requires two of the following five criteria for diagnosis [79].

1. Fever (not otherwise explained).
2. Pericardial or pleuritic chest pain.
3. Pericardial or pleural rub.
4. Pericardial effusion.
5. Pleural effusion with increased C-reactive protein.

Several treatments of PPS have been trialed, including colchicine, steroids, and nonsteroidal therapies. Colchicine is an anti-inflammatory drug used most commonly to treat gout. In recent years, it has been proposed as a way of preventing pericarditis. It may also assist in the prevention of atherosclerosis, which carries inflammation as a key part of its pathway. The mechanism of this has to do with the ability of colchicine to inhibit neutrophil entry and activation into an active inflammatory environment. By stabilizing plaque, this drug could possibly help prevent plaque rupture in an atherosclerotic vessel.

There were two specific trials, the COPPS-1 and COPPS-2 trials, that suggest that colchicine works in decreasing inflammation and mediastinal/pleural fluid drainage after cardiac surgery [80, 81]. A review of 15 randomized controlled trials using colchicine, most delivered at a dose of a dose of 1 mg/day, confirmed less pericarditis, less post-pericardiotomy syndrome, and less post-procedural atrial fibrillation [82].

Colchicine is the mainstay of treatment in the postoperative period if PPS is suspected. The most common side effect is diarrhea, which is increased by twofold in most studies [82]. However, given the association of PPS to overall mortality, treatment with colchicine is supported [83].

References

1. Lin TY, Chiu KM, Wang MJ, Chu SH. Carbon dioxide embolism during endoscopic saphenous vein harvesting in coronary artery bypass surgery. J Thorac Cardiovasc Surg. 2003;126(6):2011–5.
2. Strauss E, Taylor B, Mazzeffi M, Tanaka K, Odonkor P. The auscultation of a carbon dioxide embolization event during endoscopic vein harvest. Case Rep Anesthesiol. 2016;2016:6947679.
3. Mommerot A, Perrault LP. Carbon dioxide embolism induced by endoscopic saphenous vein harvesting during coronary artery bypass grafting. J Thorac Cardiovasc Surg. 2006;132(6):1502. author reply 1502–3.
4. Banks TA, Manetta F, Glick M, Graver LM. Carbon dioxide embolism during minimally invasive vein harvesting. Ann Thorac Surg. 2002;73(1):296–7.
5. Kypson AP, Greenville NC. Sudden cardiac arrest after coronary artery bypass grafting as a result of massive carbon dioxide embolism. J Thorac Cardiovasc Surg. 2005;130(3):936–7.
6. Suarez-Pierre A, Terasaki Y, Magruder JT, Kapoor A, Grant MC, Lawton JS. Complications of CO2 insufflation during endoscopic vein harvesting. J Card Surg. 2017;32(12):783–9.
7. Sintek M, Coverstone E, Singh J. Coronary subclavian steal syndrome. Curr Opin Cardiol. 2014;29(6):506–13.
8. Elian D, Gerniak A, Guetta V, et al. Subclavian coronary steal syndrome: an obligatory common fate between subclavian artery, internal mammary graft and coronary circulation. Cardiology. 2002;97:175–9.
9. Gaudino M, Serricchio M, Glieca F, et al. Steal phenomenon from mammary side branches: when does it occur? Ann Thorac Surg. 1998;66:2056–62.
10. Coskun I, Colkesen Y, Altay H, Ozkan U, Demirturk OS, Gulcan O, et al. Hemodynamic effects of left upper extremity arteriovenous fistula on ipsi- lateral internal mammary coronary artery bypass graft. Thorac Cardiovasc Surg. 2013;61:663–8.
11. Gaudino M, Serricchio M, Luciani N, et al. Risks of using internal thoracic artery grafts in patients in chronic hemodialysis via upper extremity arteriovenous fistula. Circulation. 2003;107(21):2653–5.
12. Kato H, Ikawa S, Hayashi A, Yokoyama K. Internal mammary artery steal in a dialysis patient. Ann Thorac Surg. 2003;75:270–1.
13. Rahbar R, McGee WR, Birdas TJ, Muluk S, Magovern J, Maher T. Upper extremity arteriovenous fistulas induce modest hemodynamic effect on the in situ internal thoracic artery. Ann Thorac Surg. 2006;81:145–7.
14. Takami Y, Tajima K, Kato W, et al. Effects of the side of arteriovenous fistula on outcomes after coronary artery bypass surgery in hemodialysis-dependent patients. J Thorac Cardiovasc Surg. 2014;147(2):619–24.
15. Shim H, Jeong DS, Kim WS, Park PW, Sung K, Jeon CS, Lee YT. Impact of arteriovenous fistula for hemodialysis on clinical outcomes of coronary artery bypass. Ann Thorac Surg. 2018;106(6):1820–6.
16. Ahn S, Han A, Kim SY, Choi C, Min SI, Ha J, Hwang HY, Kim KB, Min SK. The incidence and risk factors of coronary steal after ipsilateral AVF in patients with a coronary artery bypass graft. J Vasc Access. 2017;18(4):290–4.
17. Cuthbert GA, Kirmani BH, Muir AD. Should dialysis-dependent patients with upper limb arterio-venous fistulae undergoing coronary artery bypass grafting avoid having ipsilateral in situ mammary artery grafts? Interact Cardiovasc Thorac Surg. 2014;18(5):655–60.
18. Carpentier A, Guermonprez JL, Deloche A, Frechette C, DuBost C. The aorta-to-coronary radial artery bypass graft. A technique avoiding pathological changes in grafts. Ann Thorac Surg. 1973;16:111–21.
19. Jebara VAC, Portoghese M, Beyssen B, Pagny JY, Grare P, Chachques JC, Fabiani JN, Deloche A, Guermonprez JL. Revival of the radial artery for coronary artery bypass grafting. Ann Thorac Surg. 1992;54:652–60.
20. Lytle BW, Blackstone EH, Loop FD, Houghtaling PL, Arnold JH, Akhrass R, et al. Two internal thoracic artery grafts are better than one. J Thorac Cardiovasc Surg. 1999;117:855–72.
21. Chardigny C, Jebara VA, Acar C, Descombes JJ, Verbeuren TJ, Carpentier A, Fabiani JN. Vasoreactivity of the radial artery. Comparison with the internal mammary and gastroepiploic arteries with implications for coronary artery surgery. Circulation. 1993;88(5 Pt 2):II115–27.
22. Aldea GS, Bakaeen FG, Pal J, Fremes S, Head SJ, Sabik J, Rosengart T, Kappetein AP, Thourani VH, Firestone S, Mitchell JD, Society of Thoracic Surgeons. The Society of Thoracic Surgeons Clinical Practice Guidelines on arterial conduits for coronary artery bypass grafting. Ann Thorac Surg. 2016;101(2):801–9.
23. Denton TA, Trento L, Cohen M, Kass RM, Blanche C, Raissi S, Cheng W, Fontana GP, Trento A. Radial artery harvesting for coronary bypass operations: neurologic complications and their potential mechanisms. J Thorac Cardiovasc Surg. 2001;121(5):951–6.
24. Taggart DP, Benedetto U, Gerry S, Altman DG, Gray AM, Lees B, Gaudino M, Zamvar V, Bochenek A, Buxton B, Choong C, Clark S,

Deja M, Desai J, Hasan R, Jasinski M, O'Keefe P, Moraes F, Pepper J, Seevanayagam S, Sudarshan C, Trivedi U, Wos S, Puskas J, Flather M, Arterial Revascularization Trial Investigators. Bilateral versus single internal thoracic artery grafts at 10 years. N Engl J Med. 2019;380(5):437–46.
25. Connolly MW. Endoscopic radial artery harvesting: results of first 300 patients. Ann Thorac Surg. 2002;74(2):502–5. discussion 506.
26. Rudez I, Unic D, Sutlic Z, Biocina B, Baric D, Ivkovic M, Pavlovic M. Endoscopic radial artery harvesting reduces post-operative pain and neurologic complications. Heart Surg Forum. 2007;10(5):E363–5.
27. Tranbaugh RF, Schwann TA, Swistel DG, Dimitrova KR, Al-Shaar L, Hoffman DM, Geller CM, Engoren M, Balaram SK, Puskas JD, Habib RH. Coronary artery bypass graft surgery using the radial artery, right internal thoracic artery, or saphenous vein as the second conduit. Ann Thorac Surg. 2017;104(2):553–9.
28. Tranbaugh RF, Dimitrova KR, Lucido DJ, Hoffman DM, Dincheva GR, Geller CM, Balaram SK, Ko W, Swistel DG. The second best arterial graft: a propensity analysis of the radial artery versus the free right internal thoracic artery to bypass the circumflex coronary artery. J Thorac Cardiovasc Surg. 2014;147(1):133–40.
29. Tatoulis J, Royse A, Buxton B, et al. The radial artery in coronary surgery: a 5-year experience—clinical and angiographic results. Ann Thorac Surg. 2002;73:143–8.
30. Royse A, Royse C, Maleskar A, Garg A. Harvest of the radial artery for coronary artery surgery preserves maximal blood flow of the forearm. Ann Thorac Surg. 2004;78:539–42.
31. Brodman RF, Hirsh LE, Frame R. Effect of radial artery harvest on collateral forearm blood flow and digital perfusion. J Thorac Cardiovasc Surg. 2002;123(3):512–6.
32. Allen RH, Szabo RM, Chen JL. Outcome assessment of hand function after radial artery harvesting for coronary artery bypass. J Hand Surg Am. 2004;29(4):628–37.
33. Ruengsakulrach P, Eizenberg N, Fahrer C, Fahrer M, Buxton BF. Surgical implications of variations in hand collateral circulation: anatomy revisited. J Thorac Cardiovasc Surg. 2001;122:682–6.
34. Meharwal ZS, Trehan N. Functional status of the hand after radial artery harvesting: results in 3,977 cases. Ann Thorac Surg. 2001;72(5):1557–61.
35. Serricchio M, Gaudino M, Tondi P, Gasbarrini A, Gerardino L, Santoliquido A, Pola P, Possati G. Hemodynamic and functional consequences of radial artery removal for coronary artery bypass grafting. Am J Cardiol. 1999;84(11):1353–6. A8
36. Dick F, Hristic A, Roost-Krähenbühl E, Aymard T, Weber A, Tevaearai HT, Carrel TP. Persistent sensitivity disorders at the radial artery and saphenous vein graft harvest sites: a neglected side effect of coronary artery bypass grafting procedures. Eur J Cardiothorac Surg. 2011;40(1):221–6.
37. Royse AG, Royse CF, Shah P, Williams A, Kaushik S, Tatoulis J. Radial artery harvest technique, use and functional outcome. Eur J Cardiothorac Surg. 1999;15(2):186–93.
38. Özdemir HI, van Dijk CHB, Özdemir AB, van Straten BHM, Haanschoten M, Soliman-Hamad MA. Preventing spasm of the radial artery conduit during coronary artery bypass grafting: nicardipine versus verapamil. J Card Surg. 2019;34(12):1505–10.
39. Attaran S, John L, El-Gamel A. Clinical and potential use of pharmacological agents to reduce radial artery spasm in coronary artery surgery. Ann Thorac Surg. 2008;85(4):1483–9.
40. He GW, Yang CQ. Vasorelaxant effect of phosphodiesterase-inhibitor milrinone in the human radial artery used as coronary bypass graft. J Thorac Cardiovasc Surg. 2000;119:1039–45.
41. He GW, Yang CQ. Comparative study on calcium channel antagonists in the human radial artery: clinical implications. J Thorac Cardiovasc Surg. 2000;119:94–100.
42. Rudzinski P, Wegrzyn P, Lis GJ, Piatek J, Konstanty-Kalandyk J, Nosalski R, Mikolajczyk T, Jasinska M, Pyka-Fosciak G, Guzik T, Litwin JA, Korbut R, Sadowski J. Vasodilatory effect and endothelial integrity in papaverine- and milrinone-treated human radial arteries. J Physiol Pharmacol. 2013;64(1):41–5.
43. Lim LM, Galvin SD, Javid M, Matalanis G. Should the radial artery be used as a bypass graft following radial access coronary angiography. Interact Cardiovasc Thorac Surg. 2014;18(2):219–24.
44. Neumann FJ, Sousa-Uva M, Ahlsson A, Alfonso F, Banning AP, Benedetto U, Byrne RA, Collet JP, Falk V, Head SJ, Jüni P, Kastrati A, Koller A, Kristensen SD, Niebauer J, Richter DJ, Seferovic PM, Sibbing D, Stefanini GG, Windecker S, Yadav R, Zembala MO, ESC Scientific Document Group. 2018 ESC/EACTS Guidelines on myocardial revascularization. Eur Heart J. 2019;40(2):87–165.
45. Leriche R, Morel A. The syndrome of thrombotic obliteration of the aortic bifurcation. Ann Surg. 1948;127(2):193–206.
46. Wooten C, Hayat M, du Plessis M, Cesmebasi A, Koesterer M, Daly KP, Matusz P, Tubbs RS, Loukas M. Anatomical significance in aortoiliac occlusive disease. Clin Anat. 2014;27(8):1264–74.
47. de Heer LM, Buijsrogge MP, Lahpor JR. Acute limb ischemia after internal thoracic artery harvesting: a case report. J Thorac Cardiovasc Surg. 2011;141(1):e5–7.
48. Arnold JR, Greenberg JD, Clements S. Internal mammary artery perfusing the Leriche's syndrome. Ann Thorac Surg. 2000;69:1244–6.
49. Tsui SSL, Parry AJ, Large SR. Leg ischemia following bilateral internal thoracic artery and inferior epigastric artery harvesting. Eur J Cardiothorac Surg. 1995;9:218–20.
50. Yurdakul M, Tola M, Ozdemir E, Bayazit M, Cumhur T. Internal thoracic artery-inferior epigastric artery as a collateral pathway in aortoiliac occlusive disease. J Vasc Surg. 2006;43:707–13.
51. Hardman RL, Lopera JE, Cardan RA, Trimmer CK, Josephs SC. Common and rare collateral pathways in aortoiliac occlusive disease: a pictorial essay. AJR Am J Roentgenol. 2011;197:W519–24.
52. Ben-Dor I, Waksman R, Satler LF, Bernardo N, Torguson R, Li Y, Gonzalez MA, Maluenda G, Weissman G, Hanna NN, Monath A, Gallino R, Lindsay J, Kent KM, Pichard AD. A further word of caution before using the internal mammary artery for coronary revascularization in patients with severe peripheral vascular disease. Catheter Cardiovasc Interv. 2010;75(2):195–201.
53. Micali LR, Bonacchi M, Weigel D, Howe R, Parise O, Parise G, Gelsomino S. The use of both internal thoracic arteries for coronary revascularization increases the estimate of post-operative lower limb ischemia in patients with peripheral artery disease. J Cardiothorac Surg. 2020;15(1):266.
54. Thielmann M, Massoudy P, Jaeger BR, Neuhäuser M, Marggraf G, Sack S, Erbel R, Jakob H. Emergency re-revascularization with percutaneous coronary intervention, reoperation, or conservative treatment in patients with acute perioperative graft failure following coronary artery bypass surgery. Eur J Cardiothorac Surg. 2006;30(1):117–25.
55. Zhao DX, Leacche M, Balaguer JM, et al. Routine intraoperative completion angiography after coronary artery bypass grafting and 1-stop hybrid revascularization results from a fully integrated hybrid catheterization laboratory/operating room. J Am Coll Cardiol. 2009;53:232–41.
56. Davierwala PM, Verevkin A, Leontyev S, et al. Impact of expeditious management of perioperative myocardial ischemia in patients undergoing isolated coronary artery bypass surgery. Circulation. 2013;128(11 Suppl 1):S226–34.
57. Sef D, Szavits-Nossan J, Predrijevac M, Golubic R, Sipic T, Stambuk K, Korda Z, Meier P, Turina MI. Management of perioperative myocardial ischaemia after isolated coronary artery bypass graft surgery. Open Heart. 2019;6(1):e001027.
58. Windecker S, Kolh P, Alfonso F, Collet JP, Cremer J, Falk V, Filippatos G, Hamm C, Head SJ, Jüni P, Kappetein AP, Kastrati A, Knuuti J, Landmesser U, Laufer G, Neumann FJ, Richter DJ, Schauerte P, Sousa Uva M, Stefanini GG, Taggart DP, Torracca L, Valgimigli M, Wijns W, Witkowski A. 2014 ESC/EACTS

Guidelines on myocardial revascularization: The Task Force on Myocardial Revascularization of the European Society of Cardiology (ESC) and the European Association for Cardio-Thoracic Surgery (EACTS)Developed with the special contribution of the European Association of Percutaneous Cardiovascular Interventions (EAPCI). Eur Heart J. 2014;35(37):2541–619.

59. Fukui T. Intraoperative graft assessment during coronary artery bypass surgery. Gen Thorac Cardiovasc Surg. 2015;63:123–30.
60. Biancari F, Anttila V, Dell'Aquila AM, Airaksinen JKE, Brascia D. Control angiography for perioperative myocardial ischemia after coronary surgery: meta-analysis. J Cardiothorac Surg. 2018;13(1):24.
61. Rupprecht L, Schmid C, Debl K, Lunz D, Flörchinger B, Keyser A. Impact of coronary angiography early after CABG for suspected postoperative myocardial ischemia. J Cardiothorac Surg. 2019;14(1):54.
62. Laflamme M, DeMey N, Bouchard D, Carrier M, Demers P, Pellerin M, Couture P, Perrault LP. Management of early postoperative coronary artery bypass graft failure. Interact Cardiovasc Thorac Surg. 2012;14(4):452–6.
63. Gabriel J, Klimach S, Lang P, Hildick-Smith D. Should computed tomography angiography supersede invasive coronary angiography for the evaluation of graft patency following coronary artery bypass graft surgery? Interact Cardiovasc Thorac Surg. 2015;21(2):231–9.
64. Leung JM, O'Kelly B, Browner WS, Tubau J, Hollenberg M, Mangano DT. Prognostic importance of postbypass regional wall-motion abnormalities in patients undergoing coronary artery bypass graft surgery. SPI Research Group. Anesthesiology. 1989;71:16–25.
65. Zaidenstein R, Cohen N, Dishi V, Alon I, Golik A, Modai D. Chylothorax following median sternotomy. Clin Cardiol. 1996;19(11):910–2.
66. Pollard WM, Schuchmann GF, Bowen TE. Isolated chylopericardium after cardiac operations. J Thorac Cardiovasc Surg. 1981;81(6):943–6.
67. Medina JE. Neck dissection. In: Bailey BJ, Johnson JT, editors. Head and neck surgery: otolaryngology, vol. 2. 4th ed. Philadelphia, Pa: Lippincott Williams and Wilkins; 2006. p. 1585e609. [Chapter 113].
68. Halldorsson A. Chylothorax after coronary artery bypass grafting. Int Surg. 2009;94(2):119–29.
69. Kan CD, Wang JN, Wu JM, Yang YJ. Isolated chylopericardium after intrapericardial procedures: possible role of inadvertent right efferent lymphatic trunk injury. Tex Heart Inst J. 2007;34(1):82–7.
70. Riquet M, Le Pimpec BF, Souilamas R, Hidden G. Thoracic duct tributaries from intrathoracic organs. Ann Thorac Surg. 2002;73:892–9.
71. Ugurlucan M, Basaran M, Kocailik A, Us MH. eComment: unexpected chylopericardium and its treatment after cardiothoracic operations. Interact Cardiovasc Thorac Surg. 2009;8:177.
72. Maldonado F, Hawkins FJ, Daniels CE, Doerr CH, Decker PA, Ryu JH. Pleural fluid characteristics of chylothorax. Mayo Clin Proc. 2009;84:129–33.
73. Bender B, Murthy V, Chamberlain RS. The changing management of chylothorax in the modern era. Eur J Cardiothorac Surg. 2016;49(1):18–24.
74. Bernet-Buettiker V, Waldvogel K, Cannizzaro V, Albisetti M. Antithrombin activity in children with chylothorax. Eur J Cardiothorac Surg. 2006;29:406–9.
75. Ismail NA, Gordon J, Dunning J. The use of octreotide in the treatment of chylothorax following cardiothoracic surgery. Interact Cardiovasc Thorac Surg. 2015;20(6):848–54.
76. Boffa DJ, Sands MJ, Rice TW, Murthy SC, Mason DP, Geisinger MA, et al. A critical evaluation of a percutaneous diagnostic and treatment strategy for chylothorax after thoracic surgery. Eur J Cardiothorac Surg. 2008;33:435–9.
77. Prince SE, Cunha BA. Postpericardiotomy syndrome. Heart Lung. 1997;26:165–8.
78. Lehto J, Gunn J, Karjalainen P, Airaksinen J, Kiviniemi T. Incidence and risk factors of postpericardiotomy syndrome requiring medical attention: the Finland postpericardiotomy syndrome study. J Thorac Cardiovasc Surg. 2015;149:1324–9.
79. Adler Y, Charron P, Imazio M, Badano L, Baron-Esquivias G, Bogaert J, Brucato A, Gueret P, Klingel K, Lionis C, Maisch B, Mayosi B, Pavie A, Ristic AD, Sabate Tenas M, Seferovic P, Swedberg K, Tomkowski W, ESC Scientific Document Group. ESC Guidelines for the diagnosis and management of pericardial diseases: the Task Force for the Diagnosis and Management of Pericardial Diseases of the European Society of Cardiology (ESC) endorsed by: The European Association for Cardio-Thoracic Surgery (EACTS). Eur Heart J. 2015;2015(36):2921–64.
80. Imazio M, Trinchero R, Brucato A, Rovere ME, Gandino A, Cemin R, Ferrua S, Maestroni S, Zingarelli E, Barosi A, Simon C, Sansone F, Patrini D, Vitali E, Ferrazzi P, Spodick DH, Adler Y, COPPS Investigators. Colchicine for the prevention of the post-pericardiotomy syndrome (COPPS): a multicentre, randomized, double- blind, placebo-controlled trial. Eur Heart J. 2010;31:2749–54.
81. Imazio M, Brucato A, Ferrazzi P, Pullara A, Adler Y, Barosi A, Caforio AL, Cemin R, Chirillo F, Comoglio C, Cugola D, Cumetti D, Dyrda O, Ferrua S, Finkelstein Y, Flocco R, Gandino A, Hoit B, Innocente F, Maestroni S, Musumeci F, Oh J, Pergolini A, Polizzi V, Ristic A, Simon C, Spodick DH, Tarzia V, Trimboli S, Valenti A, Belli R, Gaita F, COPPS-2 Investigators. Colchicine for prevention of postpericardiotomy syndrome and postoperative atrial fibrillation: the COPPS- 2 randomized clinical trial. JAMA. 2014;312:1016–23.
82. Verma S, Eikelboom JW, Nidorf SM, Al-Omran M, Gupta N, Teoh H, Friedrich JO. Colchicine in cardiac disease: a systematic review and meta-analysis of randomized controlled trials. BMC Cardiovasc Disord. 2015;15:96.
83. Lehto J, Kiviniemi T. Postpericardiotomy syndrome after cardiac surgery. Ann Med. 2020;

4 Aortic and Mitral Valve Surgery

4.1 Intraoperative Aortic Dissection

A frail, thin, 85-year-old woman with emphysema and severe aortic insufficiency has just undergone a successful surgical aortic valve replacement and is weaning off CPB. The surgeon reviews the intraoperative TEE, is satisfied with the result, and prepares to decannulate. The antegrade cardioplegia catheter is to be removed first. A reduction in flow is requested, and the monitor shows a mean pressure of 55 mmHg to remove the aortic cardioplegia line from this fragile, aged aorta. The surgeon removes the catheter only to realize from finger pressure that the radial arterial line is not reading the blood pressure correctly. The mean pressure is actually much higher. The catheter is removed, and the suture is tied into an aorta that is suddenly rapidly expanding, bleeding, and developing a blue discoloration of the wall.

4.1.1 Discussion

In cardiac surgery, the overall incidence of intraoperative aortic dissection (IAAD) is 0.16% of cases [1]. According to the IRAD (International Registry of Aortic Dissections) database, IAAD accounts for 1.1% of all reported Type A dissections [2]. The mortality rate is high (25%–33%) and varies based on the timing within the operation that the dissection occurs [1, 3]. The prolonged bypass time that is required for repair clearly plays a role in this mortality rate, but cerebral malperfusion is also a key factor [4]. The brain cannot be protected until it is systemically cooled, unlike the heart, which is able to receive direct cold myocardial protection immediately if needed. If diagnosed *postoperatively*, the mortality rate for IAAD can be as high as 78% [3]. Yet if left untreated, the mortality is expected to be more than 90% [3].

Every step in establishing cardiopulmonary bypass can be life-threatening, but aortic cannulation can be the most critical. Purse-string sutures are used to secure the cannula in the largest high-pressure artery in the body. The aortic purse strings require precise and perfect placement, deep enough to hold the adventitia of the aortic wall, without entering the vessel itself. The geometry of the purse strings and an aortotomy incision of exact length, made by the scalpel held at exactly 90 degrees, are key technical features. After completion of CPB, tying sutures at 90% tension helps prevent injury to the aortic wall at this site. Sutures *must* be tied with an accurate, relatively low blood pressure. The safest blood pressure required depends on the fragility of the aorta and the age of the patient. Determining aortic pressure using finger palpation is a useful skill and can be used to confirm arterial line tracings prior to proceeding.

Aside from cannulation, there are other manipulations of the aorta which can result in acute dissection of the wall. In one study, the order of frequency of inciting points for aortic dissection was as follows: aortic cannulation, cross-clamp, partial occlusion clamp, proximal anastomosis, and direct injury [1].

Risk factors are general, as this particular complication can occur in any patient (Table 4.1) [5, 6]. The steps that include the ascending aorta in routine cardiac surgery include aortic cannulation, antegrade cardioplegia line, cross-clamping, suture lines, and proximal anastomoses. IADD can occur when initially going on bypass and is associated with sudden, persistent low systemic arterial pressure after initiation of CPB. At this point, the arch and descending aorta should be visualized by TEE.

Specific methods of prevention have been described (Table 4.2) [1, 5, 7]. As shown in one study, atheromatous plaques are associated with 71.4% of dissections near the site of cannulation [6]. To avoid these locations in accessing the aorta, epiaortic ultrasound can be helpful.

Table 4.1 Risk factors for intraoperative aortic dissection

1. Evidence of aortic wall atheroma
2. Dilated, thin-walled, or calcified aorta
3. Advanced age
4. Evidence of arteritis or genetic disorders
5. Significant hypertensive surge

S. K. Balaram, L. Bassin, *Cardiac Surgical Complications*, https://doi.org/10.1007/978-3-030-71563-2_4

Table 4.2 Preventative steps for intraoperative ascending aortic dissection

1. Cautious suture placement
2. Cannula insertions and removal and SBP <100 mmHg systolic to prevent wall stress
3. Gentle cross-clamp application and removal with low pump flow
4. Initiation of cardiopulmonary bypass slowly
5. Use of epiaortic probe to examine aortic wall
6. Additional care in minimally invasive cases (which may require more aortic manipulation)
7. Suture placement away from calcified plaques
8. Use of deep partial-thickness or full-thickness bites to secure purse-string sutures
9. Use of padded inserts on cross-clamps
10. Avoidance of partial occluding clamps
11. Careful insertion of aortic cannula such that the posterior wall is not injured by the tip
12. A thorough post-bypass TEE study that includes examination of the aorta

Table 4.3 Immediate treatment of intraoperative ascending aortic dissection

1. Decide cannulation site if the aortic cannula has been removed
2. Plan for circulatory arrest, steroids, ice, as standard Type A dissection repair
3. Replace ascending aorta, grafts, and/or root, as required
4. Use antegrade or consider retrograde perfusion

A tense blue or purple color change with increased size and tension of the ascending aorta are the classic findings. Sudden increase in bleeding at the suture lines or proximal anastomosis can also occur. If dissection occurs immediately after initiation of bypass, new acute aortic insufficiency or ischemia from shearing of the coronary artery ostia can be seen. This complication is confirmed with TEE and/or epiaortic ultrasound.

The difference between dissection and subadvential hematoma is sometimes difficult to make visually. Subadvential hematoma is typically less extensive, more focal, and soft. If incised, there is no persistent bleeding. The treatment for this hematoma is superficial incision to prevent expansion, as there has been some suggestion that these areas leave the patient with a compromised wall that could be at risk in the future. The difference between a true dissection and a superficial hematoma is best clarified with certainty using epiaortic ultrasound in the ascending aorta.

The treatment of IAAD requires peripheral arterial cannulation, cardiopulmonary bypass, circulatory arrest, and replacement of the ascending aorta (Table 4.3). The femoral artery may be preferred for ease of access and time considerations if the surgeon is at a point in the operation in which the central aortic cannula has already been removed. "Watchful waiting" has also been reported in patients who are elderly and very high risk, but this is associated with a high mortality in the postoperative period and generally not recommended.

4.2 Aortic Pseudoaneurysm

The elderly man is post-ascending aneurysm repair and AVR 1 year prior. He had a prolonged stay with persistent hypoxia, tachycardia, and pneumonia. He was never quite breathing comfortably and took in few calories. He had repeated episodes of aspiration.

On this day, the patient arrives in the office with swollen legs; he was mildly tachypneic. He lifts his shirt to reveal what he had tried to describe over the phone. The surgeon sees a large suprasternal mass that appears pulsatile. A contained disruption of the aortic wall is present.

4.2.1 Discussion

The surgical emergency of an aortic pseudoaneurysm (PSA) is one of the most complex and high risk of any cardiac surgical operations. The preoperative planning and strategy need to be clear. The incidence of aortic pseudoaneurysm is quoted as 0.5% with an operative in-hospital mortality of at least 6.7% [8]. One reason these surgeries are high risk is that a majority of patients (83%) have had at least one previous cardiac surgery and almost 43% of them present with evidence of infection, often with a staphylococcus organism [8]. The questions that are asked immediately revolve not around the need for operative intervention, but specifically when and how.

Common sites of disruption and leak are found at the anastomotic suture line, proximal grafts, aortic cannulation site, aortotomy site, or antegrade needle site. The aorta itself can be weakened from the previous procedure, and low-grade infection or underlying connective tissue diseases may be responsible for the resultant full-thickness defect in the intima and media. This pathology often follows prior to cardiac procedures that included insertion of prosthetic graft material (Fig. 4.1).

These pseudoaneurysms typically start small and are sometimes clinically silent. However, if enlarged over time, patients present with symptoms related to impingement of the pseudoaneurysm on surrounding structures. They often complain of chest pain, dysphagia, stridor, and/or a suprasternal mass. Pseudoaneurysms that are asymptomatic are discovered on routine follow-up CXR or CTA. These small asymptomatic aneurysms may be carefully followed if no change has been seen on close re-imaging [9].

Most symptomatic pseudoaneurysms have a high chance of rupture and must be operated on promptly. Many patients, depending on their underlying condition, require optimization prior to surgical intervention. The usual course is to immediately admit the patient to the hospital and proceed with full workup, including imaging. CT angiogram of the

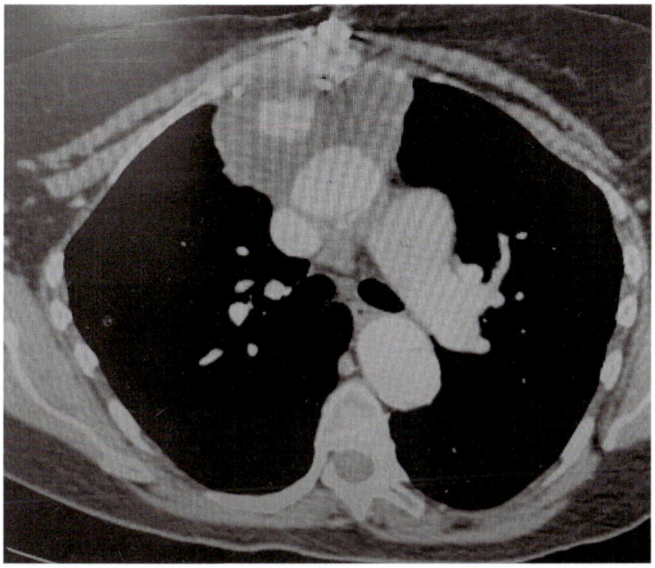

Fig. 4.1 CT scan of aortic pseudoaneurysm post-Type A dissection repair

Table 4.4 Planning for repair of aortic pseudoaneurysm

1. Review all imaging carefully: Location, critical nearby structures, extrinsic compression, distance to sternal entry
2. Obtain cultures, assume the site is infected, and consult ID service
3. Determine if the patient can tolerate a large operation immediately or needs time for optimization
4. Consider percutaneous techniques to temporize or cure, particularly if patient is not a surgical candidate
5. Require TEE and hemodynamic monitoring in the OR
6. Prepare for alternative cannulation access and assess for possibility of pseudoaneurysm rupture upon entry. Keep blood available and the CPB pump ready

chest with careful attention to the aortic replacement and anastomotic suture lines will help determine if the pseudoaneurysm is large, near the sternum, anterior, or eroding into the sternal bone itself.

Planning for sternal reentry is one of the keys to this operation (Table 4.4) [10]. Frank hemorrhage can result in immediate mortality due to rupture. It is the size and conformation of the pseudoaneurysm itself that should determine the method of entry. A significant amount of time is required to dissect out the mediastinum if adhesions are present between the graft and surrounding structures. If the rupture is free and directly under the sternum, one can consider placement of a small balloon catheter or Foley, to manage bleeding from the hole while gaining access and re-assessing options. Endovascular balloon catheters can also be used to decrease the risk of rupture during entry if the pseudoaneurysm has a small neck [11].

The surgical approach will involve cardiopulmonary bypass, usually with femoral or axillary cannulation, so that circulatory arrest followed by clear visualization of the aorta can occur. Antegrade cerebral perfusion allows for moderate hypothermia when cooling and helps shorten the total bypass time of cooling and rewarming to deep hypothermia. Going on bypass and cooling prior to opening the chest is protective against pseudoaneurysm-free rupture. However, this may result in ventricular fibrillation and aortic insufficiency with cardiac distention; the surgeon must be prepared to open the chest quickly or place a LV vent for decompression via a left thoracotomy.

The problem of active graft infection adds further complexity to these operations. Simple suture closure of the leak leaves a higher chance of recurrence due to dehiscence and continued infection [12]. Yet inserting new graft material into an infected setting is also fraught with problems. The ideal approach to aortic PSA repair in the setting of infection remains controversial. Options include debridement of infected mediastinal tissue, VAC dressings, repeated washouts, continuous irrigation, omental wrapping, or tissue flap closure at the same or delayed setting [10, 13, 14]. Partial graft excision, use of homografts, and extra-anatomic bypass are also possible treatments [15]. The excised tissue should be sent for gram stain and culture. The size and extent of infection will help guide therapy in these difficult situations and could result in a recommendation of lifelong antibiotics.

More recently, the use of minimally invasive endovascular techniques for pseudoaneurysm management has included percutaneous occlusion and embolization devices [16]. Coils, plugs, or septal occluders placed at the site of the pseudoaneurysm can occlude and thrombose the false aneurysm. This is particularly useful in patients who have already had multiple sternotomies and are at high risk of catastrophe upon sternal entry. Again, size and location determine the ability to use these methods. Critical information is obtained using CTA, MRI/A, and TEE. A narrow neck with a rim of tissue around the defect must be present for the occluder to be effective and not embolize backward into the lumen [16]. The distance from vital structures such as the coronary ostia and head vessels must be determined. These plugs can be used in patients considered inoperable or as a temporizing lifesaving measure to bridge toward a definitive operation. This additional time can then be used to treat active infection, optimize nutritional status, and stabilize acute CHF. After embolization, these patients will require close follow-up with contrast CTA chest to confirm closure and consider future operative strategies.

Another percutaneous technique is the placement of an endovascular stent graft to completely occlude the pseudoaneurysm neck. Again, imaging is critical to determine whether there is sufficient room for a graft to be placed safely in this area without impingement on surrounding structures. A suitable neck is again necessary to prevent device migration or occlusion of critical structures.

4.3 Iatrogenic Aortic Insufficiency

The patient is a slightly obese 65-year-old woman with prosthetic mitral valve endocarditis. The MRSA vegetations are extensive. The patient's anatomy makes it difficult to position the mitral retractor. The infected prosthesis is removed, and the anterior mitral annulus at A3 shows some tissue destruction. The site is debrided thoroughly and the valve is replaced. The left atrium is closed, with a small vent left across the valve for de-airing purposes.

The patient is warmed and weaning from bypass when significant blood is seen returning to the pump from the LV vent. The LV vent is clamped. LV distention and ventricular fibrillation are the result. Cardioversion is repeatedly required, despite the use of amiodarone. The TEE screen displays normal prosthetic mitral valve function with no paravalvular leak. A check of the aortic valve shows severe aortic insufficiency, with a clearly tethered non-coronary cusp.

4.3.1 Discussion

Injury to the aorto-mitral curtain can certainly occur during MV surgery. Iatrogenic AI after MVR is extremely rare but is a stark reminder of how intimately cardiac structures are related. Knowledge of anatomy here at the cardiac skeleton is vital for the surgeon. The aortic valve has been called the cornerstone of the heart, with extensions from its fibrous skeleton touching all four valves (Fig. 4.2) [17].

The possibility of this complication needs to be considered when there is known destruction or extensive involvement of the anterior MV annulus or the aorto-mitral curtain. Other surgeries have been shown to have this similar potential as well, based on anatomic or congenital factors. (Table 4.5) [18]. Injury to or restriction of the anterior mitral leaflet can occur with aortic valve surgery, when sutures are placed along the non-coronary annulus. The close relationship of these two structures is of benefit, however, when needing to upsize an aortic valve by incising the anterior leaflet and enlarging the aortic annulus with a patch.

The earliest possible detection of this injury is the key to safe treatment. Obviously, a full TEE exam both before and after surgery is mandatory to recognize abnormalities in surrounding structures. If discovered, this complication must be resolved in the same setting. A replacement of the cross-clamp and the need for repeat ischemic arrest will affect the postoperative course and potentially the long-term survival of the patient. One option available is the removal and/or replacement of the offending suture in the mitral valve if the exact location is known. If the aortic valve is significantly damaged, an aortic valve replacement will be necessary. If not discovered intraoperatively, acute AI is poorly tolerated by patients in the postoperative period, and future replacement will likely be necessary.

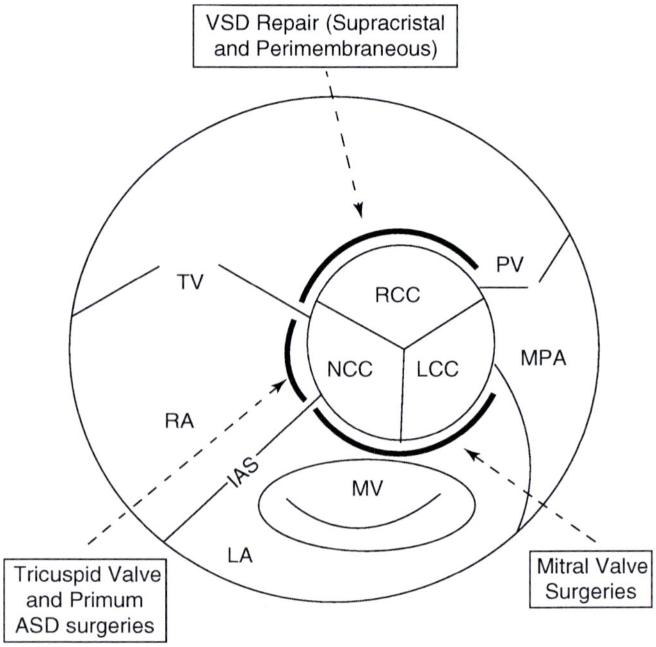

Fig. 4.2 Relationship of cardiac structures. *RCC, right coronary cusp; LCC, left coronary cusp; NCC, non-coronary cusp; TV, tricuspid valve; MV, mitral valve; PV, pulmonic valve; MPA, main pulmonary artery; LA, left atrium; RA, right atrium; IAS, interatrial septum.* From "Iatrogenic aortic insufficiency following mitral valve replacement: case report and review of the literature," by Kolakalapudi P et al., 2015, J Clin Med Res, Vol. 7, p. 488. Copyright 2015 by Elmer Press. Reprinted with permission

Table 4.5 Cardiac surgeries with potential for iatrogenic aortic insufficiency

Congenital disorders
Endocardial cushion defect
Ostium primum ASD
HCM surgery
VSD repair
Tricuspid valve surgery
Left coronary cusp: Mitral valve repair
Non-coronary cusp: Mitral valve surgery, ASD, tricuspid valve surgery
Right coronary cusp: VSD, pulmonic valve surgery

4.4 Atrioventricular Groove Rupture

An elderly female patient with rheumatic valve disease undergoes a tissue MVR and venous bypass graft to a large right posterolateral (RPL) artery. The patient is being weaned off bypass when some bleeding is seen at the distal RPL anastomosis. This requires some lifting of the heart to be fixed, and an 8-0 Prolene repair suture is placed with good result. The heart is lowered slowly, and suddenly the pericardial cavity fills with bright red blood. TEE shows large air bubbles in the left ventricle.

4.4.1 Discussion

A ruptured LV after mitral valve replacement occurs in 0.5–2% of patients and was first reported by Roberts and Morrow in 1967 [19, 20]. This disastrous complication can be classified as immediate, delayed, or chronic. Immediate rupture occurs during weaning from bypass, delayed rupture occurs hours to days later, and chronic rupture sometimes presents in the form of an LV pseudoaneurysm [21]. The majority of these injuries, two-thirds of them, occur *early* after replacement; mortality has been reported to be over 50% [21].

This devastating complication has an interesting history. Dr. Frank Spencer from NYU described a scene at the American Association of Thoracic Surgeons (AATS) meeting in 1984 in which some members argued this particular complication was due to excessive surgical trauma—a technical failure as opposed to an anatomic tendency [22]. Others disagreed and considered this problem a stretch injury due to increased afterload and "multiple endomyocardial tears" that occurred as blood dissected through the wall, causing an epicardial hematoma and eventual rupture [23]. At NYU, the excessive trauma/technical failure theory was favored until four fatal ruptures occurred with three different surgeons [22]. The idea that there was an anatomic etiology to this rupture became accepted at that point. As a result of this understanding, the group took specific care to preserve the posterior attachments to the mitral valve and reported no further ruptures in the next 365 patients who underwent MVR [22].

Critical to this understanding of anatomic structures was Cobb's theory of "the untethered loop" [24]. This concept concludes that there are muscle fibers externally that support the AV groove and that internally it is the papillary muscles and chordae that attach to the leaflet that support this continuity [24]. After mitral valve removal for replacement, the heart is volume-loaded, and increased pressure can lead to blood dissecting into the myocardium; this forms a hematoma prior to rupture. This "untethering" is the key to this complication [24]. Leaving part of the posterior mitral leaflet in place prevents stretch injury and subsequent weakening of the LV wall [24]. It is this stretch injury, after cutting the mitral chordae, that is now able to recognize as also causing a decrease in LV function after MVR by altering the anatomic features [25, 26]. A previously football-shaped ventricle will become more spherical and thus less efficient.

Over the years, different types of AV groove rupture have been described. Treasure et al. proposed two types based on location [27]. Type I is a perforation of the attachment of the atrium to the ventricle in the posterior atrioventricular groove, while Type II is a rupture that occurs in the mid-portion of the left ventricle (Fig. 4.3) [27]. This area overlies the papillary muscles. It is unknown if the etiology of these two types is different. An additional Type III is considered rupture of the LV posterior wall between the base of the papillary muscle and the AV groove [26]. Due to the anatomy involved, the ventricular tear can obviously extend, creating a mixed type of rupture.

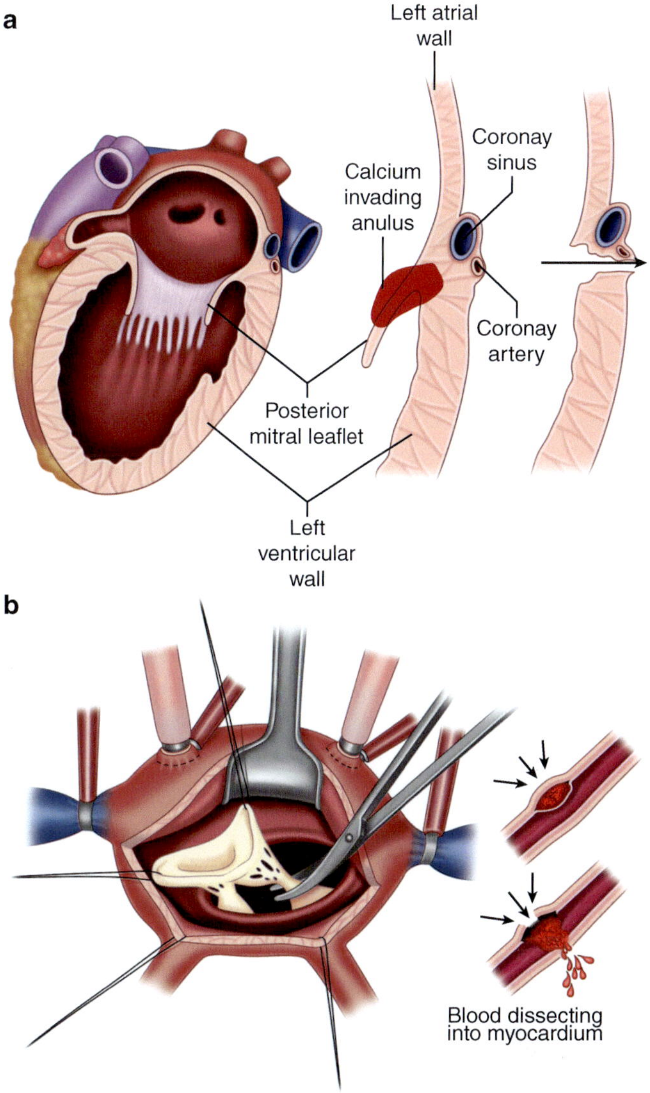

Fig. 4.3 Types of ventricular perforation after mitral valve surgery. (**a**) Location and possible mechanism of Type I perforation of ventricle. (**b**) Location and possible mechanism of Type II perforation of ventricle. From "Intraoperative left ventricular rupture associated with mitral valve replacement," by Treasure RL et al., 1974, *Chest, Vol. 66,* p. 512–3. Copyright 1974 by American College of Chest Physicians, Elsevier Inc., Reprinted with permission

Diagnosis of AV groove dissociation is heralded by epicardial hemorrhage and the subsequent visualization of large volumes of bright red blood coming from behind the heart to fill the pericardial well. Examination of the heart will demonstrate a large hematoma on the inferoposterior wall. TEE sometimes shows air within the left ventricular cavity as well as regional wall motion abnormalities. Risk factors as well as prevention techniques have been described (Tables 4.6 and 4.7) [28].

Once recognized, treatment requires a return to full CPB to decompress the heart and perform the repair. Multiple options exist for repair, which emphasize the fact that treatment of this rare condition is complex and fraught with further complications. First, it is difficult to identify the exact site of bleeding. There is an element of "tip of the iceberg" to intraoperative findings in this complication.

The most common repair technique after returning to bypass is to cross-clamp, reopen the left atrium, remove the valve, pericardial patch the entire posterior annulus, and replace the valve. Use of two strips of Teflon, one around the base of the posterior cusp of the mitral and the other in the region of the papillary muscle using deep mattress stitches in the AV groove, has been described [23]. Another choice for the treatment of a small rupture in the AV groove is to close the defect with large full-thickness pledgeted sutures into the healthy myocardium. Others have described a large patch sutured externally over the whole area [29]. Type I repairs require closure from the inside, but a Type II defect may be closed externally. If the defect involves the suture line of the mitral valve, internal repair through the left atrium is necessary. In this location, as with any repair, care must be taken to avoid the circumflex artery. A vein graft bypass to the posterior wall should be considered if there is any concern for arterial compromise. The entire AV groove needs to be reconstructed in some cases.

Resting of the heart on bypass, after what is often a long ischemic time, is prudent in these situations. IABP can be used to decrease afterload, and the judicious use of inotropes and pressors can be critical in the postoperative period. Ventricular assist devices or ECMO need to be considered, depending on cardiac function. Mechanical ventilation and prolonged sedation will also allow for healing in the postoperative period.

Table 4.6 Risk factors for AV groove rupture

1.	Calcified mitral annulus
2.	Rheumatic disease
3.	Female sex
4.	Older age
5.	Ischemic disease
6.	Endocarditis with annular abscess
7.	Hemodialysis patient
8.	Small LV size
9.	A reoperation with prosthesis in place
10.	Any reoperation

Table 4.7 Operative prevention of AV groove rupture

1.	Careful resection of mitral leaflets if needed for infection
2.	Minimize debridement of calcified annulus
3.	Accurate sizing of the valve; do not oversize
4.	Careful suture placement, avoiding sutures that are too deep
5.	Preservation of the whole posterior leaflet if possible, especially in high-risk patients
6.	Minimal retraction on the ventricle
7.	Avoidance of significant hypertension in the immediate post-CPB period
8.	Preservation of the posterior mitral leaflet-papillary muscles connection
9.	Consider the subvalvular apparatus a buttress
10.	Avoidance of excessive traction on the valve leaflets during removal
11.	Gentle insertion of the valve and careful suture tying
12.	No lifting of the heart once the prosthesis is in place
13.	Consider full adhesiolysis in a reoperative setting to prevent traction
14.	Avoid ventricular over-dilation; consider pacing

4.5 Circumflex Artery Occlusion

A complex mitral valve repair is performed for a patient with bi-leaflet prolapse. The patient is weaned slowly off bypass after partial posterior leaflet resection, placement of several chords, and a ring. Multiple inotropes are required. The TEE shows new, marked, focal inferoposterior wall hypokinesis. This is accompanied by intermittent arrhythmias and refractory hypotension. The EKG shows ST segment elevation in leads II, III, V4, and V5. The surgeon recalls previous cases of a mildly hypokinetic inferior wall after mitral valve surgery, usually attributed to the use of retrograde cardioplegia delivery. This hypokinesis always improved with reperfusion. However, the EKG changes, wall motion abnormalities, arrhythmias, and hypotension persist in this patient. One of the posterior annular sutures had occluded the circumflex artery.

4.5.1 Discussion

Likely underreported, this complication of mitral valve surgery occurs in up to 2% of patients undergoing mitral valve surgery (1.8–2.7%) [30]. Although not as devastating as atrioventricular groove disruption, it is a complex and potentially life-threatening injury, particularly if not recognized and treated expeditiously. The circumflex artery travels in the atrioventricular groove along the posterior mitral annulus and is anatomically closest to the P1 segment of the posterior leaflet. This distance has been described in rare cases to be as small as 1 mm [31]. It is the proximal part of the circumflex, close to the anterolateral commissure, that is at significant risk during mitral valve surgery.

There are a number of mechanisms by which the act of operating on a mitral valve can injure the circumflex [30]. It does not require a lacerating suture through the circumflex artery to compromise blood flow. Ischemia can also occur due to a tethering or kinking effect when the ring or valve is maneuvered into place. The manipulation of the surrounding tissue affects the artery itself. If the valve is undersized, sutures at the annulus can pull on the tissue surrounding the circumflex, potentially compromising blood flow. In addition to vascular distortion, an artery that is directly injured or perforated with a needle has ongoing bleeding or a hematoma that can compress the lumen. The artery sometimes becomes thrombosed as a result. A concomitant cryotherapy or radiofrequency MAZE procedure can also compromise the artery by a direct mechanism.

Patients at risk have been well described. In regard to coronary dominance and subsequent anatomic risk, it is critical to review the preoperative angiogram. The incidence of injury is highest with left dominant circulations [32]. Left coronary dominance is present in nearly 70% of described injuries, with only 12% of cases being right dominant. This difference can lead to injury to either the right or left system based on the proximity of the coronary arteries to the mitral valve annulus (Fig. 4.4) [33]. A distance of 3–7.3 mm between the annulus and the artery has been reported [32]. Patients with calcium that invades the annulus and surrounding tissues have anatomic distortion, putting the artery at risk. An anomalous course of the arteries has been described [31]. Indirect injury to the coronary arteries with air embolism introduced during aggressive valve testing into the LV is another mechanism that has been reported [34].

Careful suture placement directly on the mitral annulus is a way to prevent this complication. Accurate sizing of mitral valve prosthesis can also be helpful. For mitral valve repairs, avoiding wide posterior leaflet resections with small annuloplasty rings may prevent tethering. Assessing circumflex flow intraoperatively by TEE, prior to weaning from bypass, is standard in mitral valve surgery.

Early recognition is necessary. EKG changes consistent with ischemia in the setting of focal wall motion abnormalities are sufficient to address the problem intraoperatively, before major adverse outcomes such as permanent myocardial injury or arrhythmia result. Diagnosis can be made by checking circumflex flow/aliasing on TEE or by using a flow probe on the artery itself. The circumflex artery is usually readily visible on TEE views, particularly with a modified mid-esophageal long-axis view of the aortic valve and the transducer angled at 110 degrees [35]. One study reported 77% of cases showing evidence of ischemia with focal wall motion abnormalities in the operating room, with nearly 80% of patients showing EKG changes [30]. In the operating room, there is a "blind spot" in EKG with standard leads II and V5, which include the right ventricle and posterior LV wall [30].

Multiple published reports confirm that this complication is not isolated [36–39]. Options exist for treatment, but the best outcomes come from those patients for whom injury is immediately recognized and a bypass graft to the circumflex system is performed. If a specific suture is known to be causing kinking or partial vascular impingement, removal of these stitches in the left atrium is also a consideration. For suspected total occlusion, bypass surgery is obviously pre-

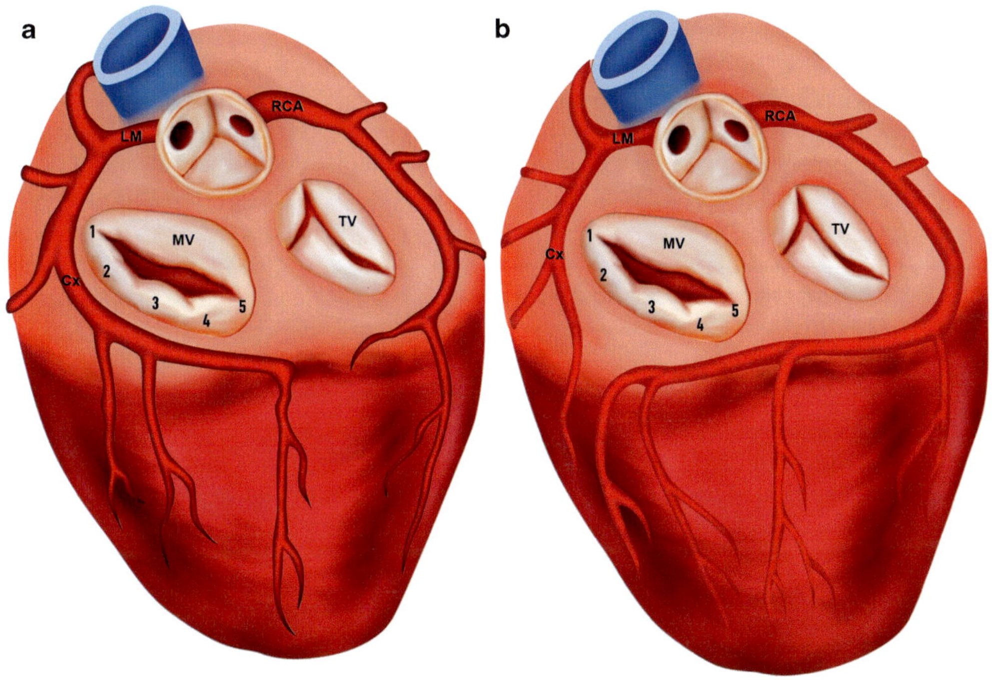

Fig. 4.4 Anatomic proximity of coronary arteries to posterior mitral valve annulus. (**a**) Left dominant system (posterior descending artery from the circumflex artery). (**b**) Right dominant system (posterior descending artery from RCA). *LM, left main; CX, circumflex; MV, mitral valve; RCA, right coronary artery; TV, tricuspid valve; 1, anterolateral commissure; 2, P1 segment; 3, P2 segment; 4, P3 segment; 5, posteromedial commissure*

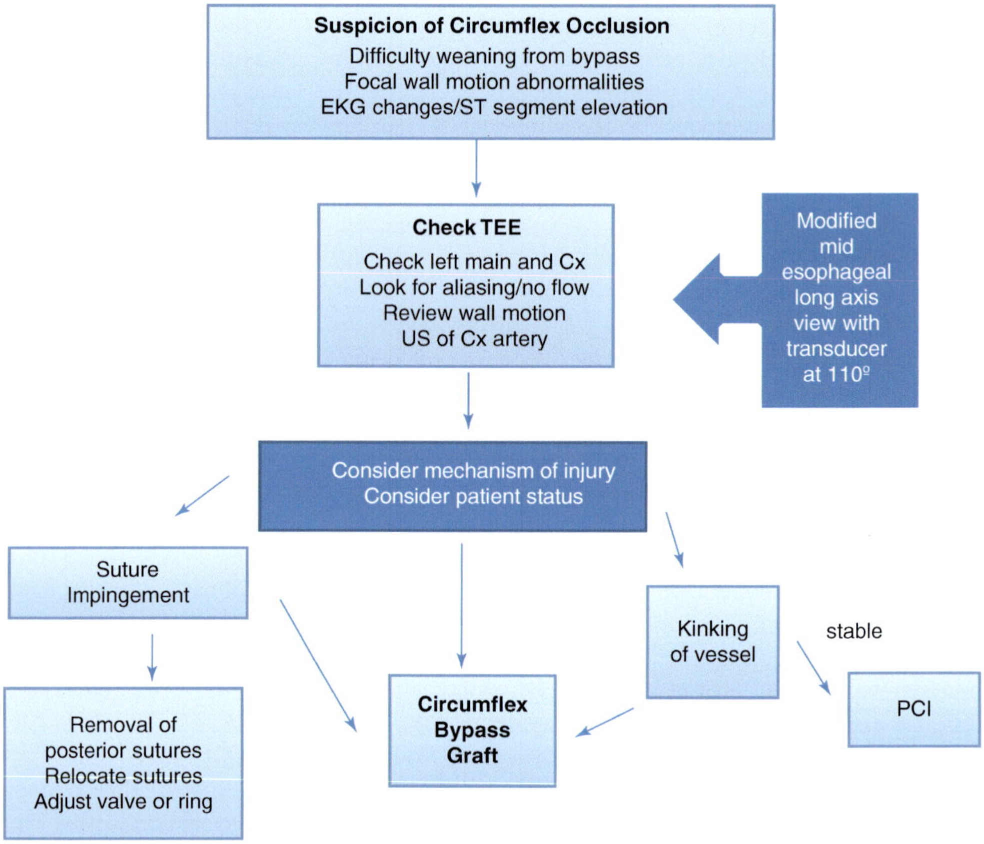

Fig. 4.5 Management of circumflex artery occlusion. *EKG, electrocardiogram; US, ultrasound; TEE, transesophageal echocardiogram; PCI, percutaneous coronary intervention*

ferred, and a vein graft to the obtuse marginal will be protective in the postoperative period.

Patients can also be treated with PCI postoperatively, with an estimated success rate of 81% [30]. Given the immediate need for return of myocardial perfusion, PCI is a better choice if the diagnosis is made when the patient is already in the ICU and then only after careful angiographic evaluation. Although kinking of the vessel can be treated with PCI, total occlusion is not amenable to this; surgical revascularization will likely be necessary. Management of this complication is highly dependent on the available resources within a specific program and hospital (Fig. 4.5).

4.6 Postoperative Systolic Anterior Motion

A complex mitral valve patient with some concerning anatomic features is scheduled for repair. These anatomic issues include (1) bi-leaflet prolapse; (2) small, tight, hyperdynamic ventricle; (3) difficult aorta-mitral angle; (4) generous anterior leaflet; and (5) small annulus. The surgeon proceeds with a P2 resection, places new chords to A2, and finishes with an appropriately sized annuloplasty band. The coaptation line appears adequate. Yet while weaning from bypass, clear LV outflow tract obstruction is seen on TEE.

4.6.1 Discussion

Systolic anterior motion (SAM) is the displacement of the anterior mitral leaflet toward the left ventricular outflow tract (LVOT) during systole. This displacement can result in significant LVOT obstruction after mitral valve repair in 4–10% of cases [40]. In the world of advanced mitral valve repair centers, this complication can occur even in the most experienced hands. Presently, with well-established techniques of repair and a better understanding of mitral valve pathology, the overall incidence in expert hands is reported as 6.6% [41]. Conservative measures are often sufficient for treatment in most cases, with only a small group (2.3%) of patients who continue to have SAM after repair [42].

The keys to treating SAM are found in fully understanding the etiology, mechanism, and effects. This flow movement of the distal anterior leaflet has been well described in the dynamics of hypertrophic cardiomyopathy, with the anterior leaflet being drawn into the outflow tract [43]. The systolic contraction of the left ventricle and movement of blood into the outflow tract can draw the tip of the distal anterior leaflet into the outflow tract, creating a turbulence that results in dynamic obstruction and mitral regurgitation [43]. Various grades have been described that characterize the severity of this physiology (Table 4.8) [42].

Detailed knowledge of the mitral valve pathology and related LV measurements are crucial in order to prevent this complication. When known risk factors for SAM are present, operative strategies sometimes need to be adjusted (Table 4.9) [43–45]. The presence of a large amount of leaflet tissue corresponding to a smaller reconstructed mitral annulus is a classic scenario [44]. Normal or increased LV contractility, particularly in younger patients with high catecholamine levels, can increase the chance of dynamic LVOT obstruction [45]. A low AL/PL ratio is essential to recognize, as a long posterior leaflet can push the anterior leaflet into the outflow tract [44]. In addition to clinical predictors such as younger age and male sex, one group found that specific echo findings such as bi-leaflet prolapse and lower LV end systolic dimension were significant [45]. The importance of annular size cannot be underestimated, and the degree of annular reduction chosen can be a crucial decision [46]. Three key risk factors for LVOT obstruction are (1) flow velocity, (2) the coaptation point and septal distance, and (3) the residual anterior leaflet length after coaptation [47]. The need for extensive posterior leaflet resection has also been associated with SAM [48].

Table 4.8 Grades of systolic anterior motion (SAM)

Grade of SAM	Anterior motion	LVOT obstruction	Mitral regurgitation
Mild	Yes	No	0-trace
Moderate	Yes	20–50 mmHg	Mild-mod
Severe	Yes	>50 mm hg	Severe

Table 4.9 Risk factors for SAM

Patient factors	Technical factors
Small hyperkinetic left ventricle Young patients with small end systolic volume	Resection of posterior leaflet height inadequate >15 mm
Anterior displaced papillary muscles	Short coaptation zone of mitral leaflet and basilar septum
Excessive leaflet tissue with tall posterior leaflet >15 mm	Small prosthetic ring
Narrow aorto-mitral plane angle <120°	Hypotension and hypovolemia
Distance between the interventricular septum and the mitral leaflet coaptation point <25 mm	
Thick basal interventricular septum (>15 mm)	
Ratio between the heights of the anterior and posterior leaflets ≤1.3	

SAM prevention requires an understanding of prophylactic techniques and may require modification of the previously chosen repair method. Confirming posterior height and the closure line are essential, as is a correctly sized annuloplasty ring [41]. A mitral band rather than a complete ring is helpful in the setting of a small annulus.

A recent study from Mayo Clinic evaluated consecutive mitral valve repair patients (761) to examine the 98 patients (13%) who developed SAM after completion of mitral valve repair [45]. Trends, echocardiographic features, and observed outcomes were reviewed. All patients received posterior annuloplasty bands, with the majority requiring a size of 28 mm. Conservative therapy alone was considered adequate in 91% of patients with postoperative SAM [45]. Treatment included use of beta blockers, volume loading, and increasing SVR [45]. Other basics of medical therapy include increasing perfusion pressure and avoiding aggressive diuresis postoperatively.

Depending on the timing of the onset of SAM, treatment and decision-making during the intraoperative and postoperative settings can vary (Figs. 4.6 and 4.7) [41]. Prior to the decision to return the cross-clamp and move to ischemic arrest for re-repair, a plan must be in place that is based on careful review of TEE findings. Changes that can be made include modifications of each of the involved parts: the AL, PL, annulus, and septum. One of the classic methods is to reduce the posterior leaflet height if >20 mm [44]. This can be done by resection or folding of the posterior leaflet, using the Carpentier sliding technique [44]. The anterior leaflet height can also be shortened with an elliptical excision [49]. The edge-to-edge technique is also considered to be effective in counteracting SAM [50–52]. Specific surgical repair techniques for SAM have been well described by Alfieri (Fig. 4.8) [52].

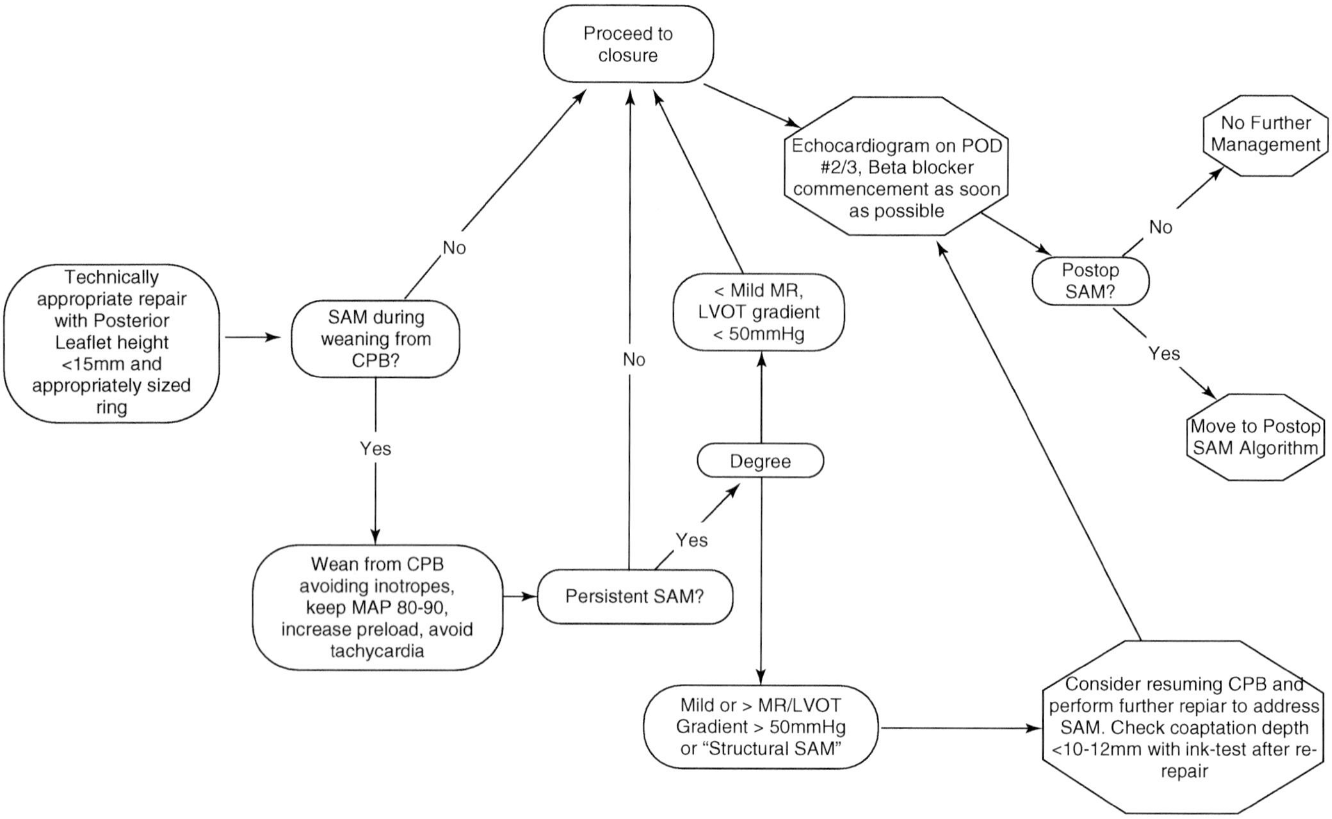

Fig. 4.6 Management of intraoperative systolic anterior motion. From "Management of systolic anterior motion after mitral valve repair: An algorithm," by Varghese et al., 2012, *J Thorac Cardiovasc Surg, Vol. 143,* p. S4. Copyright 2012 by Elsevier Science & Technology Journals. Reprinted with permission

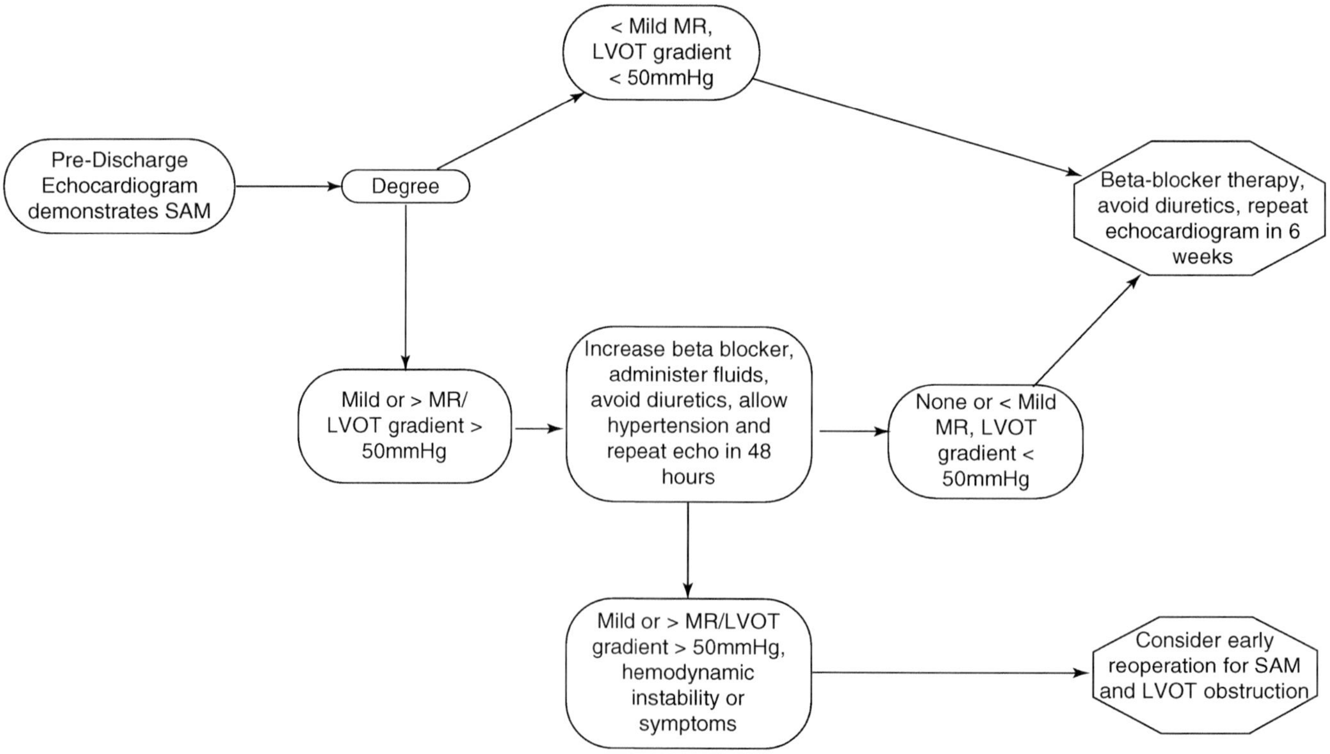

Fig. 4.7 Management of postoperative systolic anterior motion. From "Management of systolic anterior motion after mitral valve repair: An algorithm," by Varghese et al., 2012, *J Thorac Cardiovasc Surg, Vol. 143,* p. S5. Copyright 2012 by Elsevier Science & Technology Journals. Reprinted with permission

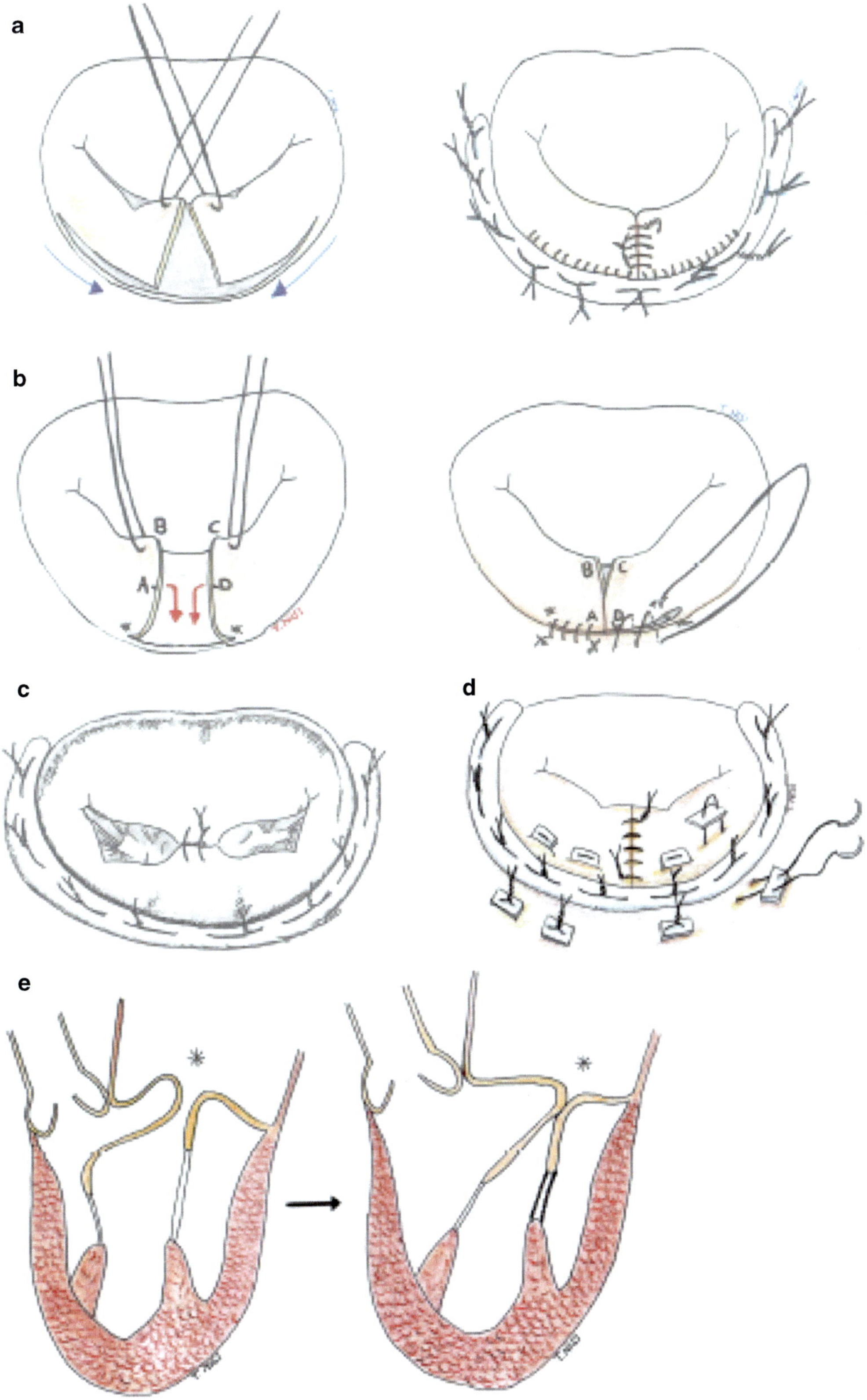

Fig. 4.8 Techniques for the treatment and prevention of postoperative systolic anterior motion: (**a**) resection and sliding plasty; (**b**) resection and folding plasty; (**c**) edge-to-edge suture; (**d**) posterior leaflet shortening technique using pledgeted interrupted horizontal mattress sutures; and (**e**) use of shortened neochordae to displace the tallest portion of the posterior leaflet into the left ventricle. From "Systolic anterior motion after mitral valve repair: where do we stand in 2015?" by Alfieri O and Lapenna E, 2015, *Eur J Cardiothorac Surg, Vol. 48*, p. 346. Copyright 2015 by Oxford University Press Journals/ Elsevier. Reprinted with permission

4.7 Hemolysis

A 60-year-old woman with a history of deep venous thrombosis, on anticoagulation, undergoes mitral valve repair for myxomatous disease. She has a posterior leaflet resection and placement of two anterior leaflet neochords. Post-surgery, her TEE shows trace mitral regurgitation. She does well initially, but returns to her physician's office 3 months later complaining of increasing fatigue, dyspnea with exertion, and dark urine. Her primary care physician is concerned about hematuria due to anticoagulation, and she is referred to a urologist. She is eventually sent back to see her cardiac surgeon. She arrives fatigued, dyspneic, mildly jaundiced, and with signs of heart failure. Her urine and bloodwork are consistent with hemolysis. The urine is positive for hemoglobinuria; she is anemic, with an elevated LDH, alkaline phosphatase, and bilirubin. Furthermore, fragmented schistocytes are present on peripheral smear. She requires a blood transfusion for profound anemia. TEE demonstrates mild-moderate MR, with a jet that is described as high velocity.

4.7.1 Discussion

Hemolysis in the clinical setting of paravalvular regurgitation for both mitral and aortic valve replacement has been well described. However rarely, patients have presented with hemolysis after mitral valve *repair* surgery. The incidence is extremely low, and therefore this complication particularly benefits from being recognized at an early stage, before further issues arise. Most patients will present within 3 months of surgery but can present earlier or up to several years later [53]. Presenting symptoms include jaundice, dark urine, heart failure, decreased kidney function, elevated liver function studies (LDH, alkaline phosphatase, bilirubin), and evidence of hemolysis on peripheral smear [54, 55].

Hemolysis can be present even in the setting of mild mitral regurgitation (MR), although it is more common with moderate to severe MR. Damage to the kidneys is a secondary complication of the hemoglobinuria deposits, and this differentiation between hematuria and hemoglobinuria, regardless of the presence of anticoagulation, must be made early to prevent kidney injury [56]. Simple injury to the RBC can result in increased phagocytosis by the liver and spleen, resulting in hepatosplenomegaly [54]. Medical treatment of this complication is generally poorly tolerated. Patients are often referred for reoperation and mitral valve replacement.

Once hemolysis is confirmed, a TEE will provide essential information. The mechanism of hemolysis in the repair setting is complex [57, 58]. Paravalvular leaks around the sewing ring or high-velocity jets through the valve injure the erythrocytes through collisions and cause fragmentation. Ring dehiscence, suture material near the offending jet, direct collision between the jet and the wall of the atrium, and poor endothelialization of the ring and sutures all can result in hemolysis [57]. Post-MV repair, increased LVOT gradients and SAM have been shown to result in hemolysis, even in the setting of a functioning mitral valve [59]. Other mechanisms for hemolysis include fragmentation of the regurgitation jet, rapid acceleration, slow deceleration, collision, and free jets [53]. Lam and colleagues studied specific locations of hemolysis jets and found posterior jets in 55%, central jets in 33%, anterior jets in 6%, and complex jets in the remainder [53]. Of importance in this study is that 77% of patients who returned with this complication had developed moderate to severe MR over time [53]. Turbulent flow or whiplash injury is more likely to develop in failed or incomplete repairs [58].

These patients do not heal spontaneously or improve with prolonged medical therapy. Irreversible injury to the kidney and multiple blood transfusions are not well tolerated. Most patients will require reoperation. Although re-repair can be considered, the majority of these patients will likely receive prosthetic valve replacements as a definitive cure for their hemolysis.

References

1. Still RJ, Hilgenberg AD, Akins CW, Daggett WM, Buckley MJ. Intraoperative aortic dissection. Ann Thorac Surg. 1992;53(3):374–9. discussion 380.
2. Januzzi JL, Sabatine MS, Eagle KA, Evangelista A, Bruckman D, Fattori R, Oh JK, Moore AG, Sechtem U, Llovet A, Gilon D, Pape L, O'Gara PT, Mehta R, Cooper JV, Hagan PG, Armstrong WF, Deeb GM, Suzuki T, Nienaber CA, Isselbacher EM. International registry of aortic dissection investigators. Iatrogenic aortic dissection. Am J Cardiol. 2002;89(5):623–6.
3. Murphy DA, Craver JM, Jones EL, Bone DK, Guyton RA, Hatcher CR Jr. Recognition and management of ascending aortic dissection complicating cardiac surgical operations. J Thorac Cardiovasc Surg. 1983;85:247–56.
4. Ruchat P, Hurni M, Stumpe F, Fischer AP, von Segesser LK. Acute ascending aortic dissection complicating open heart surgery: cerebral perfusion defines the outcome. Eur J Cardiothorac Surg. 1998;14:449–52.
5. Singh A, Mehta Y. Intraoperative aortic dissection. Ann Card Anaesth. 2015;18(4):537–42.
6. Narayan P, Angelini GD, Bryan AJ. Iatrogenic intraoperative type a aortic dissection following cardiac surgery. Asian Cardiovasc Thorac Ann. 2015;23(1):31–5.
7. Lin TY, Chen YS, Chiu KM, Hsu RB, Yu HY, Wang MJ. Eight-year experience of intraoperative aortic dissection. Asian Cardiovasc Thorac Ann. 2009;17(4):408–12.
8. Atik FA, Navia JL, Svensson LG, et al. Surgical treatment of pseudoaneurysm of the thoracic aorta. J Thorac Cardiovasc Surg. 2006;132:379–85.
9. Sabri MN, Henry D, Wechsler AS, DiSciascio G, Vetrovec GW. Late complications involving the ascending aorta after cardiac surgery: recognition and management. Am Heart J. 1991;121(6 Pt 1):1779–83.

10. Henley MD, Farber MA, Ikonomidis JS, Kolarczyk LM, Teeter EG, Barrick BP, Caranasos TG, Martinelli SM. A tale of three surgeries: management of a massive recurrent mycotic aortic pseudoaneurysm. J Cardiothorac Vasc Anesth. 2018;32(1):550–7.
11. Pettersson G, Nores M, Gillinov AM. Transfemoral control of ruptured aortic pseudoaneurysm at aortic root reoperation. Ann Thorac Surg. 2004;77(1):311–2.
12. Mohammadi S, Bonnet N, Leprince P, Kolsi M, Rama A, Pavie A, Gandjbakhch I. Reoperation for false aneurysm of the ascending aorta after its prosthetic replacement: surgical strategy. Ann Thorac Surg. 2005;79(1):147–52. discussion 152.
13. Coselli JS, Köksoy C, LeMaire SA. Management of thoracic aortic graft infections. Ann Thorac Surg. 1999;67(6):1990–3. discussion 1997–8.
14. Malvindi PG, van Putte BP, Heijmen RH, et al. Reoperations for aortic false aneurysms after cardiac surgery. Ann Thoracic Surg. 2010;90:1437–43.
15. Quevedo HC, Santiago-Trinidad R, Castellanos J, Atianzar K, Anwar A, Abi Rafeh N. Systematic review of interventions to repair ascending aortic pseudoaneurysms. Ochsner J. 2014;14(4):576–85.
16. Bashir F, Quaife R, Carroll JD. Percutaneous closure of ascending aortic pseudoaneurysm using Amplatzer septal occluder device: the first clinical case report and literature review. Catheter Cardiovasc Interv. 2005;65:547–51.
17. Kolakalapudi P, Chaudhry S, Omar B. Iatrogenic aortic insufficiency following mitral valve replacement: case report and review of the literature. J Clin Med Res. 2015;7(6):485–9.
18. Mehta AR, Hunsaker R. Iatrogenic aortic incompetence after mitral valve replacement. J Cardiothorac Vasc Anesth. 2007;21(2):276–8.
19. Roberts WC, Morrow AG. Causes of early postoperative death following cardiac valve replacement: clinicopathologic correlations in 64 patients studied at necropsy. J Thorac Cardiovasc Surg. 1967;54:422.
20. Zacharias A, Groves LK, Cheanvechai C, Loop FD, Effler DB. Rupture of the posterior wall of the left ventricle after mitral valve replacement. J Thorac Cardiovasc Surg. 1975;69(2):259–63.
21. Björk VO, Henze A, Rodriguez L. Left ventricular rupture as a complication of mitral valve replacement. J Thorac Cardiovasc Surg. 1977;73(1):14–22.
22. Spencer FC, Galloway AC, Colvin SB. A clinical evaluation of the hypothesis that rupture of the left ventricle following mitral valve replacement can be prevented by preservation of the chordae of the mural leaflet. Ann Surg. 1985;202:673–80.
23. Azariades M, Lennox SC. Rupture of the posterior wall of the left ventricle after mitral valve replacement: etiological and technical considerations. Ann Thorac Surg. 1988;46(5):491–4.
24. Cobbs BW Jr, Hatcher CR Jr, Craver JM, Jones EL, Sewell CW. Transverse midventricular disruption after mitral valve replacement. Am Heart J. 1980;99(1):33–50.
25. Lillehei CW, Levy MJ, Bonnabeau RC Jr. Mitral valve replacement with preservation of papillary muscles and chordae tendineae. J Thorac Cardiovasc Surg. 1964;47:532.
26. Miller DW Jr, Johnson DD, Ivey TD. Does preservation of the posterior chordae tendineae enhance survival during mitral valve replacement? Ann Thorac Surg. 1979;28:22–7.
27. Treasure RL, Rainer WG, Strevey TE, Sadler TR. Intraoperative left ventricular rupture associated with mitral valve replacement. Chest. 1974;66(5):511–4.
28. Zhang HJ, Ma WG, Xu JP, Hu SS, Zhu XD. Left ventricular rupture after mitral valve replacement: a report of 13 cases. Asian Cardiovasc Thorac Ann. 2006;14:26–9.
29. Celemín D, Nuñez L, Gil-Aguado M, Larrea JL. Intraventricular patch repair of left ventricular rupture following mitral valve replacement: new technique. Ann Thorac Surg. 1982;33(6):638–40.
30. Hiltrop N, Bennett J, Desmet W. Circumflex coronary artery injury after mitral valve surgery; a report of four cases and comprehensive review of the literature. Catheter Cardiovasc. 2017;89(1):78–92.
31. Grande AM, Fiore A, Massetti M, Viganò M. Iatrogenic circumflex coronary lesion in mitral valve surgery: case report and review of the literature. Tex Heart Inst J. 2008;35(2):179–83.
32. Virmani R, Chun PK, Parker J, McAlister HA. Suture obliteration of the circumflex coronary artery in three patients undergoing mitral valve operation. Role of left dominant or codominant coronary artery. J Thorac Cardiovasc Surg. 1982;84:773–8.
33. Bargagna M, Trumello C, Sala A, Blasio A, Castiglioni A, Alfieri O, De Bonis M. Left circumflex artery injury after mitral valve surgery: an algorithm management proposal. Ann Thorac Surg. 2021;111(3):899–904.
34. Obarski TP, Loop FD, Cosgrove DM, Lytle BW, Stewart WJ. Frequency of acute myocardial infarction in valve repairs versus valve replacement for pure mitral regurgitation. Am J Cardiol. 1990;65(13):887–90.
35. Ender J, Selbach M, Borger MA, Krohmer E, Falk V, Kaisers UX, Mohr FW, Mukherjee C. Echocardiographic identification of iatrogenic injury of the circumflex artery during minimally invasive mitral valve repair. Ann Thorac Surg. 2010;89(6):1866–72.
36. Danielson GK, Cooper E, Tweeddale DN. Circumflex coronary artery injury during mitral valve replacement. Ann Thorac Surg. 1967;4:53–9.
37. Dello SA, Leus SJ, Tan MES, Otterspoor LC, Botman CJ. Percutaneous coronary intervention of an iatrogenic occlusion of the circumflex coronary artery after mitral valve replacement. Eur Heart J Acute Cardiovasc Care. 2020;9(5):NP1–2.
38. Somekh NN, Haider A, Makaryus AN, Katz S, Bello S, Hartman A. Left circumflex coronary artery occlusion after mitral valve annuloplasty: "a stitch in time". Tex Heart Inst J. 2012;39(1):104–7.
39. Ahmad TA, Nudy M, Pabst D, Kozak M. Mitral valve repair complicated by left circumflex coronary artery occlusion: an underrecognized but potentially deadly complication. Cardiovasc Revasc Med. 2018;19(8S):77–81.
40. Crescenzi G, Landoni G, Zangrillo A, Guarracino F, Rosica C, La Canna G, Alfieri O. Management and decision-making strategy for systolic anterior motion after mitral valve repair. J Thorac Cardiovasc Surg. 2009;137(2):320–5.
41. Varghese R, Anyanwu AC, Itagaki S, Milla F, Castillo J, Adams DH. Management of systolic anterior motion after mitral valve repair: an algorithm. J Thorac Cardiovasc Surg. 2012;143(4 Suppl):S2–7.
42. Brown ML, Abel MD, Click RL, Morford RG, Dearani JA, Sundt TM, Orszulak TA, Schaff HV. Systolic anterior motion after mitral valve repair: is surgical intervention necessary? J Thorac Cardiovasc Surg. 2007;133(1):136–43.
43. Sherrid MV, Chaudhry FA, Swistel DG. Obstructive hypertrophic cardiomyopathy: echocardiography, pathophysiology, and the continuing evolution of surgery for obstruction. Ann Thorac Surg. 2003;75:620–32.
44. Carpentier A, Adams DH, Filsoufi F. Techniques in systolic anterior motion(SAM). In: Carpentier A, Adams DH, Filsoufi F, editors. Carpentier's reconstructive valve surgery. Philadelphia: Saunders, Elsevier; 2010. p. 157–65.
45. Ashikhmina E, Schaff HV, Daly RC, Stulak JM, Greason KL, Michelena HI, Fatima B, Lahr BD, Dearani JA. Risk factors and progression of systolic anterior motion after mitral valve repair. J Thorac Cardiovasc Surg. 2020;S0022-5223(20):30195-1.
46. Shah PM, Raney AA. Echocardiographic correlates of left ventricular outflow obstruction and systolic anterior motion following mitral valve repair. J Heart Valve Dis 2001;10:302–306.
47. Manabe S, Kasegawa H, Arai H, Takanashi S. Management of systolic anterior motion of the mitral valve: a mechanism-based approach. Gen Thorac Cardiovasc Surg. 2018;66(7):379–89.

48. Grossi EA, Galloway AC, Parish MA, Asai T, Gindea AJ, Harty S, et al. Experience with twenty-eight cases of systolic anterior motion after mitral valve reconstruction by the Carpentier technique. J Thorac Cardiovasc Surg. 1992;103:466–70.
49. Quigley RL. Prevention of systolic anterior motion after repair of the severely myxomatous mitral valve with an anterior leaflet valvuloplasty. Ann Thorac Surg. 2005;80:179–82.
50. Mascagni R, Al Attar N, Lamarra M, Calvi S, Tripodi A, Mebazaa A, et al. Edge-to-edge technique to treat post-mitral valve repair systolic anterior motion and left ventricular outflow tract obstruction. Ann Thorac Surg. 2005;79:471–4.
51. Myers PO, Khalpey Z, Maloney AM, Brinster DR, D'Ambra MN, Cohn LH. Edge-to-edge repair for prevention and treatment of mitral valve systolic anterior motion. J Thorac Cardiovasc Surg. 2013;146:836–40.
52. Alfieri O, Lapenna E. Systolic anterior motion after mitral valve repair: where do we stand in 2015? Eur J Cardiothorac Surg. 2015;48(3):344–6.
53. Lam BK, Cosgrove DM, Bhudia SK, Gillinov AM. Hemolysis after mitral valve repair: mechanisms and treatment. Ann Thorac Surg. 2004;77(1):191–5.
54. Wang J, Zhang H, Fan H, Chen K, Zhang Y, Song K, Ao H, Yu C. Intractable mechanical hemolytic anemia complicating mitral valve surgery: a case series study. BMC Cardiovasc Disord. 2020;20(1):104.
55. Hinrichs L, Rassaf T, Totzeck M. Malaise and fatigue following mitral valve repair: case report. Eur Heart J Case Rep. 2018;2(2):yty039.
56. Björkenheim A, Cha SO, Dioubanova I. Haemolytic anaemia after mitral valve repair due to recurrent mild to moderate mitral regurgitation. BMJ Case Rep. 2019;12(8):e230280.
57. Yeo TC, Freeman WK, Schaff HV, Orszulak TA. Mechanisms of hemolysis after mitral valve repair: assessment by serial echocardiography. J Am Coll Cardiol. 1998;32(3):717–23.
58. Wilson JH, Rath R, Glaser R, Panke T. Severe hemolysis after incomplete mitral valve repair [see comments]. Ann Thorac Surg. 1990;50:136–7.
59. Rabbani M, Hafiz A, Algadheeb M, Thain A, Kiaii BB. A case of systolic anterior motion after mitral valve repair causing hemolytic anemia: mechanism and treatment. Can J Cardiol. 2020;36(12):1977.e5–8.

5 Critical Postoperative Complications

5.1 Pulmonary Hypertension

Just prior to a planned CABG, a pulmonary artery catheter was placed in a 72-year-old patient with a normal EF, COPD, and tight obstructive lesions, During mammary artery harvest, the systolic PA pressures were observed to rise from 30 into the high 50 range. Review of the echo for signs of ischemia showed a newly hypokinetic lateral wall and mild-moderate mitral regurgitation. Nitroglycerin was added, and the patient's cardiac function was significantly improved. Several hours later, the surgery was completed, and the patient was sent in stable condition to the ICU. During transfer of care communications between the OR and ICU teams, the systolic PA pressures suddenly crept up from 40 to 50 to 60. The patient became hypotensive.

5.1.1 Discussion

One of the clinical focuses necessary in cardiac surgery is a clear understanding of pulmonary hypertension before, during, and after cardiopulmonary bypass. Pulmonary hypertensive crisis can lead to sudden catastrophe in what was potentially a successful operation. The line between a dramatic rise in pulmonary vascular resistance and a poorly functioning RV is small.

Cardiopulmonary bypass in itself, through multiple mechanisms, increases pulmonary vascular tone and decreases RV function and performance. The reasons behind this are both chemical and flow-related. The pulmonary vasculature cells are dynamic in nature and respond acutely to flow, pressure, circulating catecholamines, and vasopressors. Patients with pre-existing pulmonary hypertension who undergo cardiopulmonary bypass can experience pulmonary endothelial cell injury through acidosis, shear stress, hypoperfusion of the lung, and SIRS activation of the pulmonary bed [1]. All of these factors increase vasoreactivity and induce specific chemical changes: decreased nitric oxide production and increased endothelin-1 and thromboxane production [1, 2]. Furthermore, hypoxia and hypotension, the production of free radicals, and the duration of bypass all contribute to pulmonary vasospasm [3].

The principal mechanism of pulmonary endothelial dysfunction in cardiac surgery is ischemia/reperfusion, associated with an influx of inflammatory cells and complement activation [4]. The change from pulsatile flow to nonpulsatile flow alters this highly sensitive vascular bed. Blood transfusions and protamine responses can also change hemodynamics quickly during the initial post-bypass hours.

A sudden increase in PA pressures seen post-bypass should be evaluated in terms of vasoconstrictors, acidosis, atelectasis, or hypercarbia. A change in left ventricular function and increased mitral regurgitation can manifest initially with rising pulmonary hypertension. The pulmonary bed is triggered by simple mechanical stress as well as multiple biochemical changes [5] (Fig. 5.1).

Changes in pulmonary vascular resistance can occur suddenly in the operating room or in the ICU. In the OR, there is the benefit of hemodynamic monitoring, TEE, and a direct view of the right ventricular wall. Subtle changes in filling and function that are observed and treated early can prevent future problems. Pulmonary hypertension has been described as "dilator and constrictor functions out of balance" [5]. Optimizing heart rate and hematocrit can be helpful prior to weaning from bypass.

Managing of sudden pulmonary hypertension immediately post-surgery begins with re-examination of wall motion by TEE and ruling out ischemia. Bypass grafts should be re-checked with flow monitoring. If there is no evidence of ischemia and/or the patient had non-coronary surgery, other causes will need to be considered. Reversible ischemia, respiratory failure, anemia, hypoxia, and acidosis all require immediate attention [6]. The basic treatments are optimize volume status, reduce PVR by use of pulmonary vasodilators, and consider inotropes such as milrinone or dobutamine (Table 5.1) [6].

In the ICU, pulmonary hypertension and the RV failure that ensues can present rapidly but with ambiguous signs.

S. K. Balaram, L. Bassin, *Cardiac Surgical Complications*, https://doi.org/10.1007/978-3-030-71563-2_5

Fig. 5.1 Etiology of pulmonary hypertension post-cardiac surgery

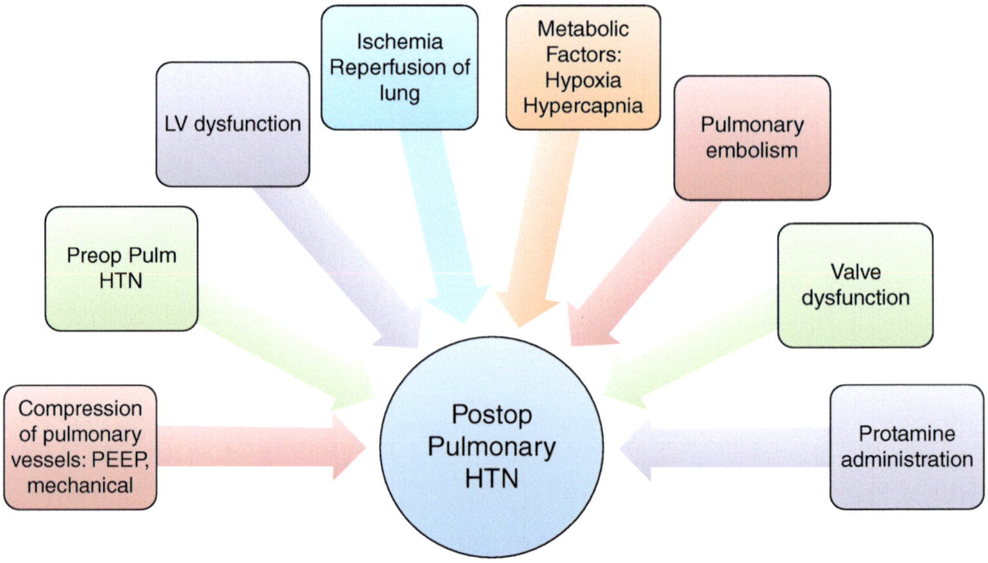

Table 5.1 Treatment goals for pulmonary hypertension

Optimize hemodynamics	Volume load, augment RV function, maintain systemic vascular resistance and RCA perfusion
Treat reversible factors	**Metabolic:** Hypoxia, hypercarbia, ischemia, acidosis **Mechanical:** Mitral regurgitation, LV dysfunction
Add vasodilators	Nitric oxide, Epoprostenol, milrinone

Again, the underlying cause needs to be determined early, before the cascade of dysfunction begins. A mechanical problem such as decreased oxygen delivery or elevated CO_2 can rapidly lead to right ventricular strain, myocardial ischemia, decreased function, and low cardiac output with oliguria. If invasive hemodynamic monitoring is not in place, low urine output is sometimes the first observed clinical change. Depending on presentation, one must also consider drug reaction and pulmonary embolism. The specific etiologies that should be ruled out include preoperative PAH, hypoxia, increased left atrial pressure due to cardiac/valve dysfunction, and pulmonary embolism [6].

The CPB run has many affects, each time in varying degrees. One such affect is decreased circulating nitric oxide (NO). When ischemia-reperfusion occurs, free radicals are produced that inactivate NO. The resulting increase in pulmonary vascular tone can be improved by the use of inhaled nitric oxide (NO). Inflammation, cytokines, and complement action depend on the baseline status of the pulmonary vasculature, the intraoperative management, and the postoperative medication environment [4].

RV ischemia can occur with both increased afterload and decreased perfusion. The RV is also sensitive to systemic afterload changes between the aorta and LV. As right ventricular perfusion occurs during both systole and diastole, the right ventricle can become ischemic even when the LV-generated pressures are not as low as to affect the left-sided perfusion [6].

5.2 Hepatic Failure

A referring physician calls to discuss a diabetic patient with a tight left main lesion who was admitted through their emergency room with a NSTEMI. The patient has a remote history of cirrhosis, GI bleeding, and thrombocytopenia. The initial response from the surgeon is that this is an appropriate patient for percutaneous therapies, not for surgical revascularization. After transfer, the GI service is consulted to assist in risk assessment. The patient has a documented MELD score, Child's score, SYNTAX score, EuroSCORE, and STS risk score. The question of whether this patient should proceed with cardiac surgery remains.

5.2.1 Discussion

Previously rarely performed due to high mortality, cardiac surgery for patients with end-stage liver disease (ESLD) is more frequently considered in the modern era. In fact, an increase of 22% was seen in this patient population from 1998 to 2006 [7]. This increase has been assisted by the use of combined heart/liver transplant as a potential treatment option for patients with acute liver failure [8]. Hepatic failure after cardiac surgery is rare, accounting for approximately 4% of gastrointestinal complications, but carrying an associated mortality of over 70% [9].

Assessment of preoperative liver dysfunction is key, and multiple methods are available for this evaluation. The Child-Turcotte-Pugh (CTP) score and now the Model End-Stage Liver Disease (MELD) score have been used to determine degrees of liver failure with associated mortality [10]. There has been some debate in the literature about which score more accurately predicts outcomes after cardiac surgery in particular, with two previous studies showing Child's class to be a better predictor [11, 12]. MELD does not account for the presence of portal hypertension and ascites which are specifically vital markers in cardiac surgery, but it does include renal function. Common practice at present utilizes the MELD score most frequently to predict mortality risk after cardiac surgery (Table 5.2) [13–15].

Patients with hepatic dysfunction are predisposed to multiple postoperative complications based on hepatic functional reserve, which is difficult to predict. This is one of the main problems of surgical intervention in liver failure patients [14]. The normal stresses of cardiac surgery include transient hypotension, fluid shifts, relative ischemia, reperfusion, SIRS, catecholamine exposure, and potential anesthetic toxins. The liver is exquisitely sensitive to these routine perioperative conditions. Although the GI service is often encouraging in their recommendations to proceed if the risk scores are low, it is critical to remember that the CPB pump is a demanding physiologic challenge for cirrhotics. Preoperative, intraoperative, and postoperative care must all be carefully controlled and monitored with even more than the usual vigilance in order to prevent the catastrophic complication of fulminant hepatic failure (Table 5.3). Causes of death in cirrhotic patients post-bypass include sepsis, multi-system organ failure, and hepatic failure [14]. More common postoperative complications include cholestasis, hemorrhage, and mesenteric ischemia [14]. Understanding the pathophysiology of dysfunction is helpful in planning for surgery.

In addition to the basics of nonpulsatile flow and low cardiac output that accompany heart surgery, catecholamines and synthetic vasoconstrictors cause a specific decrease in arterial liver perfusion. Anesthetic medications themselves can exacerbate liver failure. Furthermore, routine volume loading can lead to tissue edema, coagulopathy, and further hypoxia secondary to a hepatic venous congestion that limits oxygen delivery. Cirrhotics have low oncotic pressure based on hypoalbuminemia; they do not tolerate large volume shifts. In addition to the ischemic insult, there are circulating endotoxins and translocated bacteria that also impact the liver through the portal vein, markedly increasing the SIRS response and directly decreasing blood flow [14]. From there, it can be a short step to sepsis and multi-system organ failure. Prevention with a specific strategy is necessary (Table 5.4).

A review of multiple studies published in 2015 found that mortality rates after cardiac surgery for Childs A/B/C are 90%/37%/52%, respectively [15]. Patients with advanced-stage liver disease and/or portal hypertensive complications have difficulty surviving the stresses of cardiac surgery. Even without cardiac surgery, the expected 1-year survival for Childs A/B/C are 100%/81%/45% [15]. Keeping the CPB time to <70 minutes has been shown to be helpful in preventing complications [16]. In patients with lesser stage liver disease, close evaluation of cardiac function and a good understanding of long-term prognosis are mandatory when choosing surgical technique. These patients must also be monitored after surgery with careful fluid resuscitation and careful prevention of bleeding, infection, or other complications.

Patients with advanced liver disease and little hepatic reserve tolerate cardiac surgery poorly. Optimizing medical management and planning for percutaneous options are often

Table 5.2 ESLD risk scores and cardiac surgery

CPT A or MELD <11	CPT B or MELD 12–14	CPT C or MELD >14
Green light (relative)	Yellow light	Red light
Look for red flags: Recent GI bleed, encephalopathy, ascites, thrombocytopenia Remember that even class A has a 7x higher rate of death in 10 years [15]	Carefully examine cardiac anatomy and function in terms of complexity of surgery to make a decision Consider combined heart liver transplant Look for red flags: Portal HTN, cirrhotic complications, bleeding	Think about medical management or percutaneous therapies, not surgery Remember *any* treatment (even minimally invasive) that requires general anesthesia and/or involves hypotension will likely affect postoperative hepatic function

Table 5.3 Keys to operating on cirrhotic patients

1. Maintain splanchnic perfusion
2. Careful fluid management
3. Limit coagulopathy
4. Thermoregulation
5. Avoid/limit CPB and nonpulsatile flow
6. Optimize transfusion

Table 5.4 Preoperative and intraoperative strategy for cardiac surgery in ESLD

Preop	Intraop	Postop
Thorough history and physical. Look for stigmata of cirrhosis including encephalopathy	Avoid large volume shifts Manage coagulopathy Minimize CPB time Minimize transfusions	Maintain HD support Follow LFTs and coagulation studies closely
Tests: LFT, platelets, Creatinine, coagulation studies Liver ultrasound Abdominal CT	CPB: Low prime volume Smaller tubing High-flow rates Try to keep <70 min	Metabolic support Nutrition Prevent sepsis Prevent hemorrhage
Medical management of portal HTN with beta blockers Consider TIPS Give vitamin K and nutrition prior to surgery to improve coagulopathy (a sign of hepatic reserve)	Constant and careful monitoring of BP	

preferred. Careful consideration of cardiac function, other comorbidities, and technical options are all necessary to ensure a good outcome.

5.3 Abdominal Compartment Syndrome

A moderately obese 78-year-old patient with a low ejection fraction undergoes a difficult CABG on-pump. His targets are small and his CPB time is prolonged. He is transfused two units of packed red blood cells along with platelets and fresh frozen plasma. Despite moderate doses of inotropes, his cardiac output remains marginal. A tense abdominal wall is noted at the completion of the procedure; this is not improved with nasogastric tube decompression. Postoperatively, there is moderate postoperative bleeding and coagulopathy that requires additional blood transfusions.

Over the next several days, he does not progress well. He is extubated on day 1 and placed on high-flow oxygen, but is persistently tachypneic and tachycardic. His urine output is marginal despite a normal cardiac index. He has a poor response to fluid administration. His abdomen is distended and firm; abdominal films reveal a large ileus. Also present, and more concerning, are nonspecific changes in mental status. Eventually, the patient goes into anuric renal failure and his BUN rises over 100. The team is treating profound ileus, abdominal distention, renal failure, and altered mental status.

5.3.1 Discussion

Abdominal compartment syndrome (ACS) has been described after extensive abdominal surgery or trauma. A sudden increase in intra-abdominal pressure occurs after closure of the abdomen and results in new organ dysfunction, most commonly renal and pulmonary impairment [17]. In the setting of *non*-abdominal surgery, this effect is termed secondary abdominal compartment syndrome. Although not directly cardiac in scope, knowledge of etiology, pathology, organ dysfunction, and available treatments need to be understood by all surgical care teams.

ACS usually begins with a shock state, which in cardiac surgery patients can include a CPB run with its relative hypotension, hemodilution, vasoconstriction, decreased osmotic pressure, and nonpulsatile flow. Blood can be redistributed away from organs and lead to intestinal, renal, tissue, and cellular hypoxia. Reperfusion injury occurs in the splanchnic system. Multiple cytokines and free radicals are released, leading to capillary leak and increased edema of the intestinal walls [18]. This edema in itself can increase the intra-abdominal pressure and decrease perfusion. Bowel ischemia and hypoxia from poor flow lead to even more edema and higher pressures. CPB is the initial injury, but aggressive post-bypass fluid administration can exacerbate this effect.

This syndrome notably affects the kidneys, GI tract, and lungs. The result of increased abdominal pressure is decreased blood flow to the kidneys. This ischemic injury can be increased by the use of inotropes, increased mechanical pressure, hypotension, hemodynamic shifts, and increased renal vascular resistance [18]. Oliguria occurs, sometimes in the face of normal measured cardiac output, but always in the presence of a tense, distended abdomen [18]. CPB, particularly the cooling and rewarming, can affect the oxygen demand of the gut. Mesenteric organs can incur injury through a combination of aerobic/anaerobic conversion and endothelial cell dysfunction [19]. Changes in vascular tone and vasoconstriction at the gut mucosal level during CPB can cause tissue hypoxia of the intestinal cells, manifesting as lactate and putting the gut more at risk for edema [19, 20].

The effects of tight abdominal distention on the lung are increased restriction with decreased compliance, ventilation, lung capacity, and functional residual capacity [17]. Without clear infiltrates or pulmonary edema on CXR, these patients are hypoxic and demonstrate increased respiratory drive with poor gas exchange. This pulmonary status is accompanied by increased ventilatory pressures. The patients can appear clinically similar to those with acute heart failure or sepsis. Hypotension in the setting of a large ileus can lead to fluid resuscitation that compromises the patient further. Liver and direct bowel injury can occur in the form of ischemia, and both cardiovascular and neurologic changes occur [18].

The immediate first step for these patients is measurement of bladder pressure (Table 5.5) [21]. This simple monitoring measurement can be done intermittently and can assist in determining more pointed treatments if organ dysfunction and a tense abdominal compartment remain. This is followed by recording and interpretation of the findings (Table 5.6) [20, 21].

The clinical finding of ACS should be made early. The patients that raise concern include those who exit the operating room with a tense abdomen and/or a prolonged or difficult cardiopulmonary bypass run with aggressive intraoperative fluids and an elevated lactate. An abnormal physical exam with profound abdominal distention and an X-ray showing multiple dilated loops of bowel is seen along with new-onset renal dysfunction.

Table 5.5 Steps to measure bladder pressure

1. Fill the urinary catheter toward the bladder with 25 cc of saline
2. Clamp the tubing
3. Transmit the pressure to a manometer positioned and zeroed at the mid-axillary line
4. The measurement should be recorded at end expiration in the supine position

Table 5.6 Intra-abdominal pressure (World Society of Abdominal Compartment Syndrome)

Normal	Intra-abdominal HTN (IAH)	Abdominal compartment syndrome
IAP 5–7 mm hg	IAP > 12 mmHg	IAP > 20 mm hg
	Grade I, IAP 12–15 mmHg Grade II, IAP 16–20 mmHg	Grade III, IAP 21–25 mmHg Grade IV, IAP >25 mmHg

Some specific risk factors for IAH include age, hypertension, length of bypass, intraoperative lactate, and transfusion [22]. Other predictors include high CVP, a markedly positive fluid balance with blood transfusion, and high-dose inotropes that result in mesenteric ischemia. Age, baseline IAP, vasoactive drugs, and AKI are also significant [18].

Depending on findings on exam, some general management goals will follow. The basic themes to consider are intraluminal, extraluminal, compliance, fluids, and flow (Table 5.7) [18, 21, 23]. From a respiratory perspective, high peak airway pressures are needed to adequately ventilate the patient. Thus, sedation, analgesia, and paralysis are required in the post-cardiac surgery patient with IAH. Fluid balance is critical, as patients can be volume responsive with a large amount of third spacing, but ultimately too much volume will cause marked abdominal distention and increase IAH even further [23]. Finally, surgical invasive approaches become necessary if there is lack of improvement with conservative therapy. These treatments include percutaneous drainage of intraperitoneal fluid and decompressive laparotomy.

This syndrome needs to be carefully monitored in order to prevent major morbidity. Understanding the cause of the underlying disease and the role that IAP is playing in the overall picture of the patient will make a difference in subsequent organ dysfunction. A tense abdomen results from increased volume or decreased compliance or both. The timeline between the IAP and the organ dysfunction is crucial. Decompressive laparotomy, performed in extreme circumstances, has the potential for other serious complications, including enteric fistula and intra-abdominal sepsis [23]. Mortality rates range from 22.2 to 71.4% in individual studies recently reviewed, with an overall mortality of 49.7% [24]. The main cause of death is multi-system organ failure. The second most common cause is profound intestinal ischemia or necrosis [24].

The key to this disease is acute awareness with early recognition, early measurement of bladder pressure, conservative therapies, and constant monitoring and re-evaluation to determine the next escalation of treatment if indicated (Fig. 5.2) [20].

Table 5.7 Management of abdominal compartment syndrome

NGT decompression
NPO
Diuresis
CRRT negative fluid balance
Rectal tube, colon decompression/enemas
Sedation
Analgesia
Neuromuscular blockades
Body position (reverse Trendelenburg)
Promotability agents/gastropyrokinetics
Cisapride, Metaclopramide, Erythromycin, Neostigmine
Percutaneous drainage of ascites
Intraperitoneal fluid catheter
Decompressive laparotomy
Laparostomy (open/Bogota bag)

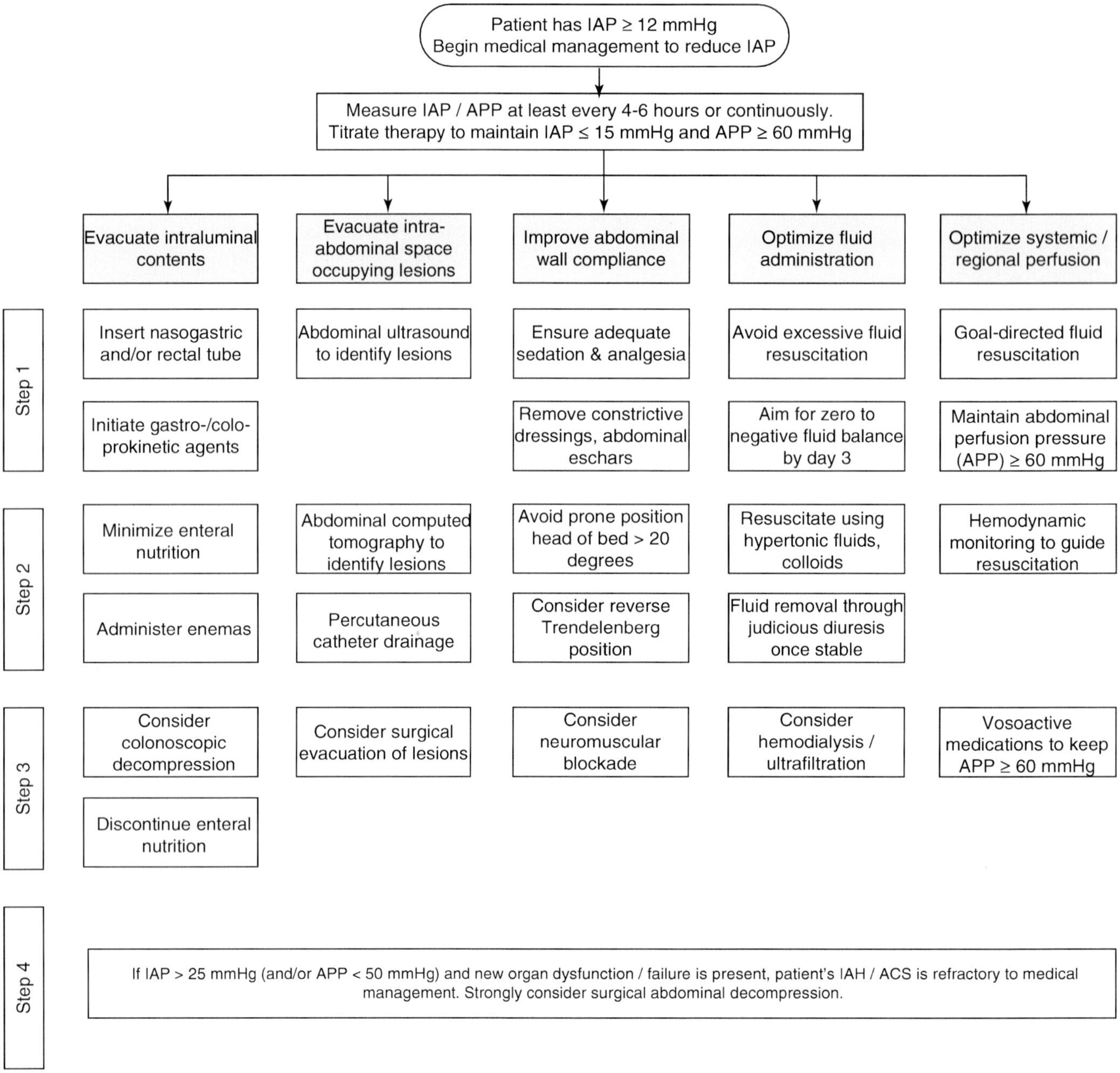

Fig. 5.2 Management of intra-abdominal hypertension. *IAP, intra-abdominal pressure; IAH, intra-abdominal hypertension; APP, abdominal perfusion pressure.* From "Nonoperative management of intraabdominal hypertension and abdominal compartment syndrome," by Cheatham ML, 2009, *World J Surg, Vol. 33,* p. 1117. Copyright 2009 by Springer Nature BV. Reprinted with permission

5.4 Pulmonary Artery Rupture

An 85-year-old female undergoes mitral valve replacement for mitral stenosis. She has known pulmonary hypertension with a systolic PAP of 80 mmHg prior to surgery. Postoperatively, the pulmonary pressure remains 60 mmHg, and she is started on inhaled nitric oxide. Two days later, the surgeon is called emergently to the bedside for a sudden drop in blood pressure that occurs during PA catheter manipulation. A large amount of blood is within the endotracheal tube, and 2 liters of blood is filling the right chest tube. The blood bank is notified and chest compressions are started.

5.4.1 Discussion

Introduced in 1970, the Swan-Ganz catheter (SGC) has a long history in cardiac surgery. Recent literature has been mixed regarding use of this invasive monitoring, citing decreased utility within the medical and surgical intensive care setting. A Cochrane Review of 13 studies and over 5000 patients showed no difference in cost of care, length of stay (LOS), or ICU LOS with use of the SGC [25].

Paradox and controversy exist regarding SGC use in cardiac surgery. There is a lack of adequate evidence for its use, but with limited data and no randomized controlled trials. It makes sense that the patients who have invasive lines in place would have longer hospital stays and higher hospital costs, based on the rationale for monitoring. Reviews of cardiac surgery patients alone show no difference in mortality, except in a small group of very-high-risk patients [26].

Others support the use of PAC in terms of its ability to provide clinically relevant, real-time information about patients who are hemodynamically unstable [27]. This information includes cardiac output and mixed venous measurements that quickly change management, particularly if echocardiography is not immediately available [27]. Overall, the consensus in the modern era is that the PAC should not be used routinely, but in complex patients with cardiac dysfunction, it can guide in management decisions [28]. Patients with low ejection fraction and severe PH may be helped with this monitoring in the setting of cardiogenic shock and decreased organ function. Specifically, the addition and weaning of inotropes, vasopressors, and pulmonary vasodilators can be tailored to the patient's condition.

Complications have been well described and have been deemed prohibitive by some, particularly when added to arguments regarding time, cost, and potential increased LOS (Table 5.8) [29]. Complete heart block can occur with passing of the PA catheter in the setting of a bundle branch block, and perforations of the RA or RV have been described.

Complications of SGC are most common at the time of insertion, but improvements in imaging and monitoring have minimized these injuries. For example, inadvertent puncture of the carotid artery rather than the internal jugular vein would sometimes result in case cancellation. This cancellation was due to the requirement for high-dose anticoagulation in cardiac surgery and thus a potential for carotid artery hematoma or dissection. In the past, this complication had an incidence of 1.9% [30]. The use of transducers to check pressure, intraoperative echo to determine correct placement of the wire, and site ultrasound for needle entry have all been used to minimize this complication further. Other injuries at the time of insertion have to do with passage of the catheter into the pulmonary artery. Right atrial perforation can occur during floating of the catheter, particularly if the balloon is not fully inflated or in the presence of a thin right atrial wall.

Certainly, the most serious complication is PA rupture, which carries an incidence of 0.031% [31]. Specific guidelines to avoid this complication have been recommended (Table 5.9) [32]. In addition to inflating the PA balloon in a small arterial branch, this injury may occur just in the setting of severe pulmonary hypertension, when the tip of the catheter migrates to smaller, more fragile PA branches. If the balloon is not inflated, it is possible for the tip to spear the wall of the pulmonary artery, particularly in elderly patients whose pulmonary arteries are not elastic.

Minor injuries to the lung that occur at the time of perforation and involve small areas of hemorrhage can self-seal by surrounding parenchyma. These injuries are treated simply with removal of the SGC, avoidance of high pressure on the pulmonary bed, and close monitoring. If a focal injury is present with ongoing hemorrhage, percutaneous treatment with coil embolization is a minimally invasive option.

Management of massive PA rupture as described above involves resuscitation with packed RBC during immediate transfer to the operating room. An emergent right anterior thoracotomy followed by placement of a large vascular clamp onto the hilum will stop the hemorrhage after initial immediate manual pressure is obtained. Continued resuscitation is necessary during this time. The lung should be carefully examined. If a clear rupture site is able to be exposed, it should be directly oversewn. However, a large bleeding site within a lobe requires lobectomy or even pneumonectomy [32].

Table 5.8 Complications of PA catheters

1. Carotid cannulation (hematoma, AV fistulas, strokes, bleeding)
2. Catheter knot
3. Arrhythmias, both atrial and ventricular, RBBB and CHB, particularly on insertion
4. Thrombus
5. Tricuspid valve rupture/injury with tricuspid regurgitation
6. Intracardiac balloon rupture
7. Insertion complications: Pneumothorax, thoracic duct injury, air embolism, perforation of RA, RV, PA with balloon tip
8. Limited local hemorrhage in lung parenchyma

Table 5.9 Methods to avoid PA rupture

1. Never inflate the balloon without monitoring in place and stop inflating when the wedge appears or the tracing dampens
2. If there is a concern about position, pull the catheter back into the RV or RA and refloat with the balloon up rather than inflating within the PA itself
3. Check the volume with every balloon inflation and make sure that if the waveform suddenly dampens at a lower volume, the balloon has not migrated
4. Check catheter position daily by CXR
5. Consider positioning of the balloon without frequent inflation or documentation of wedge pressure
6. Keep stopcock off and the syringe attached to make sure no fluid fills the balloon inadvertently

5.5 Acute Pulmonary Embolism

A 60-year-old morbidly obese diabetic male with a normal EF and good targets carries a STS mortality risk of 0.4% for CABG. Thus, four patients out of every 1000 patients with his characteristics are predicted to die within 30 days of cardiac surgery. This patient did well with surgery. The flows in all of his grafts were excellent, and he was extubated on the same day of surgery. He went home in 5 days and looked well in initial follow-up 2 weeks later. On day #24, he died suddenly at home.

Table 5.10 Etiology of prothrombotic state post-cardiac surgery

Aspirin resistance
Increased fibrinogen
Decreased fibrinolysis
Platelets normalize
Tissue factor activation
Thrombin production

Table 5.11 Pulmonary embolism risk factors

RV or LV failure
Obesity
History of DVT
Increased age
Preoperative immobility
Blood transfusion
Respiratory failure or reintubation

5.5.1 Discussion

Massive pulmonary embolism is a rare complication after cardiac surgery, although it is often the initial suspicion when unexplained sudden death occurs. DVT alone after cardiac surgery has a reported rate of 17% to 48% [33, 34]. A recent review and meta-analysis of venous thromboembolism (VTE) after cardiac surgery looked at 68 studies [35]. Of all these studies, the median incidence of asymptomatic DVT was 14.8%, asymptomatic PE 1.9%, and fatal PE 0.3% [35]. When reviewing only prospective studies and randomized controlled trials, the incidence of fatal PE is reported as 0.4% [35]. Another review found that fatal PE was present in 0.5% of patients postoperatively, occurring mostly in the first 2 weeks after surgery [36].

A certain number of VTE are asymptomatic, and the diagnosis can be difficult. In a recent study of 100 patients who received postoperative lower extremity ultrasound and CT after CABG, PE alone was seen in 13%, PE and DVT were seen in 8%, and DVT alone was seen in 4% in patients prophylaxed only with aspirin and without SQ heparin or mechanical compression devices [34]. This study, and the fact that the majority of these patients would be discharged as asymptomatic without treatment, underscores the actual presence of this complication in the postoperative period.

Earlier, standard teaching was that patients who undergo cardiac surgery, with its associated heparinization and cardiopulmonary bypass, are somehow protected from DVT and VTE. Routine aspirin use, DVT prophylaxis, inflammatory mediators, and potentially abnormal liver function all contribute to this concept. Yet studies have shown increased fibrin, thrombin, tissue factor, and aspirin resistance contribute to a *pro*thrombotic state initially after surgery [37]. These findings occur on POD 1, peak on POD 3–5, and can be present up to 30 days postoperatively [37] (Table 5.10).

Given this prothrombotic state, systemic prophylaxis during this high-risk period should be considered, particularly if a patient has multiple risk factors (Table 5.11) [35]. Previous DVT, age, obesity, bed rest, and respiratory failure are risk factors in this patient population [35]. The risk of DVT/PE with heart surgery is not the same as in general surgery. Cardiac surgery involves blood with foreign body interference, coagulopathy, platelet dysfunction, and thrombocytopenia. Each patient does have a risk of bleeding, based on the type of surgery performed. Individual risk factors need to be considered when determining choice of anticoagulation and dosing.

Patients who are treated with ultrafractionated heparin (UFH) or LMWH have not been shown to have a significantly increased rate of pericardial effusion, postoperative pericardial effusion, GI bleeding, mediastinal hemorrhage, or cardiac tamponade [35]. That being said, these complications must be monitored and the real risk-benefit ratio individualized for each patient. Not one method of anticoagulation has been found to be superior [35]. Anti-platelet agents, commonly used in CABG patients, also help in preventing VTE in the preoperative period.

Controversy continues to exist about the timing and choice of medications for prophylaxis. Guidelines are conflicting between the European Association for Cardio-Thoracic Surgery that recommends pharmacologic prophylaxis starting on POD 1 [38] and the American College of Chest Physicians that has concerns about risk for bleeding in the immediate postoperative period and favors mechanical compression stockings alone unless the patient is at high risk for VTE [39]. A more recent study of 331,950 CABG patients found the incidence of VTE with PE to be 1.3% and with DVT alone to be 0.9% [40]. Despite this low number, these patients had significantly higher mortality, morbidity, and cost [40]. The mortality in particular was striking: 6.8% vs 1.7% [40]. Specific risk factors included coagulopathy, obesity, COPD, prior DVT, heart failure, and increased age [40].

Sudden unexplained death in the postoperative period is often thought to be PE. Autopsy studies reviewing sudden death in cardiac surgery report fatal PE occurs in 11–20% of studied cases [41, 42]. In actuality, nearly 50% of deaths are of cardiac origin [41]. These numbers are not helpful in any individual case unless an autopsy has been performed. In summary and practice, risks and benefits must be weighed to protect each patient from VTE/PE carefully in the postoperative period after cardiac surgery.

5.6 Postoperative Bleeding

A 78-year-old male undergoes an uneventful coronary bypass X 3 and is extubated at 3 pm. He remains stable but at 11 pm is found to have 300 ml of bleeding in the previous 5 minutes. An additional 300 ml is recorded in the drains, and the vasopressor requirements are increasing. On the surgeon's arrival into the ICU, the patient is peri-arrest with 2 liters of the blood in the drainage chamber. The surgeon opens the chest rapidly and there is relief of some degree of tamponade. The source of bleeding is unclear as there is a large volume of clotted and fresh blood present. Careful inspection in the operating room reveals that a clip from a venous graft to the RCA was dislodged, leading to immediate blood loss and tamponade.

5.6.1 Discussion

Postoperative bleeding presents in two categories—catastrophic acute bleeding after many hours of stability or continuous bleeding after leaving the OR. The former is mechanical in nature, while the latter is often related to coagulopathy. All bleeding is considered to be from a surgical source until proven otherwise.

Definitions of significant bleeding in the literature relate to drain output in the first 4–12 hours or a threshold value of red cell transfusion. The risk of major bleeding increases with the complexity of surgery including multiple valves, aortic/arch replacement, aortic dissections, redo operations, renal or liver dysfunction, and the use of preoperative antiplatelets or anticoagulants. Postoperative bleeding requires careful assessment, management, and treatments (Tables 5.12, 5.13, 5.14 and 5.15).

Table 5.12 Initial assessment and management of postoperative bleeding

Drain output per hour
Hemodynamics—Heart rate, rhythm, arterial pressure, venous pressures, inotropic support, vasopressors
Respiratory—Lung volumes and pressures, gas exchange
End organ function—pH, base excess, urine output, lactate
Hemoglobin: Absolute value, trend, red cell transfusion required
Basic coagulation profile: INR, APTT, platelet count, fibrinogen
Advanced coagulation profile: ACT, TEG/ROTEM
Temperature
Remainder of lab values (including calcium)

Table 5.13 Initial treatment of postoperative bleeding with sudden increase in drain output

Urgent return to the OR
Activate surgical team—Anesthesia, nursing, assistant, operating room available
Ensure perfusionist and cell salvage available
Prepare for possible reopening in ICU
4 units of red cells X-matched, plus platelets and clotting factors
Delegate where possible
Inform the family

Table 5.14 Correction of reversible non-surgical causes for bleeding

Correct coagulopathy	Non-surgical options
Low fibrinogen: Fibrinogen concentrate or cryoprecipitate	Correct hypocalcemia
High INR/PR: Clotting factors	Maintain normothermia
High ACT: Protamine (25–50 mg at a time)	Consider prothrombin-X
Recent anti-platelets: Platelet transfusion	Consider recombinant activated factor VII

Table 5.15 Systematic assessment of potential bleeding sites

Common sites of bleeding	Less common sites of bleeding
Mammary harvest bed	Coronary sinus tear
Pericardial and thymic branches	Left atrial appendage
Cannulation sites—Arterial, venous, vent sites, retrograde cardioplegia site	Drain entry sites
Aortotomy or atriotomy	LV apex from vent
Proximal/distal anastomoses	Pacing wires
Conduit side branches	
Sternum—Wire holes, periosteum, marrow, fractures	

Ongoing high-volume bleeding does not always mandate urgent return to the OR, particularly in the setting of known and documented coagulopathy. However, this scenario involves very close attention to the patient with correction of all reversible causes of coagulopathy and keeping the option open for an urgent return to the OR. Monitoring includes:

Every 15 min—review drain output and hemodynamics.
Every 60 min—arterial blood gas—Hb drop, pH, BE, lactate.
Every 4 h—CXR for retained intrapleural clot.
Maintain Hb > 90 in a bleeding patient.
Resuscitate and replace volume loss.
Be ready to rapidly reopen in the ICU.

Return to the OR is indicated if there is a rapid increase in drain output, signs of tamponade, or ongoing bleeding despite correction of a coagulopathy (Table 5.14). The surgical team should be activated and a decision made whether a perfusionist is required. At all times, the patient's volume status and hemodynamics should be maintained, ideally with volume resuscitation and early use of packed red cell transfusion.

Once in the OR, the anesthesia team must resuscitate the patient and correct any coagulopathy. The surgeon needs to open the chest and reopen the pericardium as soon as possible—there is often a degree of cardiac tamponade that accompanies any major bleeding episode. Catastrophic bleeding sources are usually identified quickly and at least temporarily managed with finger pressure and then definitive management. For most bleeding patients, the source is not immediately clear, and a systematic approach to all potential bleeding sites should be used (Table 5.15).

The lateral wall of the left ventricle can be difficult to visualize in the bleeding patient without cardiopulmonary bypass. The best approach to inspecting the lateral wall is to use the techniques used in off-pump. Opening the right pleura widely is a simple technique to improve exposure. Once the bleeding has been identified and corrected, the patient should be packed temporarily, while resuscitation is continued. The volume status should be restored, coagulopathy corrected, and the patient warmed before contemplating sternal closure. If there is a profound coagulopathy or hemodynamic instability, the chest can be packed for 24 hours until some degree of stability has been achieved.

5.7 Cardiac Tamponade

A 55-year-old female undergoes an elective mitral valve repair and initially has high drain output which settled after transfusion of clotting factors. At the end of day, the patient is noted to have sudden escalating vasoconstrictor requirements and a fall in urine output. An urgent bedside transthoracic echocardiogram was ordered, but the views were suboptimal. The report shows good ventricular function and possible intrapericardial clot.

5.7.1 Discussion

Cardiac tamponade in the postoperative setting is relatively common and usually occurs in the first few hours after returning to the ICU.

Findings include:

- Increasing vasoconstrictor and inotrope requirements.
- Raised filling pressures (CVP and PA pressures).
- Low cardiac output state.
- Low urine output.
- Decreased mixed venous saturation.
- Metabolic acidosis.
- Pericardial effusion and features of tamponade visible on echocardiography.

Not all tamponade patients have such a classic presentation. The mentioned findings are typical for a circumferential pericardial effusion. However, in the postoperative setting, there can also be a localized tamponade of the right atrium, the right ventricular free wall, or the left atrium (clot in the oblique sinus). Localized tamponade will present with a low cardiac output state, rising inotrope requirements, and low urine output. The filling pressures and echocardiogram are sometimes difficult to interpret. Any postoperative patient with falling cardiac output and oliguria should have tamponade excluded.

The treatment for a genuine suspicion of tamponade should be an urgent return to the operating room for surgical drainage. For any cardiac arrest in the postoperative period, tamponade is the diagnosis of exclusion, and the chest and pericardium must be opened immediately at the bedside. Emergency reopening of the lower aspect of the wound and entering the pericardium inferiorly (as per a subxiphoid window) can often provide excellent initial drainage if the patient is a week or 2 post-surgery. For very early tamponade, however, complete sternal reopening is required to drain the effusion and correct the bleeding source.

References

1. Wessel DL, Adatia I, Giglia TM, Thompson JE, Kulik TJ. Use of inhaled nitric oxide and acetylcholine in the evaluation of pulmonary hypertension and endothelial function after cardiopulmonary bypass. Circulation. 1993;88(Pt 1):2128–38.
2. Komai H, Adatia IT, Elliott MJ, de Leval MR, Haworth SG. Increased plasma levels of endothelin-1 after cardiopulmonary bypass in patients with pulmonary hypertension and congenital heart disease. J Thorac Cardiovasc Surg. 1993;106:473–8.
3. Fattouch K, Sbraga F, Sampognaro R, Bianco G, Gucciardo M, Lavalle C, Vizza CD, Fedele F, Ruvolo G. Treatment of pulmonary hypertension in patients undergoing cardiac surgery with cardiopulmonary bypass: a randomized, prospective, double-blind study. J Cardiovasc Med (Hagerstown). 2006;7(2):119–23.
4. Riedel B. The pathophysiology and management of perioperative pulmonary hypertension with specific emphasis on the period following cardiac surgery. Int Anesthesiol Clin. 1999; 37(2):55–79.
5. Denault A, Deschamps A, Tardif JC, Lambert J, Perrault L. Pulmonary hypertension in cardiac surgery. Curr Cardiol Rev. 2010;6(1):1–14.
6. Price LC, Wort SJ, Finney SJ, Marino PS, Brett SJ. Pulmonary vascular and right ventricular dysfunction in adult critical care: current and emerging options for management: a systematic literature review. Crit Care. 2010;14(5):R169.
7. Modi A, Vohra HA, Barlow CW. Do patients with liver cirrhosis undergoing cardiac surgery have acceptable outcomes? Interact Cardiovasc Thorac Surg. 2010;11(5):630–4.
8. Axelrod D, Koffron A, DeWolf A, et al. Safety and efficacy of combined orthotopic liver transplantation and coronary artery bypass grafting. Liver Transpl. 2004;10:1386–90.
9. Hessel EA. Abdominal organ injury after cardiac surgery. Semin Cardiothorac Vasc Anesth. 2004;8:243–63.
10. Wiesner RH, McDiarmid SV, Kamath PS, et al. MELD and PELD: application of survival models to liver allocation. Liver Transpl. 2001;7:567–80.

11. Suman A, Barnes DS, Zein NN, et al. Predicting outcome after cardiac surgery in patients with cirrhosis: a comparison of Child-Pugh and MELD scores. Clin Gastroenterol Hepatol. 2004;2:719–23.
12. Filsoufi F, Salzberg SP, Rahmanian PB, et al. Early and late outcome of cardiac surgery in patients with liver cirrhosis. Liver Transpl. 2007;13:990–5.
13. Thielmann M, Mechmet A, Neuhäuser M, et al. Risk prediction and outcomes in patients with liver cirrhosis undergoing open-heart surgery. Eur J Cardiothorac Surg. 2010;38:592–9.
14. Diaz GC, Renz JF. Cardiac surgery in patients with end-stage liver disease. J Cardiothorac Vasc Anesth. 2014;28(1):155–62.
15. Jacob KA, Hjortnaes J, Kranenburg G, de Heer F, Kluin J. Mortality after cardiac surgery in patients with liver cirrhosis classified by the child-Pugh score. Interact Cardiovasc Thorac Surg. 2015;20(4):520–30.
16. Kumle B, Boldt J, Suttner SW, et al. Influence of prolonged cardiopulmonary bypass times on splanchnic perfusion and markers of splanchnic organ function. Ann Thorac Surg. 2003;75:1558–64.
17. De Wolf A, Poelaert J, Herck I, De Waele JJ. Surgical decompression for abdominal compartment syndrome after emergency cardiac surgery. Ann Thorac Surg. 2008;85:2133–5.
18. Dalfino L, Sicolo A, Paparella D, Mongelli M, Rubino G, Brienza N. Intra-abdominal hypertension in cardiac surgery. Interact Cardiovasc Thorac Surg. 2013;17:644–51.
19. Andrási TB, Buhmann V, Soós P, Juhász-Nagy A, Szabó G. Mesenteric complications after hypothermic cardiopulmonary bypass with cardiac arrest: underlying mechanisms. Artif Organs. 2002;26:943–6.
20. Cheatham ML. Nonoperative management of intraabdominal hypertension and abdominal compartment syndrome. World J Surg. 2009;33(6):1116–22.
21. Kirkpatrick AW, Roberts DJ, De Waele J, et al. Intra-abdominal hypertension and the abdominal compartment syndrome: updated consensus definitions and clinical practice guidelines from the World Society of the Abdominal Compartment Syndrome. Intensive Care Med. 2013;39:1190–206.
22. Kılıç B, Yapıcı N, Yapıcı F, Kavaklı AS, Kudsioğlu T, Kılıç A, Aykaç Z. Factors associated with increased intra-abdominal pressure in patients undergoing cardiac surgery. Turk Gogus Kalp Damar Cerrahisi Derg. 2020;28(1):134–42.
23. De Laet IE, Malbrain MLNG, De Waele JJ. A clinician's guide to management of intra-abdominal hypertension and abdominal compartment syndrome in critically ill patients. Crit Care. 2020;24(1):97.
24. Van Damme L, De Waele JJ. Effect of decompressive laparotomy on organ function in patients with abdominal compartment syndrome: a systematic review and meta-analysis. Crit Care. 2018;22(1):179.
25. Rajaram SS, Desai NK, Kalra A, Gajera M, Cavanaugh SK, Brampton W, Young D, Harvey S, Rowan K. Pulmonary artery catheters for adult patients in intensive care. Cochrane Database Syst Rev. 2013;2:CD003408.
26. Joseph C, Garrubba M, Smith JA, Melder A. Does the use of a pulmonary artery catheter make a difference during or after cardiac surgery? Heart Lung Circ. 2018;27(8):952–60.
27. Lee M, Curley GF, Mustard M, Mazer CD. The Swan-Ganz catheter remains a critically important component of monitoring in cardiovascular critical care. Can J Cardiol. 2017;33(1):142–7.
28. De Backer D, Vincent JL. The pulmonary artery catheter: is it still alive? Curr Opin Crit Care. 2018;24(3):204–8.
29. Asteri T, Tsagaropoulou I, Vasiliadis K, Fessatidis I, Papavasi-Liou E, Spyrou P. Beware Swan-Ganz complications. Perioperative management. J Cardiovasc Surg (Torino). 2002;43(4):467–70.
30. Shah KB, Rao TL, Laughlin S, El-Etr AA. A review of pulmonary artery catheterization in 6,245 patients. Anesthesiology. 1984;61(3):271–5.
31. Tayoro J, Dequin PF, Delhommais A, Alison D, Perrotin D. Rupture of pulmonary artery induced by Swan-Ganz catheter: success of coil embolization. Intensive Care Med. 1997;23(2):198–200.
32. Kelly TF Jr, Morris GC Jr, Crawford ES, Espada R, Howell JF. Perforation of the pulmonary artery with Swan-Ganz catheters: diagnosis and surgical management. Ann Surg. 1981;193(6):686–92.
33. Ambrosetti M, Salerno M, Zambelli M, Mastropasqua F, Tramarin R, Pedretti RF. Deep vein thrombosis among patients entering cardiac rehabilitation after coronary artery bypass surgery. Chest. 2004;125:191–6.
34. Viana VB, Melo ER, Terra-Filho M, Dallan LA, Gonzalez MM, Hajjar LA, et al. Frequency of deep vein thrombosis and/or pulmonary embolism after coronary artery bypass grafting investigation regardless of clinical suspicion. Am J Cardiol. 2017;119(2):237–42.
35. Ho KM, Bham E, Pavey W. Incidence of venous thromboembolism and benefits and risks of thromboprophylaxis after cardiac surgery: a systematic review and meta-analysis. J Am Heart Assoc. 2015;4(10):e002652.
36. Shammas NW. Pulmonary embolus after coronary artery bypass surgery: a review of the literature. Clin Cardiol. 2000;23(9):637–44.
37. Parolari A, Mussoni L, Frigerio M, Naliato M, Alamanni F, Galanti A, Fiore G, Veglia F, Tremoli E, Biglioli P, Camera M. Increased prothrombotic state lasting as long as one month after on-pump and off-pump coronary surgery. J Thorac Cardiovasc Surg. 2005;130:303–8.
38. Dunning J, Versteegh M, Fabbri A, Pavie A, Kolh P, Lockowandt U, Nashef SA, EACTS Audit and Guidelines Committee. Guideline on antiplatelet and anticoagulation management in cardiac surgery. Eur J Cardiothorac Surg. 2008;34:73–92.
39. Gould MK, Garcia DA, Wren SM, Karanicolas PJ, Arcelus JI, Heit JA, Samama CM, American College of Chest Physicians. Prevention of VTE in nonorthopedic surgical patients: antithrombotic therapy and prevention of thrombosis, American College of Chest Physicians evidence-based clinical practice guidelines. Chest. 2012;141:e227S–77S.
40. Panhwar MS, Ginwalla M, Kalra A, Gupta T, Kolte D, Khera S, Bhatt DL, Sabik JF 3rd. Association of acute venous thromboembolism with in-hospital outcomes of coronary artery bypass graft surgery. J Am Heart Assoc. 2019;8(19):e013246.
41. Rastan AJ, Gummert JF, Lachmann N, Walther T, Schmitt DV, Falk V, Doll N, Caffier P, Richter MM, Wittekind C, Mohr FW. Significant value of autopsy for quality management in cardiac surgery. J Thorac Cardiovasc Surg. 2005;129(6):1292–300.
42. Zehr KJ, Liddicoat JR, Salazar JD, Gillinov AM, Hruban RH, Hutchins GM, Cameron DE. The autopsy: still important in cardiac surgery. Ann Thorac Surg. 1997;64(2):380–3.

6 Neurologic Complications

6.1 Stroke after CPB

The patient is a 55-year-old male transferred in from an outside hospital for a 95% left main lesion and a NSTEMI. He appears otherwise well, except for a long history of diabetes. A noncontrast CT of the chest shows no arterial calcifications, and the patient undergoes an off-pump CABG x 3, clampless, with a heartstring device utilized for the proximal anastomosis. Epiaortic ultrasound shows only Grade 1 disease. He is extubated and doing well postoperatively.

During his second walk around the ICU, the patient suffers acute left hemiparesis. CTA head shows an acute right-sided stroke. The next day, he is aphasic, and repeat CT shows an acute left-sided CVA. TEE demonstrates no intracardiac source of emboli. There are no significant carotid lesions or evidence of large, retrievable thrombus. MRI confirms the stroke and also shows significant intracerebral atherosclerotic changes including an occluded vertebral artery.

6.1.1 Discussion

Stroke is a devastating complication in cardiac surgery, increasing morbidity, length of stay, mortality, and long-term survival [1, 2]. Nearly all strokes after heart surgery are thought to be embolic. Stroke is not surprising in a patient with severe atheroma in the ascending aorta or arch. But sometimes stroke patients have none of those findings and undergo surgery with minimal aortic manipulation. A strategy to prevent stroke has to move beyond diseased carotid arteries and visualized plaque in the aorta by epiaortic ultrasound/CT.

A critical factor for intraoperative stroke is not always embolizing atheroma, but may also involve cerebral hypoperfusion. A recent functional evaluation of cerebral perfusion used SPECT to find that adequate pulsatile flows with off-pump CABG or IABP with conventional CABG made a significant impact on cerebral hypoperfusion [3]. What we learn from this is something we already suspected; stroke can be a systemic problem. Lack of pulsatile flow during cardiopulmonary bypass and episodes of hypotension can play critical roles in this process. Aside from significant stenosis or occlusion, patients can have other diagnoses that predispose them to this complication. A history of stroke is one key factor. A high degree of peripheral vascular disease is another marker [3].

The main causes of stroke after cardiac surgery are thromboembolism and/or intraoperative hypoperfusion [4]. The incidence rates of these mechanisms have been reported as embolism (62.1%), multiple etiologies (10.1%), and hypoperfusion (8.8%) [5]. Hemorrhage or unknown etiologies are other mechanisms [5].

The majority of strokes are thought to be embolic, due to the vascular manipulation required in cardiac surgery; one study cited embolic disease in 83% and watershed infarcts in 24% [1]. Most strokes occur on the first postoperative day (41.7%), and an additional 20.4% occurred on the second postoperative day [5]. The instantaneous risk of postoperative stroke has been reported to peak at 40 hours after surgery [6].

There are areas in the brain that can have chronic hypoperfusion due to lack of collaterals. Multiple severe stenoses or occlusions in the same system can also predispose to hypoperfusion in the carotid or vertebral arteries. The presence of these intracranial occlusions also play a role in perioperative stroke [7, 8]. In the smaller regions of the brain, tiny emboli can cause both ischemic and hemodynamic changes. Low BP can cause prothrombotic conditions and prevent washout.

Some of the data from TAVR has been revelatory regarding the importance of anatomy and the location of cerebrovascular disease; the absence of significant carotid stenosis is not necessarily protective in terms of CVA. One study in TAVR patients found vertebral artery stenosis to be a significant risk factor with 60% of new CVA lesions to be located in the posterior circulation by MRI [9].

Specific predictors for stroke include advanced age, previous CVA or TIA, carotid artery stenosis, PVD, postop atrial

S. K. Balaram, L. Bassin, *Cardiac Surgical Complications*, https://doi.org/10.1007/978-3-030-71563-2_6

fibrillation, unstable angina, and prolonged CPB [10]. It is critical to maintain BP and stable hemodynamics to manage these patients during and after cardiac surgery. Patients with cerebrovascular disease who undergo CABG have a documented increased risk of stroke for many years afterward; they should be followed for existing extracranial disease [11].

Hypoperfusion as a reason for stroke in CABG has always been considered secondary. But in fact, the usual autoregulation that works with cerebral blood flow is not always compensatory at pressures of 50 mmHg and in the setting of intrinsic cerebrovascular disease. Techniques including off-pump surgery and the use of IABP to increase perfusion can be applied to consistently keep SBP >90. Having a planned strategy for all patients but varying this based on suspicion (previous stroke, even moderate disease bilaterally, TIA symptoms, etc.) is good prevention. If patients have a history of multiple strokes, it is prudent to consider minimally invasive options. In some cases, this includes percutaneous revascularization. These patients will otherwise need careful monitoring in the perioperative period to keep systemic blood pressures relatively high. In most cases, it is beneficial to choose and perform a surgery that minimizes operative time and hypotension.

6.2 Stroke after MIS

A mitral valve P2 prolapse repair case is scheduled for a new minimally invasive program. No other radiologic imaging other than a chest X-ray is performed. The patient is a vague historian, but denies other significant comorbidities. The case goes forward using standard femoral arterial cannulation and endoaortic balloon occlusion. All goes according to plan, and the repair is straightforward. Two days later, the patient still has not woken up appropriately, and brain imaging demonstrates small bilateral embolic infarcts. After another 3 days of supportive care, he has made a full recovery with postoperative delirium having contributed significantly to his neurologic dysfunction.

6.2.1 Discussion

The risk of stroke in minimally invasive mitral surgery remains a controversial topic with cannulation strategy and method of aortic occlusion at its core. Several large studies have been published that have specifically considered the problem of stroke after minimally invasive surgery. A 2010 review of the STS database comprising 28,143 patients undergoing first-time isolated mitral valve surgery demonstrated an *increased* risk of permanent stroke with less invasive mitral valve surgery compared to sternotomy, in both the unadjusted and adjusted analysis (1.87% vs 0.93%, $p = 0.0002$ and OR 1.96, 95% CI 1.46 to 2.64, $p < 0.0001$), thus triggering concern within the cardiac surgery community [12]. Interestingly, however, it was shown that femoral cannulation was not independently related to the increased stroke risk, whereas beating or fibrillating heart surgery without aortic occlusion demonstrated an odds ratio of 3.03 (95% CI 1.66 to 5.51, $p = 0.0003$) [12]. Nevertheless, when this group was excluded, there still remained an increased rate in the less invasive group, inferring that not all the risk could be accounted for by a non-aortic occlusion technique.

In fact, further analysis in the STS data revealed that endoaortic balloon occlusion (EABO) was also not an independent predictor of stroke when compared to transthoracic cross-clamping; it did increase the risk of CVA in non-robotic cases [12]. De-airing is more difficult to perform in less invasive mitral surgery, particularly in the setting of a lack of control of the ascending aorta. Adequate de-airing maneuvers are essential. The Leipzig group recommends Trendelenburg position with a slight leftward rotation of the table in order to elevate the left atrium relative to the left ventricle, and to vent the aorta whenever possible, particularly when using a non-aortic occlusion technique [13]

However, a 2019 systematic review and meta-analysis of 11 observational studies comprising 4181 patients demonstrated no difference in the risk of death or stroke when comparing EABO to transthoracic cross-clamping [14]. There was a suggestion that EABO via central aortic cannulation has a lower risk of stroke than either transthoracic cross-clamp or EABO via femoral cannulation [14]. Retrograde passage of the endoaortic balloon in association with retrograde perfusion via the femoral artery may cause dislodgement of aortic plaque or thrombus and subsequent cerebral emboli. Additionally, the endoaortic balloon has a risk of slipping during the case and occluding the innominate artery; the use of bilateral radial arterial lines and cerebral oximetry should alert the surgeon to this before any neurological insult arises. This risk may be lowered using central cannulation, which prevents distal balloon migration. EABO via central cannulation has an additional advantage over transthoracic clamping in terms of aortic root venting.

Another main issue with stroke is that of central vs peripheral cannulation. A significantly increased risk of stroke and postoperative delirium was seen with retrograde perfusion in one study in which EOBO was used for the majority of these patients [15]. A follow-up series including only transthoracic clamping patients also showed a 3.3-fold increase in stroke rate between retrograde and antegrade perfusion (3.5% vs 1.1%, $p = 0.005$) [16]. However, the only significant risk factor for a neurological event was found to be retrograde perfusion in patients >70 yrs. old with an atherosclerotic burden; it appeared safe in younger patients [16]. Nonetheless, they now advocate for central cannulation whenever possible [17].

Similarly, the NYU group were concerned about treating the elderly population and developed a strategy that avoided femoral perfusion in those with significant atherosclerotic burden, switching to a predominantly central cannulation strategy for all patients as a result [18]. They had previously demonstrated an increased neurologic risk with retrograde perfusion in patients >50 years old [19]. Their current strategy attempts to restrict retrograde perfusion to situations where there is very limited central aortic access [18]. In these cases, CT angiography of the descending aorta with distal runoff as well as intraoperative TEE assessment of the thoracic aorta is performed (Table 6.1) [18, 19].

Routine preoperative imaging is a critical part of patient evaluation in order to avoid stroke. A study from the Cleveland Clinic using preoperative CT imaging found that approximately one in five patients had significant atherosclerosis of the aorta or its pelvic branches [20]. Moreover, this information was useful in that 93% of these patients had an alteration in surgical approach [20]. Conversely, Lamelas and colleagues studied 2645 patients undergoing minimally invasive cardiac surgery, including 13.5% with previous history of stroke, 10.4% with previous cardiac surgery, and 10.4% with peripheral vascular disease, but showed only a 1.17% rate of clinical stroke despite using femoral cannulation preferentially (>90% of cases), without routine preoperative imaging [21]. Their strategy focused on identifying those with significant peripheral vascular disease via history and examination as well as at the direct assessment at the time of femoral cannulation [21]. If disease was significant, or if descending aorta atherosclerosis was seen on TEE, then central cannulation was performed [21].

The concept of the combination of adequate pre- and intraoperative assessment as well as a patient-specific approach is favored. This tailored approach, using a preoperative vascular screen (either CT angiography or aortoiliac/femoral angiography at the time of coronary angiography), demonstrated a reduction in stroke rate from 5.2% to 1% over 10 years, despite an increase in case numbers and complexity (Table 6.2) [22].

Table 6.1 Summary: stoke and peripheral cannulation

Retrograde perfusion appears safe in properly selected patients (approx. 90% of all patients)
Good preoperative preparation will allow identification of the other 10% and alter cannulation and clamping strategies accordingly
Aortic clamping or occlusion appears safer than beating/fibrillating heart if feasible
EABO (central) is possibly better than transthoracic cross-clamping which is possibly better than EABO (femoral)
Consider using bilateral radial arterial lines and cerebral oximetry if using EABO

Table 6.2 Stroke prevention imaging

Preoperative CT angiography of aorta with femoral runoff
Careful intraoperative TEE assessment of the thoracic aorta
Intraoperative ultrasound of the femoral artery (if using peripheral cannulation)

6.3 Brachial Plexus Injury

The 50-year-old diabetic man was morbidly obese and presented with acute chest pain with a finding of multi-vessel coronary artery disease. He had a history of a left shoulder injury and was a chronic smoker. His surgery was planned as an all-arterial bypass using bilateral internal mammary arteries. Given his size, he had a difficult exposure of his grossly enlarged heart within the mediastinum, as well as visualization of the mammary arteries. He complained of symptoms immediately upon awakening. Specifically, he stated that he had constant left shoulder pain, inability to lift his left arm, persistent numbness, and paresthesias of the left fourth and fifth fingers.

6.3.1 Discussion

The incidence of brachial plexus injury (BPI), the most common peripheral nerve injury after cardiac surgery, can be as high as 5–10% [23]. It is accepted that this injury can be transient and therefore underreported. Sensory deficits involving the ulnar nerve have an incidence of 1.9–18.3% [24]. Paresthesias, loss of two-point discrimination, and numbness in the ring and little fingers are the most common presentation of ulnar neuropathy [25]. Traction stretch, direct injury, compression, and ischemia can all contribute to injury.

The anatomy of these nerve fibers is key to this injury. The brachial plexus is composed of the anterior rami of C5–8 plus T1 [26]. They pass through the middle and anterior scalene muscles, over the first rib, and under the clavicle, where they split into divisions and recombine before passing into the axilla. They split after they pass under the clavicle, but the space between the first rib and clavicle is very small. The nerve roots of the brachial plexus are fixed proximally at the origin and distally onto fascia and muscle of the insertion points (Fig. 6.1) [23].

Thus, the specific anatomical problems with the brachial plexus during sternotomy for cardiac surgery include:

1. A *superficial location* after they pass through the scalenes, over the rib, and under the clavicle. Here, the plexus can sustain direct pressure damage.
2. *Fixed* nerve roots proximally and distally, conferring a susceptibility to tension injury.
3. A *tight space* between the first rib and clavicle.

The mechanism of injury can occur directly from pressure at the superficial location, tension/stretch due to the fixed nerve roots, and compression or direct damage from the first rib as it is pushed upward [23, 26]. The first rib is fixed medially and is obviously inflexible. The ulnar nerve can also be injured much lower: at the elbow (ulnar groove), below the elbow (edge of the supinator muscle) and the cubital tunnel, and at the palm [23].

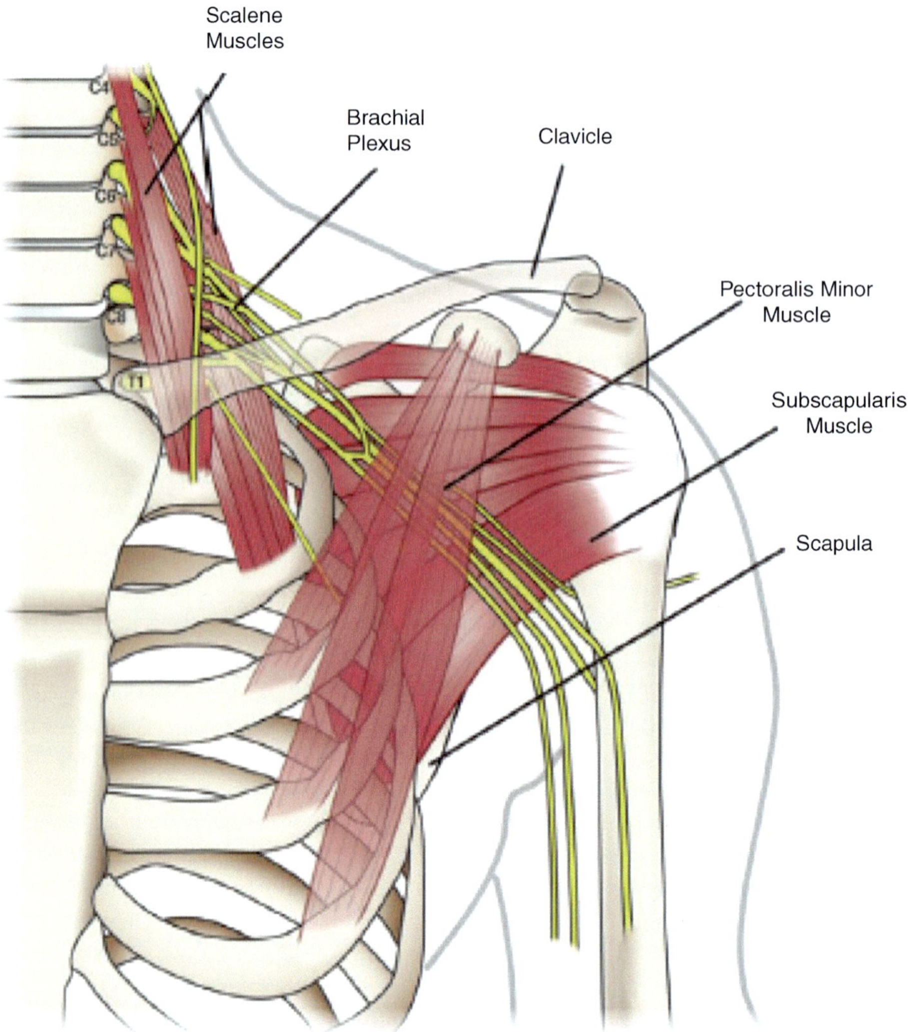

Fig. 6.1 Anatomy of the brachial plexus. From "Peripheral Nerve Injury in Cardiac Surgery," by Jellish WS and Oftadeh M, 2018, *J Cardiothorac Vasc Anesth, Vol. 323*, p. 498.Copyright 2017 by Elsevier Inc. Reprinted with permission

BPI has been described since the early days of cardiac surgery and the understanding of its pathology began with cadaver studies. The initial studies, performed by Kirsch et al., found that opening a sternal retractor halfway did not stretch the brachial plexus, but that the rotation of the first rib superiorly as the clavicles were pushed posteriorly increased the tension on the points of fixation [26]. The sternal retractor causes the lateral first rib to move upward toward the brachial plexus. The lateral clavicle pinches the plexus between the first rib and the middle third of the clavicle (Fig. 6.2) [26]. When the retractor was fully opened, the clavicles are pushed into the retroclavicular space, and the first rib moves superiorly, which can also cause a stretch injury given that both ends of the brachial plexus are fixed. This can be direct and indirect injury based on anatomy.

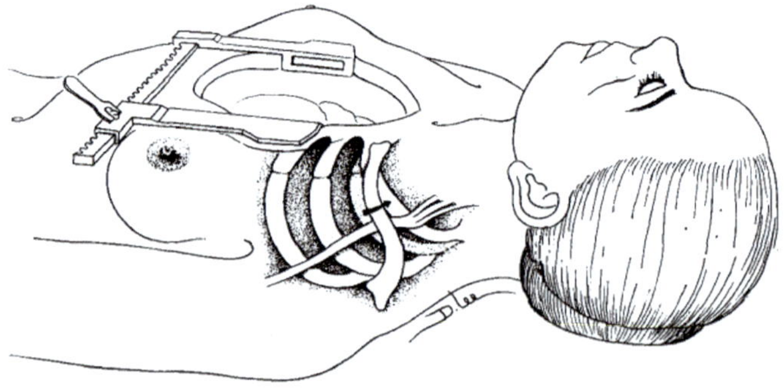

Fig. 6.2 Wide retraction of the sternum: first rib elevation and stretch of brachial plexus. From "Brachial plexus injury following median sternotomy incision," by Kirsh MM et al., 1971, *Ann Thorac Surg, Vol. 11*, p. 319. Copyright 1965 by Elsevier Science & Technology Journals. Reprinted with permission

In 1973, the "double crush hypothesis" was developed that explained how relatively minor nerve injuries combine to form a more significant injury [27]. BPI can be caused by the arm positioning and sternal retraction combination itself, leading to compression, direct pressure, stretch, and ultimately an ischemia of the nerve that arises from the disruption of intramural vessels (Fig. 6.3) [23, 26]. The anatomic mechanisms are the key to prevention of this injury.

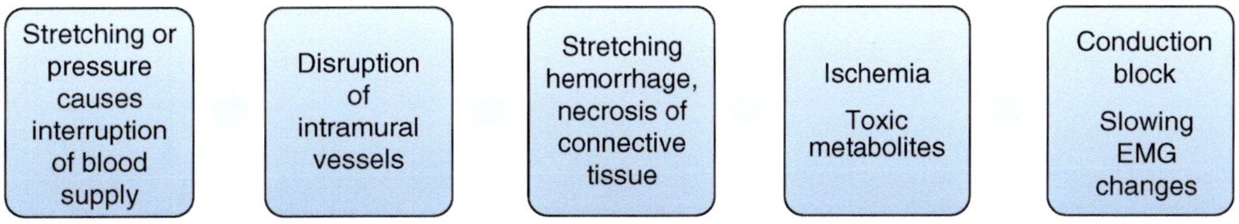

Fig. 6.3 Mechanism of peripheral nerve injury

Table 6.3 Protection and prevention in brachial plexus injury

Precise midline sternotomy to avoid asymmetric retraction
Caudally placed sternal retractor
Maintain head in neutral position
Beware of patients with extremes of body size when positioning arms and retracting
Monitor arm positioning on all patients
Consider less retraction during IMA harvesting
Put sternal retractor low, do not open too wide, and prefer fourth intercostal space
Watch for direct pressure on the site of the BP
Avoid prolonged hypotension, hypothermia, operative time

Specific patient anatomic factors have been associated with this injury. Body extremes (obese and extremely thin) seem to be more at risk with traction/stretch and compression injury [23, 28]. Thin patients can have more direct pressure on the site; obese patients may have their chests opened wider for adequate exposure of the entire heart. Other common risk factors include diabetes mellitus, pre-existing neuropathy, malnutrition, smoking, and HTN [23, 28]. The harvest of the LIMA is in itself a risk factor due to retraction required of the left sternal wall. One study of over 1000 patients showed an increase in incidence of BPI from 2.7% to over 10% by the addition of LIMA harvest [29]. Other risk factors include sternal retraction, first rib fracture, positioning, and advanced age [30].

Components of prevention of BPI in cardiac surgery include careful planning regarding the use of the sternal retractor, the IMA retractor, and arm positioning. Patients who are at risk based on individual anatomy must have more attention paid to positioning, placement, and degree of retraction. Placing the retractor more caudally, below the manubrium and around the site of the fourth rib, has been shown to be helpful in the prevention of injury. Wider retraction and longer bypass time correlate with an increased incidence of injury [28]. The arms should be tucked to the patient's side carefully, with appropriate padding, and in a position of function. The ulnar nerve can be damaged directly with poor arm positioning, either at the elbow with external compression at the cubital tunnel or at the medial base of the palm (Table 6.3) [28, 30].

Most commonly, diagnosis of BPI is made by history and physical exam. Patients complain of shoulder, arm, or hand pain, as well motor weakness and numbness. Pre-existing neuropathy can be additive, especially when considering the double crush hypothesis. Formal EMG studies show conduction slowing or conduction block. The patient should be referred immediately to physical and occupational therapy, as the mainstay of treatment is conservative. After cardiac surgery, BPI carries an excellent prognosis, with many patients showing complete recovery even by the time of discharge [31]. Recovery involves the growth of axons in the area of injury at 1 mm/day, and some injuries resolve in 3–4 weeks [23]. In one study, 79% of patients demonstrated a complete recovery, although half of those patients required 5 months to 1 year for this improvement [31]. Data suggests that prognosis may be worse in patients with pre-existing neuropathy or motor dysfunction in addition to a sensory deficit. Neurologic consultation and EMG studies are helpful in determining options for treatment for these patients.

6.4 Phrenic Nerve Injury

The patient is a 60-year-old with severe COPD and multi-vessel coronary disease who is referred for CABG. Bilateral internal mammary arteries and bilateral radial arteries are chosen as conduits to provide all-arterial bypasses. The patient looks well post-surgery and his X-ray is clear; he is extubated within 6 h. He requires high-flow oxygen and positive pressure ventilation the very day of extubation. He complains of dyspnea, particularly at night, and cannot lay flat. His echo is normal by report, and his oxygen saturations are adequate. There are no signs of pulmonary embolism, and his legs are not swollen. There is no evidence of pneumonia or signs of infection. The CXR reveals poor inspiratory effort, small bilateral effusions, and an elevated left hemidiaphragm.

6.4.1 Discussion

The phrenic nerve, comprised of the anterior horns of cervical nerves 3–5, crosses the scalene fat pad to enter the chest. It then passes medially and posteriorly behind the IMA down along the lateral sides of the pericardial sac, within the layers of the fibrous pericardium [32]. Its importance lies in its stimulation of the diaphragm to contract and flatten, thereby increasing intrathoracic volume, ventilation, and intra-abdominal pressure.

The location of the phrenic nerve is a key consideration during cardiac surgery. The phrenic nerve is rarely seen near the site of the proximal left internal mammary dissection

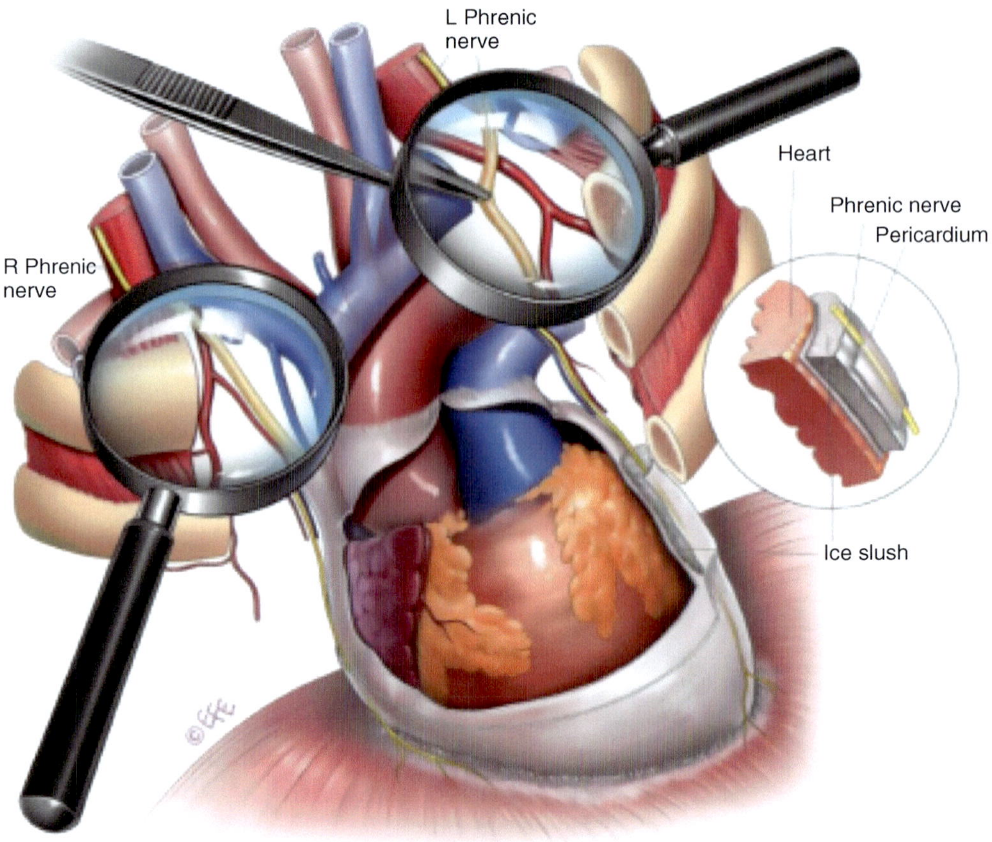

Fig. 6.4 Location of the phrenic nerve in relation to the right and left internal mammary artery. From "Peripheral Nerve Injury in Cardiac Surgery," by Jellish WS and Oftadeh M, 2018, *J Cardiothorac Vasc Anesth, Vol. 323*, p. 502.Copyright 2017 by Elsevier Inc. Reprinted with permission

(Fig. 6.4) [23]. It is at risk when ligating the first intercostal artery. In rare instances, electrocautery near the proximal IMA dissection results in sudden diaphragmatic contraction. The phrenic nerve is better visualized lower on the left lateral pericardium. A defect in the pericardium is often placed just anterior to the phrenic nerve, in order to provide the mammary artery with a safe path to enter the middle mediastinum, prior to anastomosis.

On the right side, specific cannulation of the SVC places the surgeon close to the right phrenic nerve as it traverses the lateral pericardium. Injury to the left side is more common than the right, perhaps because of the more common left IMA dissection, exposure to ice slush, and potential loss of blood supply to the nerve. However, on the right, the phrenic nerve near the proximal RIMA is under particular risk due to its variable and superficial course at this site [23].

Specific precautions should be taken to protect the phrenic nerve. Keeping the pleural space closed, using clips and less cautery at the proximal IMA, and limiting medial dissection at the phrenic entry point into the mediastinum are technical methods to avoid the nerve [33]. Not only can the nerve itself can be damaged directly from electrocautery or retraction, but it can also be subject to ischemic injury from a small branch of the pericardiophrenic artery that is ligated during dissection [32].

The term "frostbitten phrenic nerve" originated with Scannell in 1963 as a description of the effect topical cooling has on the left diaphragm [34]. The details of this effect were further described by Robiscek in 1990, who found in conduction studies that the defect of nerve function was directly related to the depth and time exposure of ice used for topical hypothermia and myocardial protection [35]. The simple use of phrenic protective nerve pads has been shown to decrease the incidence of diaphragm paralysis from over 45% to 17% [36].

Symptoms from this injury are related to the function of the nerve. Dyspnea, both at rest and with exertion, is the most common complaint. The sensation of "inability to catch my breath," particularly while laying flat, is reported by these patients. The intercostal and abdominal accessory muscles will attempt to compensate for phrenic nerve injury. These muscles can minimize the symptoms of phrenic nerve injury, but nocturnal symptoms and dyspnea with activity are common.

The early diagnosis of this injury is central to the prognosis and treatment of symptomatic patients. The most common diagnostic test is a "sniff test" performed during fluoroscopy of the diaphragm. This shows immobility or paradoxical movement of the diaphragm. Fluoroscopy performed in patients with an elevated hemidiaphragm on chest

Table 6.4 Diagnosis of phrenic nerve injury

Diagnostic test	Findings
Sniff test with fluoroscopy	Immobile/paradoxical diaphragm
Ultrasound	Diaphragm hypokinesis
EMG	Quantification of injury
Spirometry	Restrictive pattern

X-ray will confirm diaphragmatic paralysis in approximately 10–15% of cases [33]. EMG can be performed to quantify injury. An ultrasound can assist in the direct visualization of hypokinetic, immobile, or paradoxical muscle movement [30]. Direct spirometry will show a restrictive pattern as the diaphragm does not flatten and lung volumes do not increase with inhalation (Table 6.4).

The prognosis of phrenic neuropathy improves with time; most (90%) of patients with early diaphragm elevation normalize their diaphragm on CXR within 1 year [37]. However, if patients have poor pulmonary function preoperatively, their postoperative course can be greatly affected by their limited reserve [30]. This would include patients with restrictive lung disease, reduced oxygen diffusing capacity, and interstitial lung disease [33]. Interestingly, some literature suggests that patients with severe COPD and increased lung volumes actually garner some protection from the reduction in lung volumes that occurs with this injury [33]. In a small percentage of patients, bilateral phrenic injury can occur. This serious injury results in prolonged mechanical ventilation with the need for tracheostomy and the potential for ventilator-associated complications [38].

Options for treatment are few for those patients who manifest significant morbidity. Prolonged intubation is required if phrenic nerve dysfunction is severe. For those patients with profound atelectasis and tachypnea, an early tracheostomy is sometimes necessary. In the setting of significant symptoms or bilateral phrenic nerve injury, implantation of diaphragmatic pacemaker, thoracoscopic diaphragm plication, and phrenic nerve reconstruction have all been described [39].

6.5 Ischemic Optic Neuropathy

A 64-year-old poorly controlled diabetic female undergoes a difficult CABG x 4. She presented with a NSTEMI and history of severe peripheral vascular disease, stroke, carotid stenosis, and a low ejection fraction. She does not have a straightforward operation: her distal targets are small, and her cross-clamp and CPB times are long. She requires two units of blood transfusion. There are problems with venous drainage, and a left ventricular vent is placed. Two of the targets are intramuscular, and with massive cardiomegaly, the exposure is difficult. The transit time flows at completion are adequate, but the ejection fraction remained decreased. She left the operating room on moderate vasopressors and inotropes. She is seen postoperatively after the sedation was lifted and the endotracheal tube is removed. The first words out of her mouth were clear and slow, "I'm blind."

6.5.1 Discussion

The unexpected nature of ischemic optic neuropathy (ION) is devastating, but can happen after any cardiac surgery. The actual incidence is quite rare, with initial studies ranging from 0.05% to 0.1% [30]. A national inpatient study showed an incidence of 1.43/10,000 (1998–2013), without a significant change over time [40]. However, a more recent study from 1995 to 2017 included 44,568 patients on CPB, with a reported incidence of 0.22 per 10,000 cases [40]. This improvement in incidence is likely due to less invasive procedures (PCI, off-pump techniques) for high-risk patients and basic surgical improvements such as membrane rather than bubble oxygenators [41, 42].

Spontaneous ION has been shown to occur in 1.03 per 10,000 patients, with comorbid conditions in those groups similar to most cardiac surgery patients [43]. These common comorbidities include coronary artery disease, stroke, diabetes, HTN, OSA, and high cholesterol [44].

The etiology of ION is based on anatomy. There are three main mechanisms of retinal artery occlusion: external compression causing ischemia, decreased arterial flow causing ischemia, and venous occlusion [45]. The optic nerve traverses the optic canal and is supplied by multiple posterior ciliary arteries, branches of the ophthalmic artery. These form an incomplete anastomotic ring, the circle of Zinn and Haller [46]. This special anatomy creates an area of watershed and, similar to the brain, ends in ischemic injury and disc edema in certain clinical situations [45]. In surgeries where patients are positioned prone, this complication occurs as a result of persistent compression of the globe over time, which causes ischemia due to central retinal artery or vein occlusion.

In supine cardiac surgery, this problem is more likely multifactorial and directly related to those watershed zones that surround the retinal globe. Retinal embolization and thrombosis are more likely the culprits [45]. Both systemic and local factors can result in ischemic optic neuropathy: inflammation, vasospasm from CPB, circulating catecholamines, medications, hypotension, and low oxygen levels [46]. The use of bubble oxygenators and embolic disease in CPB patients could also affect these patients [42].

Clinically, there is a difference between visual loss in cortical blindness and optic neuropathy; the difference is the pupillary reflex [30]. The two types of visual loss are:

1. *Preserved pupillary reflex* with homonymous vision loss. This occurs in the setting of ischemia to the occipital cortex as the occipital lobe is supplied by the watershed between the middle and posterior cerebral arteries.
2. *Loss of pupillary reflex* with optic neuropathy or retinal injury occurs because of inadequate arterial supply to two sections of the optic nerve: anterior and posterior. The anterior optic nerve is supplied by posterior ciliary end arteries, branches of the ophthalmic artery. The posterior optic nerve is supplied by pial vessels that are more susceptible to decreased oxygen delivery from hypotension, anemia, or facial edema [47].

The US national inpatient study of 5.8 million cardiac surgery procedures from 1998 to 2013 found 4564 cases of specific retinal artery occlusion after cardiac surgery; this is 7.7 per 10,000 cases [48]. Preoperative risk factors included cerebrovascular disease in the form of TIA, carotid artery stenosis, embolic stroke, and hypercoagulability [48]. Diabetic retinopathy was also a risk factor. Bleeding, valve surgery, and septal surgery were significant procedural factors [48]. Another group found severe PVD, recent angiogram, severe anemia, RBC transfusion, and long CPB time as predictive risk factors [45].

Just the use of CPB alone leads to an increase in circulating catecholamines, which act as alpha agonists and can decrease arterial perfusion. Such an inflammatory process can lead to edema which can further decrease perfusion. Anemia decreases the oxygen-carrying capacity of blood flow in the watershed zones. The combination of use of vasoconstrictors and the production of complement and inflammatory mediators can cause tissue edema and affect organ perfusion throughout the body. Degenerative eye diseases, hypertensive retinopathy, macular degeneration, glaucoma, and cataracts also play a local role in these effects [41].

Given the rarity of ION, prevention techniques are very difficult to implement. Certainly, this complication could be seen in patients with severe PVD, carotid or intracerebral atherosclerosis, older age, diabetic retinopathy, or other retinal pathology. However, many cardiac surgery patients have this set of risk factors, and yet the incidence of ION is very low. Maintaining good perfusion, preventing air or embolic phenomena with choice of operation and technique, and maximizing oxygen-carrying capacity are the main tenets in every operation. Severe anemia should be avoided, both while on bypass and afterward. General considerations such as minimizing CPB time can decrease inflammation and complement activation. It is also helpful to prevent hypotensive episodes, particularly in patients with a history of cerebrovascular disease.

Although some cases of spontaneous improvement have been reported, there is no effective treatment for perioperative ischemic optic neuropathy [46]. There have been reports of using acetazolamide to decrease aqueous humor and lower intraocular pressure to improve perfusion [49]. Basic increases in perfusion and hematocrit are always helpful in increasing oxygen delivery. However, perioperative treatments have not shown an improvement in the course of disease [50]. Corticosteroids have been utilized, but poor wound healing and infection may limit their use in the perioperative period [51].

6.6 Recurrent Laryngeal Nerve Injury

The 55-year-old patient presents for a reoperative total arch replacement after a previous Type A dissection repair. All the tissues are densely adherent, and the use of circulatory arrest leaves the surgeon with a long period of necessary hemostasis at the end. Surprisingly, the patient does not bleed excessively and is extubated safely on POD #2. She complains of hoarseness and some shortness of breath after extubation. Although she is not struggling significantly, she does have some stridor and is placed on high-flow oxygen. The working diagnosis of pharyngeal edema is made, in the setting of a long operation with an endotracheal tube and TEE in place. The patient is reassured. However, the next several days show no improvement. The patient is frustrated at her inability to speak. The ENT surgeon performs direct laryngoscopy: the left vocal cord is completely paralyzed. The patient has hoarseness, voice fatigue, dyspnea, and aspiration risk.

6.6.1 Discussion

Iatrogenic injury to the recurrent laryngeal nerve (RLN) can be debilitating and occurs in <2% of thoracic surgery cases [52, 53]. Although phrenic nerve injury is well described, RLN injury can be overlooked as a cause of respiratory insufficiency after cardiac surgery. Yet it can incur great morbidity, particularly in the setting of bilateral RLN injury or multiple reintubations. Many cardiac surgeries do not involve the area surrounding the RLN. The course of the right and left RLNs is non-symmetrical [52]. After the vagus nerve crosses the first part of the subclavian artery, the *right* RLN arises and hooks backward and up behind the artery. It is found in the tracheoesophageal groove. The *left* RLN branches off the vagus just as it crosses the aortic arch [53]. A number of anatomic variables can be present, however. More than half of left RLN arise from the vagal nerve just at or above the aortic arch [54].

Functionally, paralysis to the ipsilateral RLN branch of the vagus nerve places the vocal cord into adduction, which affects both phonation and respiration. The symptoms typically present with a change in voice quality, one that can be described as hoarseness or breathiness. Patients complain that their voice becomes fatigued as the day proceeds. Both the vocal intensity and range is lost. On the respiratory side, difficulties arise with an inability to clear secretions; aspiration and pneumonia have been observed. The patients can also have the sensation of choking while eating. This warrants immediate investigation.

Similar symptoms are seen immediately after extubation and can improve with time. Typically, unilateral vocal cord paralysis will appear immediately and days following extubation with a breathy hoarse voice, inefficient weak cough, and normal respirations. Bilateral vocal cord paralysis occurs just hours following extubation with an often normal voice and insufficient breathing [52]. Symptoms can escalate to stridor, hoarseness, aspiration, and respiratory distress.

As always, prevention is key to this neurologic injury. Direct visualization, identifying and avoiding the nerve, is a standard step in any aortic arch surgery. However, RLN injury can occur at multiple points throughout the operation. Ice slush can cause RLN neuropathy. Sometimes required for operative exposure, even simple retraction near the nerve or of the heart itself can result in neuropathy [55]. The left RLN is in the operative field during a thoracotomy for descending aortic surgery and is often clearly exposed. Thermal electrocautery should be avoided near the location of the nerve. Other, lesser known mechanisms of injury have been identified (Fig. 6.5) [52]. For instance, the inflated cuff of an endotracheal tube can paralyze the RLN anterior branch with pressure against the thyroid cartilage [56].

Once an injury is suspected, the immediate next step is to ensure safety by temporarily ceasing oral intake and directly visualizing the vocal cords with laryngoscopy. In 85% of patients, idiopathic injury of the RLN will spontaneously recover within 6 months [57]. In pediatric congenital cardiac surgery, the recovery can be more problematic. Studies have shown a recovery rate of 61% with up to 39% of patients still showing airway symptoms after 6 months [58]. Patients should be closely monitored over the several weeks, and if no improvement is seen, an injection by ENT to lateralize the vocal cords has been shown to improve symptoms of hoarseness and voice changes. External laryngoplasty implants have also been used for persistent vocal cord paralysis. Unfortunately, permanent symptoms can remain despite treatment.

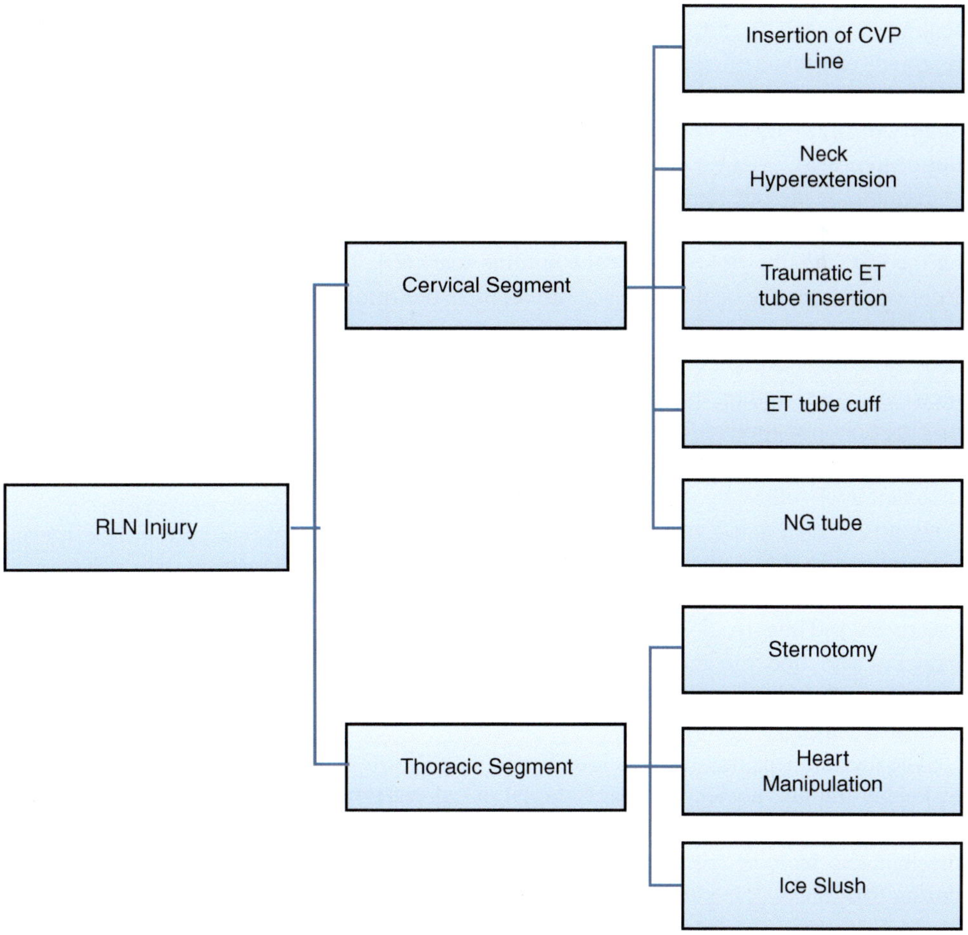

Fig. 6.5 Location and mechanisms of recurrent laryngeal nerve neuropraxia. *RLN*, recurrent laryngeal nerve, *CVP*, central venous pressure, *ET*, endotracheal, *NG*, nasogastric

6.7 Chronic Sternal Pain

A patient 1 year post-uncomplicated CABG is still complaining of incisional pain to both her primary care physician and cardiologist. She has formed a clear keloid, and the skin at and around the surgical site has always been exquisitely tender to touch. In addition, she often reports right-sided parasternal pain and occasional numbness/tingling. There is real anguish in her face; she jumps just with the lightest palpation at the incision. Her physical exam is unhelpful except for tenderness at the incision and surrounding chest wall. The stable sternum shows no palpable wires or erythema. A chest CT scan is ordered. Again, the wires and sternum look intact, with well-opposed and well-healed bone. There is no surrounding inflammatory tissue, evidence of stitch abscess, or wire erosion.

6.7.1 Discussion

Chronic post-sternotomy pain, strictly defined as continued pain 3 months after surgery, does occur. Most sternal pain is described as mild to moderate, but approximately 15%–25% of patients will continue to describe it as severe at 3 months, scaling it at >7/10 [59, 60]. This pain, not always well reported, can result in sleep disorders, depression, anxiety, chronic drug use, decreased productivity, and changes in quality of life. Patients report interference with daily activities of living.

Symptoms include pain, palpation tenderness, numbness, and allodynia. There can be pain from the anterior chest wall on the left side, the right side, or the midline scar [60]. The location and quality of the pain are varied, as is the timing. Some have sternal pain with deep breathing, movement, and activity. Others complain of tenderness from touching the incision or surrounding tissues. Some complain of pain from just the pressure applied from loose-fitting clothing.

A large multicenter study looking at post-sternotomy pain included 97% follow-up and found the prevalence of pain at 3 months, 1 year, and 3 years to be 35.4%, 26.8, and 19.8%, respectively [61]. Another study found that 23% of patients complained of pain at >1 year after surgery, non-anginal in nature [62]. Time from surgery is an essential factor for patients who undergo sternotomy. Overall, most patients who report pain at 3 months will improve at 6 months or a year after surgery.

The mechanisms involved in chronic pain include musculoskeletal trauma, nociceptive pain, sternal pseudoarthrosis, instability, or intercostal neuralgia [60]. Other possible mechanisms include anterior intercostal nerve entrapment during IMA harvest, hypersensitivity with sternal wire placement, sternal fracture, incomplete bone healing, or wound infection (Table 6.5) [63].

When the sternum is spread, soft tissues and nerves can be damaged; the intercostal and brachial nerves can cause chronic pain. Immediately after surgery, inflammation and activation of nociceptor and damaged nerve structures account for the majority of discomfort. Acute pain postoperatively sometimes activates and upregulates neural pathways, thereby leading to a chronic pain syndrome [63]. Release of cytokines, histamine, bradykinin, and prostaglandins also affects the experience of pain [64]. Neuropathic pain includes signs of nerve injury and has been shown to play a central role in chronic pain [65].

Risk factors can be both patient and case specific (Table 6.6) [60, 63]. Some studies have shown that patients who undergo IMA harvest have more pain after surgery. In addition to more pain in the shoulder, neck, and arms, the IMA group showed more pain in all different positions compared with non-IMA group (88.3% versus 75.5%) [66]. Diabetes and gender were not significant risk factors for pain in either group [66].

Technical aspects of surgery are always important, but not always controllable based on the patient's anatomy. The use of sternal retractor and the necessary degree of retraction for adequate exposure can place different stresses on the sternum and lead to more patient discomfort. Rib fractures should obviously be avoided. Retractors should be placed more caudal, to decrease pain and protect the brachial plexus. Extensive cauterization of the chest wall can cause significant pain in the postoperative period, particularly around the mammary bed if a pedicled mammary was harvested. IMA exposure in itself should ideally not include extensive retraction, and intercostal nerves should always be avoided.

Table 6.5 Possible etiologies of chronic sternal pain

Trapped nerve
Sternal wires
Musculoskeletal injury/trauma
Sternal fracture
Incomplete bone healing
Wound infection/mediastinitis
Instability
Intercostal neuralgia
Pseudoarthrosis

Table 6.6 Risk factors for chronic post-sternotomy pain

Case-specific factors	Patient-specific factors
Reoperative surgery	Female sex
Severe acute postoperative pain	Anxiety/depression
Wound infection	Preop chronic pain
Sternal nonunion	Hypothyroidism
Rib fractures	Advanced NYHA class

Table 6.7 Treatments for chronic sternal pain

Gabapentin
Pregabalin
Ketamine
Clonidine
Lidocaine patches
Nerve block
Infiltration of local anesthetic
Thoracic epidural anesthesia
Paravertebral block
Intercostal nerve block
Intrathecal opioid injection
Removal of sternal wires
Multimodal anesthesia
Antidepressants
Psychotherapy

Well-opposed, accurate closure of the chest is the basis of preventing sternal nonunion, instability, and the incidence of infection. Strict perioperative glucose control is also a preventive technique for wound infection. If a patient is morbidly obese and has other risk factors, sternal plates for stability can be considered. Obviously, the use of minimally invasive alternative approaches to cardiac surgery is helpful if the patient is felt to be at high risk for sternotomy. Anesthesiology consultation to discuss multimodal drug combinations is helpful in the immediate- and long-term postoperative period (Table 6.7). Treatment modalities are varied and require referral to dedicated pain services [60, 65].

References

1. Salazar JD, Wityk RJ, Grega MA, Borowicz LM, Doty JR, Petrofski JA, Baumgartner WA. Stroke after cardiac surgery: short- and long-term outcomes. Ann Thorac Surg. 2001;72(4):1195–201.
2. Puskas JD, Winston AD, Wright CE, Gott JP, Brown WM 3rd, Craver JM, Jones EL, Guyton RA, Weintraub WS. Stroke after coronary artery operation: incidence, correlates, outcome, and cost. Ann Thorac Surg. 2000;69(4):1053–6.
3. Teramoto C, Mutsuga M, Kawaguchi O, Araki Y, Matsuda J, Usui A. A functional evaluation of cerebral perfusion for coronary artery bypass grafting patients. Heart Vessel. 2019;34(7):1122–31.
4. Caplan LR, Wong KS, Gao S, Hennerici MG. Is hypoperfusion an important cause of strokes? If so, how? Cerebrovasc Dis. 2006;21(3):145–53.
5. Likosky DS, Marrin CA, Caplan LR, Baribeau YR, Morton JR, Weintraub RM, Hartman GS, Hernandez F Jr, Braff SP, Charlesworth DC, Malenka DJ, Ross CS, O'Connor GT, Northern New England Cardiovascular Disease Study Group. Determination of etiologic mechanisms of strokes secondary to coronary artery bypass graft surgery. Stroke. 2003;34(12):2830–4.
6. Tarakji KG, Sabik JF 3rd, Bhudia SK, Batizy LH, Blackstone EH. Temporal onset, risk factors, and outcomes associated with stroke after coronary artery bypass grafting. JAMA. 2011;305(4):381–90.
7. Yoon BW, Bae HJ, Kang DW, et al. Intracranial cerebral artery disease as a risk factor for central nervous system complications of coronary artery bypass graft surgery. Stroke. 2001;32:94–9.
8. Lee EJ, Choi KH, Ryu JS, et al. Stroke risk after coronary artery bypass graft surgery and extent of cerebral artery atherosclerosis. J Am Coll Cardiol. 2011;57:1811–8.
9. Arnold M, Schulz-Heise S, Achenbach S, et al. Embolic cerebral insults after transapical aortic valve implantation detected by magnetic resonance imaging. JACC Cardiovasc Interv. 2010; 3:1126–32.
10. Mao Z, Zhong X, Yin J, Zhao Z, Hu X, Hackett ML. Predictors associated with stroke after coronary artery bypass grafting: a systematic review. J Neurol Sci. 2015;357(1–2):1–7.
11. Imura H, Maruyama Y, Amitani R, Maeda M, Shirakawa M, Nitta T. Long-term impact of critical silent cerebrovascular disease in patients undergoing coronary artery bypass surgery: a propensity score and multivariate analyses. Perfusion. 2019;34(2):147–53.
12. Gammie JS, Zhao Y, Peterson ED, O'Brien SM, Rankin JS, Griffith BP. Less-invasive mitral valve operations: trends and outcomes from the Society of Thoracic Surgeons Adult Cardiac Surgery Database. Ann Thorac Surg. 2010;90(5):1401–10.
13. Garbade J, Davierwala P, Seeburger J, Pfannmueller B, Misfeld M, Borger MA, Mohr FW. Myocardial protection during minimally invasive mitral valve surgery: strategies and cardioplegic solutions. Ann Cardiothorac Surg. 2013;2(6):803–8.
14. Rival PM, Moore THM, McAleenan A, Hamilton H, Du Toit Z, Akowuah E, Angelini GD, Vohra HA. Transthoracic clamp versus endoaortic balloon occlusion in minimally invasive mitral valve surgery: a systematic review and meta-analysis. Eur J Cardiothorac Surg. 2019;56(4):643–53.
15. Murzi M, Cerillo AG, Miceli A, Bevilacqua S, Kallushi E, Farneti P, Solinas M, Glauber M. Antegrade and retrograde arterial perfusion strategy in minimally invasive mitral-valve surgery: a propensity score analysis on 1280 patients. Eur J Cardiothorac Surg. 2013;43(6):e167–72.
16. Murzi M, Cerillo AG, Gasbarri T, Margaryan R, Kallushi E, Farneti P, Solinas M. Antegrade and retrograde perfusion in minimally invasive mitral valve surgery with transthoracic aortic clamping: a single-institution experience with 1632 patients over 12 years. Interact Cardiovasc Thorac Surg. 2017;24(3):363–8.
17. Murzi M, Glauber M. Central versus femoral cannulation during minimally invasive aortic valve replacement. Ann Cardiothorac Surg. 2015;4(1):59–61.
18. Yaffee DW, Galloway AC, Grossi EA. Editorial analysis: impact of perfusion strategy on stroke risk for minimally invasive cardiac surgery. Eur J Cardiothorac Surg. 2012;41(6):1223–4.
19. Grossi EA, Loulmet DF, Schwartz CF, Solomon B, Dellis SL, Culliford AT, Zias E, Galloway AC. Minimally invasive valve surgery with antegrade perfusion strategy is not associated with increased neurologic complications. Ann Thorac Surg. 2011;92(4):1346–9. discussion 1349–50.
20. Moodley S, Schoenhagen P, Gillinov AM, Mihaljevic T, Flamm SD, Griffin BP, Desai MY. Preoperative multidetector computed tomography angiography for planning of minimally invasive robotic mitral valve surgery: impact on decision making. J Thorac Cardiovasc Surg. 2013;146(2):262–8.e1.
21. Lamelas J, Williams RF, Mawad M, LaPietra A. Complications associated with femoral cannulation during minimally invasive cardiac surgery. Ann Thorac Surg. 2017;103(6):1927–32.
22. Barbero C, Marchetto G, Ricci D, Cura Stura E, Clerici A, El Qarra S, Filippini C, Boffini M, Rinaldi M. Steps forward in minimally invasive cardiac surgery: 10-year experience. Ann Thorac Surg. 2019;108(6):1822–9.
23. Jellish WS, Oftadeh M. Peripheral nerve injury in cardiac surgery. J Cardiothorac Vasc Anesth. 2018;32(1):495–511.
24. Warner MA, Warner ME, Martin JT. Ulnar neuropathy: incidence, outcomes and risk factors in sedated or anesthetized patients. Anesthesiology. 1994;81:1332–40.
25. Casscells CD, Lindsey RW, Ebersole J, Li B. Ulnar neuropathy after median sternotomy. Clin Orthop Relat Res. 1993;291:259–65.

26. Kirsh MM, Magee KR, Gago O, Kahn DR, Sloan H. Brachial plexus injury following median sternotomy incision. Ann Thorac Surg. 1971;11(4):315–9.
27. Upton AR, McComas AJ. The double crush in nerve entrapment syndromes. Lancet. 1973;2(7825):359–62.
28. Healey S, O'Neill B, Bilal H, Waterworth P. Does retraction of the sternum during median sternotomy result in brachial plexus injuries? Interact Cardiovasc Thorac Surg. 2013;17(1):151–7.
29. Vahl CF, Carl I, Muller-Vahl H, et al. Brachial plexus injury after cardiac surgery. The role of internal mammary artery preparation: a prospective study on 1000 consecutive patients. J Thorac Cardiovasc Surg. 1991;102:724–9.
30. Grocott HP, Clark JA, Homi HM, Sharma A. "Other" neurologic complications after cardiac surgery. Semin Cardiothorac Vasc Anesth. 2004;8(3):213–26.
31. Ben-David B, Stahl S. Prognosis of intraoperative brachial plexus injury: a review of 22 cases. Br J Anaesth. 1997;79:440–5.
32. O'Brien JW, Johnson SH, VanSteyn SJ, et al. Effects of internal mammary artery dissection on phrenic nerve perfusion and function. Ann Thorac Surg. 1991;52:182–188.
33. Tripp HF, Bolton JW. Phrenic nerve injury following cardiac surgery: a review. J Card Surg. 1998;13(3):218–23.
34. Scannell JG, Discussion of, McGoon DC, Mankin HT, Kirklin JW. Results of open heart operations for acquired aortic valve disease. J Thorac Cardiovasc Surg. 1963;45:47–66.
35. Robicsek F, Duncan GD, Hawes AC, et al. Biological thresholds of cold induced phrenic nerve injury. J Thorac Cardiovasc Surg. 1990;99:167-I 70.
36. Esposito RA, Spencer FC. The effect of pericardial insulation on hypothermic phrenic nerve injury during open-heart surgery. Ann Thorac Surg. 1987;43(3):303–8.
37. DeVita JA, Robinson LR, Rehder J, et al. Incidence and natural history of phrenic neuropathy occurring during open heart surgery. Chest. 1993;103:850–6.
38. Werner RA, Geiringer SR. Bilateral phrenic nerve palsy associated with open-heart surgery. Arch Phys Med Rehabil. 1990;71(12):1000–2.
39. Kaufman MR, Elkwood AI, Colicchio AR, CeCe J, Jarrahy R, Willekes LJ, Rose MI, Brown D. Functional restoration of diaphragmatic paralysis: an evaluation of phrenic nerve reconstruction. Ann Thorac Surg. 2014;97(1):260–6.
40. Assawakawintip C, Nuttall GA, Garrity JA, Smith MM, Dearani JA. Incidence of ischemic optic neuropathy after cardiopulmonary bypass: 20-year experience. J Cardiothorac Vasc Anesth. 2021;35(1):35–8.
41. Rubin DS, Matsumoto MM, Moss HE, et al. Ischemic optic neuropathy in cardiac surgery: incidence and risk factors in the United States from the National Inpatient Sample 1998 to 2013. Anesthesiology. 2017;126:810–21.
42. Blauth CI, Smith PL, Arnold JV, et al. Influence of oxygenator type on the prevalence and extent of microembolic retinal ischemia during cardiopulmonary bypass. Assessment by digital image analysis. J Thorac Cardiovasc Surg. 1990;99:61–9.
43. Hattenhauer MG, Leavitt JA, Hodge DO, et al. Incidence of nonarteritic anterior ischemic optic neuropathy. Am J Ophthalmol. 1997;123:103–7.
44. Biousse V, Newman NJ. Ischemic optic neuropathies. New Engl J Med. 2015;373:1677.
45. Nuttall G, Garrity J, Dearani J, et al. Risk factors for ischemic optic neuropathy after cardiopulmonary bypass: a matched case/control study. Anesth Analg. 2001;93:1410–6.
46. Raphael J, Moss HE, Roth S. Perioperative visual loss in cardiac surgery. J Cardiothorac Vasc Anesth. 2019;33:1420–9.
47. Dunker S, Hsu HY, Sebag J, Sadun AA. Perioperative risk factors for posterior ischemic optic neuropathy. J Am Coll Surg. 2002;194(6):705–10.
48. Calway T, Rubin DS, Moss HE, et al. Perioperative retinal artery occlusion: risk factors in cardiac surgery from the United States National Inpatient Sample 1998-2013. Ophthalmology. 2017;124:189–96.
49. Hayreh SS. Anterior ischaemic optic neuropathy. III. Treatment, prophylaxis, and differential diagnosis. Br J Ophthalmol. 1974;58:981–9.
50. Buono LM, Foroozan R. Perioperative posterior ischemic optic neuropathy: review of the literature. Surv Ophthalmol. 2005;50:15–26.
51. Saxena R, Singh D, Sharma M, et al. Steroids versus no steroids in nonarteritic anterior ischemic optic neuropathy: a randomized controlled trial. Ophthalmology. 2018;125:1623–7.
52. Hamdan AL, Moukarbel RV, Farhat F, Obeid M. Vocal cord paralysis after open-heart surgery. Eur J Cardiothorac Surg. 2002;21(4):671–4.
53. Dimarakis I, Protopaps AD. Vocal cord palsy as a complication of adult cardiac surgery: surgical correlations and analysis. Eur J Cardiothorac Surg. 2004;26(4):773–5.
54. Ng C, Woess C, Maier H, Schmidt VM, Lucciarini P, Öfner D, Rabl W, Augustin F. Nerve at risk: anatomical variations of the left recurrent laryngeal nerve and implications for thoracic surgeons. Eur J Cardiothorac Surg. 2020;58(6):1201–5.
55. Tewari P, Aggarwal SK. Combined left-sided recurrent laryngeal and phrenic nerve palsy after coronary artery operation. Ann Thorac Surg. 1996;61(6):1721–2. discussion 1722–3.
56. Ellis PD, Pallister WK. Recurrent laryngeal nerve palsy and endotracheal intubation. J Laryngol Otol. 1975;89:823–6.
57. Holinger LD, Wolter RK. Neurologic disorders of the larynx. In: English GM, editor. Otolaryngology, revised ed., vol. 3, ch. 42. Philadelphia: J.B. Lippincott; 1992. p. 1–15.
58. Alfares FA, Hynes CF, Ansari G, Chounoune R, Ramadan M, Shaughnessy C, Reilly BK, Zurakowski D, Jonas RA, Nath DS. Outcomes of recurrent laryngeal nerve injury following congenital heart surgery: a contemporary experience. J Saudi Heart Assoc. 2016;28(1):1–6.
59. Ochroch EA, Gottschalk A, Augostides J, et al. Long-term pain and activity during recovery from major thoracotomy using thoracic epidural analgesia. Anesthesiology. 2002;97:1234–44.
60. Kleiman AM, Sanders DT, Nemergut EC, Huffmyer JL. Chronic Poststernotomy pain: incidence, risk factors, treatment, prevention, and the Anesthesiologist's role. Reg Anesth Pain Med. 2017;42(6):698–708.
61. Marcassa C, Faggiano P, Greco C, Ambrosetti M, Temporelli PL, Italian Association of Cardiovascular Prevention, Rehabilitation (GICR-IACPR). A retrospective multicenter study on long-term prevalence of chronic pain after cardiac surgery. J Cardiovasc Med (Hagerstown). 2015;16(11):768–74.
62. Taillefer MC, Carrier M, Bélisle S, et al. Prevalence, characteristics, and predictors of chronic nonanginal postoperative pain after a cardiac operation: a cross-sectional study. J Thorac Cardiovasc Surg. 2006;131:1274–80.
63. Kehlet H, Jensen TS, Woolf CJ. Persistent postsurgical pain: risk factors and prevention. Lancet. 2006;367:1618–25.
64. Mazzeffi M, Khelemsky Y. Poststernotomy pain: a clinical review. J Cardiothorac Vasc Anesth. 2011;25:1163–78.
65. Kraychete DC, Sakata RK, Lannes Lde O, Bandeira ID, Sadatsune EJ. Postoperative persistent chronic pain: what do we know about prevention, risk factors, and treatment. Braz J Anesthesiol. 2016;66(5):505–12.
66. Kamalipour H, Vafaei A, Parviz Kazemi A, Khademi S. Comparing the prevalence of chronic pain after sternotomy in patients undergoing coronary artery bypass grafting using the internal mammary artery and other open heart surgeries. Anesth Pain Med. 2014;4(3):e17969.

Peripheral Cannulation

7

7.1 Introduction to Peripheral Access

Complications in minimally invasive cardiac surgery occasionally originate from peripheral cannulation as opposed to the cardiac procedure itself (Table 7.1) [1]. These procedures require obtaining indirect access in a remote location to initiate cardiopulmonary bypass while operating on the heart through a small incision. Careful and thorough preoperative imaging is essential in planning these complex operations. Small, calcified, tortuous, or diseased peripheral vessels lead to complications, so alternative strategies need to be considered. Having a thorough working knowledge of the patient's anatomy and technical equipment options helps mitigate these risks. In addition to providing minimally invasive access for small access incisions, peripheral cannulation is also used to facilitate simple and safe reoperative cardiac surgery.

Table 7.1 Access-related vascular injuries

Pseudoaneurysm
Hematoma
Nerve injury
Dissection
Stenosis
Rupture
AVF
Compartment syndrome
Percutaneous device closure failure
Perforation
Lower extremity ischemia
Visceral ischemia
Hemorrhage

7.2 Acute Femoral Artery Bleed

Successful deployment of a TAVR in a frail 91-year-old woman resulted in a new left bundle branch block that is being monitored. She is hemodynamically stable. The 14 Fr right femoral artery sheath is removed, and a soft 0.035 inch, 150 cm, J-tipped wire is left in for emergencies. As the first Perclose ProGlide™ (Abbott Vascular) suture is tied, it pulls out. The second ProGlide does tie down, but there is now persistent bleeding from the femoral site. Firm pressure is held over the femoral head to gain control and review options.

7.2.1 Discussion

Access complications are a large source of the morbidity and mortality from TAVR [2, 3]. Prevention is always better than cure, and so too for femoral access. Careful evaluation of the CT and ultrasound will assist in choosing the optimal location for puncture of the common femoral artery. The necessary conditions are suitable diameter (>5 mm), freedom from calcium at the puncture site (ProGlides cannot grasp a calcified arterial wall), and minimal tortuosity. If these conditions are not met, then surgical cutdown should be performed to minimize the risks of vascular injury and blood loss [4].

Leaving the long 0.035 inch wire in, as the large sheath is removed, is the safest way to regain control and buy time. This allows replacement of the sheath (with the dilator) back through the puncture site to stop the bleeding in the artery temporarily. The other arterial access site (usually contralat-

S. K. Balaram, L. Bassin, *Cardiac Surgical Complications*, https://doi.org/10.1007/978-3-030-71563-2_7

Table 7.2 Management of acute femoral arterial bleeding

Prolonged manual compression
Placement of covered stent across injury site
Balloon angioplasty
Aortic balloon occlusion and manual compression
Surgical exposure and repair

eral femoral) should also always be available, increasing the options for bailout in the setting of uncontrolled bleeding (Table 7.2).

That said, there are both percutaneous and open options for the treatment of femoral arterial bleeding. Placement of a covered stent involves crossing the injured site from the contralateral side, and this is somewhat difficult with a 14 Fr sheath in situ. The more significant issue is that a stent in this location is likely to be crushed with mobilizing, as it is adjacent to the hinge point of the hip. Furthermore, if the puncture is quite distal from insertion, close to the femoral artery bifurcation, a covered stent may occlude the profunda femoris artery and alter perfusion to the leg. If the stent option is chosen, a self-expanding stent is recommended; these are far less prone to crush injuries than balloon-expandable stents [5].

Surgical repair is the most durable treatment if the patient can tolerate it. There are two ways of gaining vascular control to perform the repair: (1) place an iliac occlusion balloon from the contralateral side or (2) perform the cutdown with the large 14 Fr sheath in place, provided it is achieving hemostasis. Following cutdown, distal control is achieved by clamping the distal common femoral artery. For proximal control, a nylon tape is passed around the proximal femoral artery (and sheath) and then snared. The large sheath is then removed and the proximal femoral artery clamped. Usually, the intima is destroyed, and there is a small dissection flap. The key to surgical repair is to minimize the chance of stenosis with either a transverse direct closure or patch repair if the tissues are too friable. A completion digital subtraction angiogram (DSA) of the right iliofemoral system should be performed with a pigtail catheter placed in the lower abdominal aorta via the left femoral artery.

7.3 Iliac Vein Perforation

A 70-year-old man with severe aortic stenosis has a bicuspid valve with a very large annulus that is not suitable for TAVR. Coronary angiography is normal. CT of the chest shows an ascending aorta to the right of midline with a maximal dimension of 3.8 cm. The patient is offered a minimally invasive AVR via a right thoracotomy. Planned cannulation is direct aorta and percutaneous femoral venous for drainage. Using ultrasound, a 0.035 inch J-tipped soft wire is guided from the femoral vein into the SVC, and serial dilatations are performed. The cannula is advanced over the wire but feels "stuck" at the 20 cm mark. It eventually is guided into the planned 50 cm mark. The anesthesiologist cannot quite see the cannula in the right atrium on TEE, and when opened, there is no blood return. The patient becomes hypotensive over the next few minutes, and iliac vein injury is suspected.

7.3.1 Discussion

Perforation of the iliac vein is a rare but often deadly complication of femoral venous cannulation. The course of the iliac veins is quite tortuous, traveling posteriorly once over the pelvic brim and then changing direction again to become more vertical in the retroperitoneum. This injury occurs from guiding a very stiff cannula through a soft vein that changes course quite abruptly. Anatomically, this is more likely to occur when cannulating the *left* femoral vein as there is a further sharp change in direction when the left iliac vein crosses the midline to meet the IVC, a right-sided structure.

The risk of iliac injury can be mitigated by guiding the cannula over a very stiff guidewire, rather than the soft wire than comes with the cannula or cannulation kit. The stiff guidewire comes with its own risks of perforation which must also be controlled.

A safe technique to reduce the risk of perforation starts with cannulation of the vein with a relatively soft wire and guiding it into the right atrium. A long 6 Fr sheath (45–60 cm) is then advanced over this wire and visualized in the right atrium. It is now safe to pass a stiff wire (Lunderquist) through this long sheath with no risk of iliac vein perforation. Once the tip of stiff wire is confirmed in the SVC, the long 6 Fr sheath is removed, and the venous cannula guided over it safely. Because the wire is now stiffer than the cannula, it is far less likely that the cannula will perforate the femoral vein.

Treatment options for iliac vein perforation depend on the size and site of injury. Endovascular balloon occlusion is often sufficient to control the bleeding and allow time for the vessel to seal [6]. Alternate reported open techniques include (1) replacement of the injured segment with a pediatric descending aorta homograft, (2) custom-made bovine pericardial tubular graft, (3) autologous spiral saphenous vein, or (4) repair with PTFE graft [7, 8]. Emergency access to the iliac vein is achieved through superior extension of the femoral incision across the inguinal ligament [7]. Another less invasive alternative, when occlusion techniques are insufficient, includes the use of endovascular stenting [9].

7.4 Groin Complications

A 75-year-old male presents with tearing chest pain and is found to have a Type A dissection. He was very unstable and arrests in the CT scanner. Resuscitation was successful, and he was transferred to the operating theater; he maintains his airway but in extremis. The patient arrests again on induction due to hemopericardium/tamponade, and the decision is made to cannulate the femoral artery via cutdown, while another surgical colleague opens the sternum to release the tamponade (titrating to the blood pressure). After a successful procedure, the patient is ready to go home on day 7 when his examination reveals his femoral incision is swollen and leaking clear fluid. An ultrasound is performed to rule out a femoral artery pseudoaneurysm and confirms a simple fluid-filled cavity.

7.4.1 Discussion

Lymph leaks occur in more than 5% of femoral cutdowns [10] and are sometimes difficult to manage. The draining lymphatic vessels from the legs traverse the cribriform fascia (along with the saphenous vein) and can be injured during a femoral cutdown. Risk factors for a lymph leak include obesity, chronic limb swelling, immunosuppression, and diabetes [10].

Technical choices used to reduce the risk of a lymph leak include (1) meticulous layered closure, (2) the use of topical agents including sclerosants such as doxycycline or talc, (3) minimizing the incision size, and (4) avoiding an incision altogether with complete percutaneous cannulation. Percutaneous cannulation with the use of arterial closure devices has been reported to reduce overall groin complications by almost 3/4 (2.3 vs 8.7%) [11].

However, the types of complications seen between open and percutaneous techniques are quite different. While complications of femoral cutdown include lymph leaks and infections, which are not common in percutaneous cannulation, complications in the percutaneous group were predominantly of a vascular nature and included local dissections, stenoses, and vessel injuries requiring repair [11]. Advocates of the procedure propose that the end scenario is the same open cutdown technique used to perform direct vascular repair. Yet intraoperative recognition of complications is subtle and sometimes life-threatening in this setting.

Another approach that can be taken is to minimize the incision size by performing a very limited cutdown of the femoral artery only and to cannulate the femoral vein percutaneously on the other side. Using purse-string sutures rather than a conventional closure technique also helps reduce the required incision size; this approach was found to decrease the rate of hematoma formation and the incidence of delayed discharge secondary to groin complications [12]. However, use of purse strings produced similar rates of seromas, ischemic events, pseudoaneurysms, need for direct repair, surgical reintervention, and local luminal narrowing >25% [12]. The routine use of a small drain minimizes the collection of a seroma and potentially obviates the need for further intervention. In most cases of postoperative groin seroma, conservative management is appropriate. In problematic cases, negative pressure dressings or muscle flaps have been utilized to good effect.

7.5 Arteriovenous Fistula

In preparation for a mini-AVR, the femoral vein of a 45-year-old woman is cannulated percutaneously, and all appears well. Bypass is established, and the aortic valve is replaced successfully with normal ventricular function and a well-seated prosthetic valve. Upon removal of the venous cannula in the groin, there is a sudden large spray of bright red blood from the site. Firm pressure is used to control the bleeding. It is clear that the venous cannula traversed the femoral artery on its way into the venous circulation.

7.5.1 Discussion

It is often the case that the femoral artery lies directly anterior to the femoral vein. If this anatomy is seen with ultrasound guidance, percutaneous cannulation should be abandoned. A direct cutdown or an evaluation of the contralateral side is often warranted. The femoral artery is more superficial and overlaps the vein more within the femoral triangle than is suggested in traditional anatomy textbooks [13].

The majority of patients have some degree of overlap, but this decreases as the vessels move proximally to approach the inguinal ligament. Cannulation should be performed as close to the inguinal ligament as possible to minimize the risk of inadvertent arterial puncture [13]. Ultrasound visualization is mandatory, and guided puncture of vessels should show the exact point where the tip of the needle enters the vessel. This is difficult when the ultrasound is used in the short axis as it is very easy to be "out of plane." Longitudinal use of the ultrasound will more clearly show the entire course of the needle including the puncture site.

If an AV fistula occurs, intermittent pressure is initially used as a large incision is quickly made over the femoral artery. The basic tenets of obtaining proximal and distal control are observed. The inguinal ligament often requires partial division to visualize this area. Depending on the degree of injury, direct suture repair is often adequate. Unfortunately, if there is both an anterior and posterior large defect, the segment will need to be excised, and an end-end anastomosis will be required.

If a through and through injury is not discovered initially and repaired, chronic AV fistulas develop at this site. Although open surgical repair is required for a large defect, there have been reports of repair of small fistulas using ultrasound compression or thrombin injection [14, 15].

7.6 Limb Ischemia

A 55-year-old male presents for a minimally invasive mitral repair through a right mini-thoracotomy. Percutaneous SVC and IVC cannulation through the right internal jugular vein and right femoral vein are planned. Arterial access is obtained through the right common femoral artery. The mitral repair, a complex Barlow's bi-leaflet prolapse, takes longer than expected. The repair is not successful, and the valve requires replacement. The patient tolerates the long pump run surprisingly well and is on minimal inotropic support in the ICU. A few hours later, the right lower leg is edematous and very firm. The patient develops a compartment syndrome. Angiography reveals the initial cannulation site to be in the superficial femoral artery, now severely stenosed.

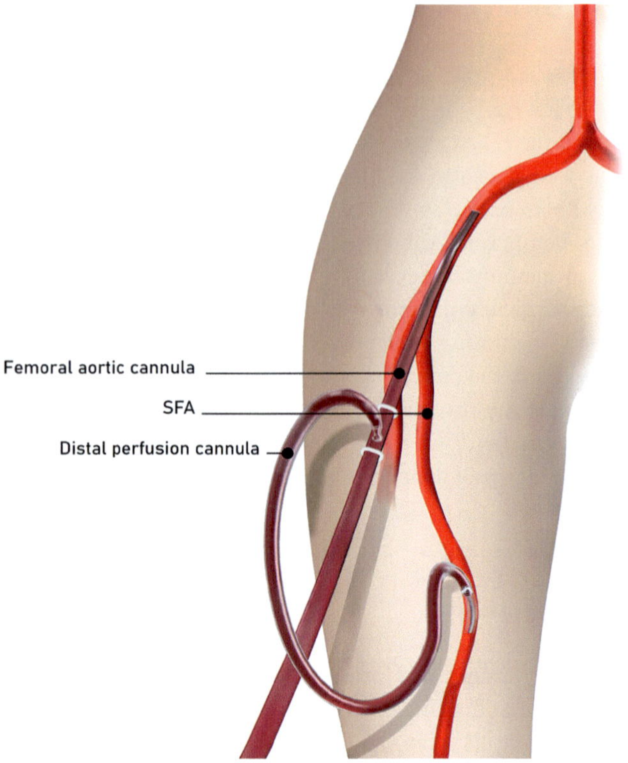

Fig. 7.1 Common femoral artery cannulation with distal perfusion catheter in place

7.6.1 Discussion

Limb ischemia can occur for a number of reasons following femoral artery cannulation. Intraoperative ischemia occurs when the superficial femoral artery is occluded for a prolonged period without distal perfusion. This manifests as a compartment syndrome after removal of the cannula and restoration of antegrade flow. The major risk factors for this include small femoral artery, long bypass run, increased BMI, large cannula, and cannulation of the superficial femoral rather than the common femoral artery [16].

It is imperative to cannulate the *common* femoral artery. Using ultrasound guidance, the skin site should be marked prior to draping. Making a relatively high skin incision and dissecting down onto the inguinal ligament is another method to secure this insertion site. An open incision to visually identify the femoral bifurcation will allow the artery to be cannulated at least 2 cm proximally. Acute vascular injuries are less likely to occur with open cutdown techniques [11].

To ensure adequate distal perfusion in patients at high risk, a distally directed cannula, placed with ultrasound, is also utilized (Fig. 7.1). Another option is an end-to-side graft sewn to the femoral artery for indirect cannulation [17]. A more recent development is the arrival of a specially designed bidirectional cannula; this has shown promise in early clinical trials [18].

Femoral artery occlusion can occur from a local dissection flap or thrombosis at the cannulation site. This often presents as a cold, pale, and pulseless limb following surgery. It is good practice to always examine the leg in the operating theater as soon as the drapes are removed. Near-infrared reflectance spectroscopy (NIRS) has been shown to identify early ischemia intraoperatively [18]. In the postoperative period, measurement of compartment pressures of the lower leg should be obtained immediately if clinical findings suggest ischemic changes.

Claudication symptoms from femoral artery stenosis usually present after the patient has been discharged. At risk are patients with cannulation of a small femoral artery or SFA closed with a purse-string or longitudinal repair. The safest cannulation technique for a small femoral artery is a cutdown, a transverse arteriotomy, and closure with interrupted sutures. Any suspicion of stenosis requires a duplex or angiogram. For persistent claudication, the patient will require an open repair to patch and thus augment the artery.

7.7 Perforation of Right Ventricle

During percutaneous access of the right femoral vein for a mini-mitral procedure, the cannula is opened, and dark blood flows out as expected. All seems well until the initiation of CPB when the perfusionist states that the drainage is suddenly quite poor; they are experiencing significant venous air in the circuit. The visual exposure of the right ventricle is poor through the small right mini-thoracotomy, but the assistant manages to lift up the sternum for the surgeon to see the tip of the cannula protruding from the RV free wall.

7.7.1 Discussion

Right atrial or right ventricular perforation occurs when the tip of the guidewire is not well positioned in the superior vena cava (SVC), far from its origin in the femoral vein. It is of vital importance that cannulation does not commence until both the surgeon and anesthesiologist are confident that the guidewire is placed well within the SVC. This can be hard to achieve, especially with a very prominent limbus of the fossa ovalis, which helps direct the wire into the right atrial appendage or right ventricle. Using a long 6 Fr cannula can facilitate passage of a stiff wire more directly into the SVC. Sometimes, this simply requires patience to guide the wire into the correct position.

Despite visualization of the wire in the SVC, one must also ensure that there are no loops in the wire, as reported in one case of right ventricular perforation [19]. Estimating the length of wire needed to reach the SVC and marking it prior to insertion decrease the risk of this complication, as does routine assessment of the tricuspid valve and right ventricle during wire placement [19]. The cannula itself is often be more easily visualized once the introducer is withdrawn slightly. In the event of injury, repair is generally relatively straightforward, once appropriate access is gained. RV injury may be fixed with a pledgeted 3–0 polypropylene suture on a large MH needle.

7.8 Retrograde Aortic Dissection

A 72-year-old, otherwise healthy female with severe symptomatic mitral regurgitation is referred for a minimally invasive operation. Her coronary anatomy is normal. She is 150 cm tall and weighs 55 kg. A right femoral artery cutdown is performed, and the artery is extremely small, about 5 mm in diameter. A 5-0 Prolene purse string is placed, and Seldinger technique is used to guide the cannula into position. When the cannula is de-aired, it is noted that the backflow is not brisk. Going on bypass, the line pressures are high, hovering around 300 mm Hg. Just prior to cross-clamp, the line pressures increase to 400 mmHg, and the pump stops intermittently. The TEE shows an aortic dissection involving the entire aorta, including the arch and ascending segments.

7.8.1 Discussion

Retrograde aortic dissection is a well-described complication of femoral artery cannulation, with multiple series reporting a 1–2% incidence during the early years of cardiopulmonary bypass when femoral cannulation was routinely used [20]. The incidence was noted to be significantly less with central aortic cannulation, which has ultimately become the preferred method in routine cases. The advent of minimally invasive surgery has brought cannulation techniques full circle. Iatrogenic dissection currently accounts for approximately 5% of aortic dissections [21], has a reported incidence of 0.06–0.6% [22], and results in approximately 0.7% of deaths post-cardiac surgery according to the STS database from 1996 to 2007 [20].

Most reports highlight an increased mortality risk associated with iatrogenic injury when compared to spontaneous dissections. This trend is changing due to earlier diagnosis and routine use of intraoperative TEE [23]. Early recognition is critical, as aortic dissection discovered in the postoperative period doubles the mortality risk [22]. Intraoperatively recognized dissections have the best survival overall when compared to early, late acute, or late chronic dissections [24]. Iatrogenic dissections also carry a high rate of stroke (9%) and renal failure (13.5%) [25]. Undiagnosed aortic dissection account for up to 5% of sudden deaths after cardiac surgery [24]. Autopsy examination is not routinely performed, but this correlates with the high mortality rate observed in those dissections discovered in the early postoperative phase [24].

Femoral cannulation has been identified as a risk factor with an incidence of 0.6% compared to 0.06% with central aortic cannulation [20]. It is difficult to ascertain cause vs effect, however, as most studies are retrospective, and femoral entry techniques are often reserved for those with hostile ascending aortas or requiring alternative strategies [25]. Axillary cannulation carries a reported incidence of dissection (0.5%), likely for similar reasons [20].

Patients considered for peripheral cannulation in minimally invasive cardiac surgery, in particular mitral surgery, generally lack the risk factors typically associated with aortic dissection (atherosclerosis, hypertension, advanced age, dilated ascending aorta) [25, 26]. In fact, a large series of 2645 patients was found to have a 0% incidence of dissection, despite a relatively high-risk cohort [10]. Diabetes has been variously identified as both a risk factor and a protective factor, highlighting the inherent difficulties with identification of risk factors for these rare complications [25, 27]. There also appears to be a reduced risk of aortic dissection with placement of a transthoracic clamp when compared to use of femoral endoaortic balloon occlusion (EABO) [28]. There is a significant learning curve associated with the use of EABO, with 1 group reporting a 3% risk of aortic dissection in their first 100 cases which reduced to 0.6% in their last 500 [28].

An acute dissection in the operating room is observed as a sudden change in line or systemic pressure, reduction in venous return, or evidence of organ malperfusion. With femoral artery retrograde dissection, the findings can be more subtle. Sometimes, the patient will only have abdominal distention or reduced urine output. These patients typically have a more benign course than that documented in the literature, as long they are immediately identified. It is critical to monitor the descending aorta with TEE when initiating cardiopulmonary bypass so that the pump may be turned off at any sign of aortic dissection. The benefits of TEE in this setting cannot be overemphasized. Usually, the injury occurs at the time of commencement of CPB, and if identified early enough, the aim should be to discontinue bypass after resuming ventilation. The dissection sometimes resolves with the resumption of antegrade flow. Intervention will be required, of course, if the ascending aorta is involved.

The treatment strategy is to determine the extent of the dissection, assess for organ malperfusion, reduce CPB flow and blood pressure, and aim to come off bypass. Depending on the timing and extent of dissection, the patient should be cooled through alternative cannulation, particularly if the true lumen is not perfused through the original arterial access. Replacement of the ascending aorta under circulatory arrest is the next step. The originally planned surgery is typically completed during the time of rewarming.

7.9 Axillary Artery Malperfusion

This 52-year-old woman presents with a dilated ascending aorta and arch and requires replacement from the sinotubular junction to the proximal descending aorta. The right axillary artery is cannulated directly with an 18 Fr straight cannula. This provides unilateral antegrade cerebral perfusion, used throughout the operation. The case proceeds well and there is minimal blood loss. The ICU team plans to keep the patient intubated and sedated overnight, based on case complexity and a difficult airway.

Sedation is weaned the next morning, and it is noted that there is no spontaneous respiratory effort. The patient's pupils are fixed and dilated, with no brainstem reflexes. An urgent CT brain scan does not show a CVA but does note a very dominant right vertebral artery with an atretic left vertebral. There are no filling defects suggestive of embolus. MRI brain scan shows complete infarction of the brainstem and cerebellum. There was significant ischemia in the right vertebral system; the patient is brain dead.

7.9.1 Discussion

The axillary artery is an excellent choice for peripheral arterial cannulation as it provides antegrade flow in the aortic arch and is usually disease-free. It facilitates continuous and safe cerebral perfusion during the circulatory arrest time of moderate hypothermia. This is done by reducing flows and placing a clamp on the innominate artery to facilitate suturing in a bloodless field at the aortic arch. Studies have shown improved results in aortic dissection and aneurysm surgery with right axillary flow, particularly for atherosclerotic aneurysms [29]. The risks of axillary cannulation include dissection which propagates into the aorta, limb ischemia, and injury to the pectoral nerves, brachial plexus, or axillary artery [30].

In this tragic case of brain death, the most likely cause was direct cannulation of the right axillary artery with a long cannula that prevented flow into the right vertebral artery. This may result from direct occlusion of the vertebral ostium by the cannula itself or from flow reversal within the vertebral artery [31]. This has also been demonstrated to occur if the cannula tip is positioned too close to the ostium [31].

The two main techniques of cannulating the axillary artery include the direct method (inserting a straight cannula through an arteriotomy) or the side-arm technique (sewing a 6–10 mm Dacron graft onto the artery in an end-to-side fashion) (Fig. 7.2) [29, 30]. The side-arm technique was shown in one study to decrease the rate of complications from axillary cannulation from 7% to 2.1% [31]. Use of this technique causes less direct injury to the vessel and allows for perfu-

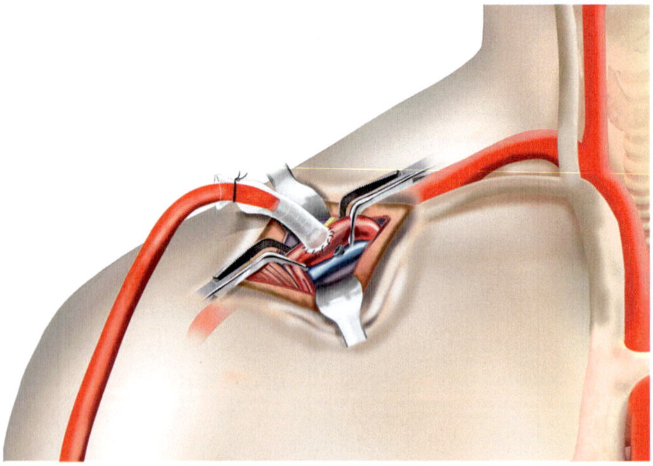

Fig. 7.2 Right axillary cannulation with side-arm graft

sion of the arm during CPB [32]. The side-arm technique has also been shown in some studies to reduce the rate of cannulation-related complications such as axillary artery dissection and stroke, but at the expense of an increased blood transfusion requirement [33]. Smaller axillary artery diameter increases the overall risk for these complications [33].

The side-arm technique avoids the risk of malperfusion of the right arm and right vertebral, but does so at increased risk of hyperperfusion of the right arm. As systemic flows increase into the axillary artery graft to provide adequate flow to the body, the right arm may possibly become hyperperfused and edematous. This hyperperfusion should be prevented with the use of bilateral invasive radial artery blood pressure monitoring to guide gentle snaring of the right axillary artery distal to the graft. The right radial blood pressure is carefully titrated, preventing hyperperfusion of the arm and ensuring adequate perfusion to the brain and body.

References

1. Kappetein AP, Head SJ, Généreux P, Piazza N, van Mieghem NM, Blackstone EH, Brott TG, Cohen DJ, Cutlip DE, van Es GA, Hahn RT, Kirtane AJ, Krucoff MW, Kodali S, Mack MJ, Mehran R, Rodés-Cabau J, Vranckx P, Webb JG, Windecker S, Serruys PW, Leon MB, Valve Academic Research Consortium-2. Updated standardized endpoint definitions for transcatheter aortic valve implantation: the Valve Academic Research Consortium-2 consensus document. J Thorac Cardiovasc Surg. 2013;145(1):6–23.
2. Winter M-P, Bartko P, Hofer F, Zbiral M, Burger A, Ghanim B, et al. Evolution of outcome and complications in TAVR: a meta-analysis of observational and randomized studies. Sci Rep. 2020;10(1):15568.
3. Raju S, Eisenberg N, Montbriand J, Cusimano RJ, Feindel C, Ouzounian M, et al. Vascular complications and procedures following transcatheter aortic valve implantation. Eur J Vasc Endovasc Surg. 2019;58(3):437–44.
4. Burri M, Ruge H, Erlebach M, Lange R. Surgical cutdown avoids vascular complications in transcatheter aortic valve replacement in calcified and small femoral arteries. Thorac Cardiovasc Surg. 2021; https://doi.org/10.1055/s-0041-1725202.
5. Patel SM, Li J, Parikh SA. Design and comparison of large vessel stents: balloon expandable and self-expanding peripheral arterial stents. Interv Cardiol Clin. 2016;5(3):365–80.
6. Yokoyama H, Mizuno S, Shishido K, Saito S. Iliac vein rupture during transcatheter edge-to-edge mitral valve repair with MitraClip. JACC Cardiovasc Interv. 2020;13(13):e117–e9.
7. Handa A, Frcs SS, Moorjani N, Frcs WSF. Iliac vein replacement with a descending aortic homograft. Ann Thorac Surg. 2001;72(3):957–8.
8. Del Campo C. Replacement of the iliac vein. Ann Thorac Surg. 2002;73(5):1692.
9. Demir D, Smeets R, Schurink GWH, Mees BME. Use of thoracic stent grafts to control major iliac vein bleeding. J Vasc Surg Venous Lymphat Disord. 2021;9(4):1058–61.
10. Lamelas J, Williams RF, Mawad M, LaPietra A. Complications associated with femoral cannulation during minimally invasive cardiac surgery. Ann Thorac Surg. 2017;103(6):1927–32.
11. Moschovas A, Amorim PA, Nold M, Faerber G, Diab M, Buenger T, Doenst T. Percutaneous cannulation for cardiopulmonary bypass in minimally invasive surgery is associated with reduced groin complications. Interact Cardiovasc Thorac Surg. 2017;25(3):377–83.
12. Cakici M, Yazicioglu L, Baran C, Ozcinar E, Ozgur A, Soykan C, Eryilmaz S, Bilgic S, Kaya B, Akar AR. A retrospective analysis of surgical femoral artery closure techniques: conventional versus purse suture technique. Ann Vasc Surg. 2017;44:103–12.
13. Hughes P, Scott C, Bodenham A. Ultrasonography of the femoral vessels in the groin: implications for vascular access. Anaesthesia. 2000;55(12):1198–202.
14. Mittleider D, Cicuto K, Dykes T. Percutaneous thrombin injection of a femoral artery pseudoaneurysm with simultaneous venous balloon occlusion of a communicating arteriovenous fistula. Cardiovasc Intervent Radiol. 2008;31(Suppl 2):S115–9.
15. Steppich B, Schürmann F, Bruskina O, Hadamitzky M, Kastrati A, Schunkert H, Fusaro M, Ott I. Ultrasound-guided thrombin injection for treatment of femoral artery pseudoaneurysm with concomitant AV-fistula—a retrospective single Centre experience. Vasa. 2018;47(6):507–12.
16. Tarui T, Miyata K, Shigematsu S, Watanabe G. Risk factors to predict leg ischemia in patients undergoing single femoral artery cannulation in minimally invasive cardiac surgery. Perfusion. 2018;33(7):533–7.
17. Vander Salm TJ. Prevention of lower extremity ischemia during cardiopulmonary bypass via femoral cannulation. Ann Thorac Surg. 1997;63(1):251–2.
18. Marasco SF, Tutungi E, Vallance SA, Udy AA, Negri JC, Zimmet AD, DC MG, Pellegrino VA, Moshinsky RA. A phase 1 study of a novel bidirectional perfusion cannula in patients undergoing femoral cannulation for cardiac surgery. Innovations (Phila). 2018;13(2):97–103.
19. Passage J, Joshi P. Right ventricular perforation during femoral venous cannulation—lessons learned. Heart Lung Circ. 2013;22(9):772–4.
20. Ram H, Dwarakanath S, Green AE, Steyn J, Hessel EA 2nd. Iatrogenic aortic dissection associated with cardiac surgery: a narrative review. J Cardiothorac Vasc Anesth. 2020;S1053-0770(20):30789-8.
21. Rylski B, Hoffmann I, Beyersdorf F, Suedkamp M, Siepe M, Nitsch B, et al. Iatrogenic acute aortic dissection type A: insight from the German Registry for Acute Aortic Dissection Type A (GERAADA)†. Eur J Cardiothorac Surg. 2013;44(2):353–93.
22. Leontyev S, Borger MA, Legare JF, Merk D, Hahn J, Seeburger J, Lehmann S, Mohr FW. Iatrogenic type A aortic dissection during cardiac procedures: early and late outcome in 48 patients. Eur J Cardiothorac Surg. 2012;41(3):641–6.
23. Hajj-Chahine J. Post-cardiotomy type A aortic dissection. Eur J Cardiothorac Surg. 2014;46(3):508.
24. Stanger O, Schachner T, Gahl B, Oberwalder P, Englberger L, Thalmann M, Harrington D, Wiedemann D, Südkamp M, Sheppard MN, Field M, Rylski B, Petrou M, Carrel T, Bonatti J, Pepper J. Type A aortic dissection after nonaortic cardiac surgery. Circulation. 2013;128(15):1602–11.
25. Williams ML, Sheng S, Gammie JS, Rankin JS, Smith PK, Hughes GC, Richard E. Clark Award. Aortic dissection as a complication of cardiac surgery: report from the Society of Thoracic Surgeons database. Ann Thorac Surg. 2010;90(6):1812–6. discussion 6–7.
26. Hwang HY, Jeong DS, Kim KH, Kim KB, Ahn H. Iatrogenic type a aortic dissection during cardiac surgery. Interact Cardiovasc Thorac Surg. 2010;10(6):896–9.
27. Januzzi JL, Sabatine MS, Eagle KA, Evangelista A, Bruckman D, Fattori R, Oh JK, Moore AG, Sechtem U, Llovet A, Gilon D, Pape L, O'Gara PT, Mehta R, Cooper JV, Hagan PG, Armstrong WF, Deeb GM, Suzuki T, Nienaber CA, Isselbacher EM, International Registry of Aortic Dissection Investigators. Iatrogenic aortic dissection. Am J Cardiol. 2002;89(5):623–6.
28. Rival PM, Moore THM, McAleenan A, Hamilton H, Du Toit Z, Akowuah E, Angelini GD, Vohra HA. Transthoracic clamp versus endoaortic balloon occlusion in minimally invasive mitral valve

surgery: a systematic review and meta-analysis. Eur J Cardiothorac Surg. 2019;56(4):643–53.
29. Etz CD, Plestis KA, Kari FA, Silovitz D, Bodian CA, Spielvogel D, Griepp RB. Axillary cannulation significantly improves survival and neurologic outcome after atherosclerotic aneurysm repair of the aortic root and ascending aorta. Ann Thorac Surg. 2008;86(2):441–6. discussion 446–7.
30. Sabik JF, Nemeh H, Lytle BW, Blackstone EH, Gillinov AM, Rajeswaran J, Cosgrove DM. Cannulation of the axillary artery with a side graft reduces morbidity. Ann Thorac Surg. 2004;77(4):1315–20.
31. Hillebrand J, Konerding MA, Koch M, Kaufmann T, Steinseifer U, Moritz A, Dzemali O. Anatomic and flow dynamic considerations for safe right axillary artery cannulation. J Thorac Cardiovasc Surg. 2013;146(2):467–71.
32. Schachner T, Nagiller J, Zimmer A, Laufer G, Bonatti J. Technical problems and complications of axillary artery cannulation. Eur J Cardiothorac Surg. 2005;27(4):634–7.
33. Puiu PC, Pingpoh C, Beyersdorf F, Czerny M, Keyl C, Kreibich M, Kondov S, Rylski B, Zimmer E, Siepe M. Direct versus side graft cannulation from the right axillary artery in thoracic aortic surgery. Ann Thorac Surg. 2021;

8 Transcatheter Aortic Valve Replacement

8.1 Iliac Artery Rupture

A 92-year-old woman presents for transfemoral TAVR. Planning CT shows an extremely calcified and tortuous aorta with multiple curves in the descending and abdominal aorta. The access vessels are small with a minimal luminal diameter of 5.0 mm and non-circumferential calcification. The right femoral artery appears to be more accessible, but there is extensive circumferential calcification at the origin of the right iliac artery. Bilateral common femoral artery access with 6 Fr sheaths is obtained, and the valve was crossed with a JR4 catheter and exchanged for a Safari wire. A 6 Fr pigtail catheter is placed into the non-coronary sinus via the left femoral artery for aortography.

The 14 Fr sheathless valve is inserted into the right femoral artery over the Safari wire, but is not progressing smoothly. The nose cone appears to be stuck at the origin of the common iliac, and with pushing, it seems as if the entire aorta moves to the left. Unfortunately, the Safari wire is not stiff enough to straighten the path. The valve is removed, and the femoral entry site is plugged with a standard 14 Fr sheath prior to replacing the Safari wire with a stiffer wire into the LV for more support. Again, the valve will not traverse the origin of the iliac artery. A stiffer Lunderquist wire is placed through the pigtail in the left femoral artery. The valve now passes into the aorta and is deployed appropriately. The in-line sheath is removed and replaced with a standard 14 Fr sheath with the proximal portion sitting in the abdominal aorta. The sheath is pulled back into the external iliac artery. Unfortunately, the patient becomes unstable, and aortography confirms an injury to the proximal right common iliac.

8.1.1 Discussion

Vascular complications after transfemoral TAVR most commonly occur in the iliofemoral vessels. The initial incidence of 15.3% of major vascular injuries as reported in the PARTNER trial using first-generation devices has decreased significantly over the past several years [1]. This is due to technical improvements in the devices themselves, patient selection, and the use of alternative access sites based on high-quality imaging and preoperative planning. A recent study suggests a current major vascular complication rate of 6.1% [2].

The risk factors for iliofemoral injury include both patient and technical factors that should be carefully evaluated preoperatively. The luminal diameter of the vessels along with the degree of calcification and tortuosity is measured. The sheath-to-femoral artery ratio can be a key indicator for vascular injury [3]. An "iliac morphology score" comprising minimal luminal diameter and degree of calcification has also been shown to be predictive of major vascular complications and 30-day mortality [4]. Tortuosity alone has not consistently been found to be predictive of vessel injury. Gender is a risk factor, with females more likely to have a major vascular complication [1, 2, 4].

During transfemoral sheath removal, it is reasonable to pull the sheath back into the distal external iliac artery and pause for a few minutes. This will "uncover" any injury in the iliac system and allow the sheath to be advanced with the dilator back into the abdominal aorta for hemostatic control if needed. Others have advocated routine use of the cross-over balloon occlusion technique for temporary control during sheath removal [5]. Ultimately, definitive control usually requires a covered stent in the iliac or, in the case of an ostial rupture, a covered stent into the abdominal aorta.

If there is significant concern about the iliofemoral vessels, alternative insertion sites such as the subclavian artery should be considered. This access is performed directly with a sheath or through a Dacron graft sewn onto the subclavian artery for better control. Ultimately, as more centers gain experience and with careful surgical planning, these vascular complications will likely continue to decrease.

S. K. Balaram, L. Bassin, *Cardiac Surgical Complications*, https://doi.org/10.1007/978-3-030-71563-2_8

8.2 Torrential Aortic Regurgitation

The TAVR patient is a small and frail 75-year-old with severe aortic valve stenosis and a small LV cavity. The aortic valve is crossed and the standard wire is exchanged for a stiff Safari wire. Suddenly, the patient's hemodynamics collapse without an obvious cause. The rhythm is sinus at 80, but the blood pressure is 50/30 mmHg, and she is in profound respiratory distress. The anesthesiologists perform a rapid sequence induction and intubation. They are pushing IV epinephrine with little effect. The decision is made to inject 25 mcg of epinephrine directly into the aortic root via the pigtail catheter in the non-coronary sinus which results in a mild improvement in the blood pressure. Echocardiography shows severe aortic regurgitation thought to be secondary to the wire crossing. The LV wire is hugging the outer curve of the aorta and jammed against the non-coronary sinus; it is splinting the aortic valve open. The valve is quickly deployed into position, and the patient's hemodynamics improve.

8.2.1 Discussion

Torrential aortic regurgitation (AR) can occur prior to TAVR deployment due to wire splinting or aggressive pre-TAVR balloon valvuloplasty. This sometimes also occurs during TAVR deployment with the use of self-expanding valves. Wire splinting is often simply corrected by centering the wire within the aortic valve annulus. The other causes of predeployment AR are really only treated by rapid valve deployment. The hemodynamic effects of torrential AR are compounded in patients without pre-existing insufficiency and in those with a very small and non-compliant LV chambers. These patients typically have severe left ventricular hypertrophy.

The effects of sudden AR are mitigated with the addition of a mild tachycardia using either pacing or chronotropes. This will reduce the time spent in diastole, and thus reduce the regurgitant time, all while awaiting valve deployment. Severe AR is poorly tolerated in the setting of a sick ventricle. Once ventricular fibrillation ensues, it is very difficult to resuscitate these patients. A dilated LV with severe AR is difficult to defibrillate. Initiation of ECMO may be required. In these cases of AR due to wire placement, with ventricular arrhythmias and hemodynamic instability, the valve should be deployed safely and immediately.

8.3 Left Ventricular Wire Perforation

An 82-year-old man with critical aortic stenosis presents for transfemoral TAVR. He has preserved LV function and severe ventricular hypertrophy. Vascular access is straightforward, but the valve is crossed with some difficulty. After measuring the intracavitary pressures, the surgeon places a Safari wire into the LV cavity with a normal curve at the apex. The systolic blood pressure immediately drops, and the patient becomes anxious and tachypneic. Aortic root injection shows torrential aortic insufficiency. The wire is hugging the outer curvature of the arch and ascending aorta and seems to be pressed against the non-coronary sinus. The wire is pulled back so that it is centered through the aortic valve—the AI reduces, and the hemodynamics return to normal. The TAVR is deployed but the wire looks compressed. It is removed using a pigtail catheter to protect the vascular structures. The blood pressure starts to fall and TTE shows an enlarging pericardial effusion.

8.3.1 Discussion

Perforation of the LV after TAVR is a rare but catastrophic complication. It is a leading cause of emergency cardiac surgery in these patients [6]. The current incidence of emergency surgery is 0.7%, with LV perforation from guidewires and annular rupture at deployment accounting for more than 50% of the cases [6]. A stiff wire is needed to guide the TAVR from the femoral artery into the LV. At the dawn of TAVR, this was achieved by manually creating a pigtail curve in a stiff wire (usually Amplatz Super Stiff), but there are now commercially made stiff wires with a cushioned soft spiral "pigtail" at the tip for use in TAVR [6, 7].

Risk factors for LV perforation include a small LV cavity, a hypercontractile ventricle, and a narrow aorto-mitral angle directing the wire toward the ventricular free wall [8]. In addition, a tortuous aorta builds a significant amount of tension into the delivery system with forward movement of the valve. The Safari wires come in a range of sizes and should be selected based on the size of the LV cavity. A careful watch for the tip of the wire is required to not allow any tension to build up and push the pigtail wire against the wall of the LV apex.

If in an injury does occur, it can be fatal, as the wire macerates the LV muscle in patients that are inherently frail [6]. These patients carry an in-hospital mortality rate of greater than 50% [6]. A percutaneous pericardial drain should be placed, and once aortic root injection shows no evidence of rupture, the patient needs to be intubated and resuscitated, in preparation for surgery. Urgent sternotomy is often required. It is helpful to use ventricular pacing to control any severe hypertension that results after the pericardium is opened. Pledgeted sutures (3/0 polypropylene on an SH needle) are used to control the bleeding. The location of the injury sometimes involves or is in proximity to a coronary artery. If the team is unable to control hemodynamics, cardiopulmonary bypass is needed. An IABP is helpful postoperatively to reduce LV systolic pressures and allow time for hemostatic healing.

8.4 Coronary Obstruction

A few minutes following deployment of a valve-in-valve TAVR, the patient becomes hypotensive and requires large doses of adrenaline. Echocardiography shows marked LV dysfunction, and there is widespread ST segment elevation. A high-dose contrast aortic root angiogram shows poor flow into the left main. The patient's condition remains unstable and is intubated, and the team prepares for peripheral V-A ECMO. The left femoral artery 6 Fr sheath is replaced with an 18 Fr ECMO cannula over a stiff wire. Similarly, the 6 Fr sheath in the right femoral vein that was used for temporary pacing is removed, and a stiff wire is guided into the SVC. A 25 Fr venous cannula is placed into the SVC with imaging guidance.

Flows of >4 L/min are obtained. Interventional colleagues attempt to wire the left main coronary artery and place a stent, "snorkeling" out above the sinotubular junction.

8.4.1 Discussion

Occlusion of the coronary arteries during TAVR is a dreaded complication, with a 30-day mortality of 35% to 40% [9, 10]. It most commonly affects the left main coronary artery (80%) and occurs within 1 hour of the procedure in the vast majority of cases [9, 10].

The risk factors for coronary occlusion are low coronary height (<10 mm), tall leaflet height, small aortic root (< 28 mm), and low sinotubular junction (STJ) [10]. The surgeon must be able to assess the anatomy clearly on CT and provide an accurate risk assessment. If there is indeed a high likelihood of obstruction, surgical AVR should be considered if the patient can tolerate surgery.

For inoperable patients, there are two main techniques to mitigate the risk of obstruction: (1) leaving a stent in the coronary, ready for deployment [11], or (2) lacerating the valve leaflet—the BASILICA (Bioprosthetic or native Aortic Scallop Intentional Laceration to prevent Iatrogenic Coronary Artery obstruction) procedure [12]. The first method involves placing a coronary wire into the at-risk coronary artery and leaving a long stent "parked" in the mid-coronary artery. The guide catheter is placed just proximal to the coronary ostia. If the coronary becomes obstructed during valve deployment, the stent is withdrawn so that it starts above the STJ and ends within the coronary. It is expanded, thus providing a "snorkel" for blood to enter the coronary.

A multicenter retrospective review of patients at risk for coronary obstruction showed that the 3-year cardiac mortality was lower in those that had a stent placed vs those that just had a prophylactic wire placed in the coronary artery (7.8% vs 15.7%) [13]. There were two cases of delayed stent thrombosis, both in patients that had a self-expanding valve-in-valve (ViV) TAVR [13]. Of the 93 patients *without* stent placement (i.e., no evidence of coronary occlusion intraoperatively), there were 4 delayed coronary occlusions, with 3 ViV cases [13].

The second method of prevention is the BASILICA technique, for which a microcatheter is positioned above the leaflet in question and a snare is placed in the LV just underneath the leaflet [12]. A straight wire is then inserted via the microcatheter and placed in contact with the aortic valve leaflet. The wire is then electrified using radiofrequency to perforate the leaflet and is snared below. This creates a loop from the microcatheter—through the leaflet and then into the snare. The wire is again electrified and pulled through to completely lacerate the leaflet to the free edge. The leaflet then splays apart during systole, but coapts during diastole. In the initial trial of 30 patients, 13 underwent native valve TAVR and the other 17 valve-in-valve (ViV) TAVR [12]. The study reported one death, three strokes, and no cases of coronary obstruction during the study period following leaflet laceration [12]. A larger registry of 214 patients undergoing leaflet laceration (72.8% of which were ViV TAVR) showed a 30-day mortality of 2.8% and 30-day stroke rate of 2.8% [14]. In this series, 4.7% suffered partial or complete coronary obstruction [14].

8.5 Paravalvular Leak

An 84-year-old woman presents with severe aortic stenosis and previous coronary bypass surgery. All grafts are patent. The aortic root and annulus are severely calcified, with a large bar of calcium located in the aorto-mitral curtain. After heart team discussion, a self -expanding valve is selected to reduce the chance of aortic root rupture. The valve deployment height is satisfactory, but there is a severe paravalvular leak at the non-coronary sinus.

8.5.1 Discussion

Paravalvular leak (PVL) is a known complication of TAVR, though the results are far better with the third-generation TAVR valves [15]. The fourth-generation balloon-expandable valves appear to portend even further improvement, but are still associated with a 10–20% rate of mild paravalvular leak [16, 17].

A common risk factor for PVL in self-expanding valves is increased calcium of the valve and aortic root [18]. A thick calcium ridge prevents the valve from expanding fully and becoming tightly fixed against the surrounding tissue. Another risk factor is incorrect deployment height relative to the annulus. Bicuspid valves have a higher rate of paravalvular leak in most registries, although this may continue to decrease with newer-generation devices [19].

If the deployment height is satisfactory, then a careful examination of the prosthesis stent should be conducted with TTE and fluoroscopy to determine whether the valve is fully expanded or if a focal collapse is present due to immobile calcium. In the first instance, a balloon aortic valvuloplasty (BAV) is performed, being aware of the increased risk of annular rupture and cognizant of any additional risk factors for this complication. Performing a post-TAVR balloon dilation aortic valvuloplasty is associated with an increased risk of stroke [20]. This balloon also has the potential to injure the porcine valve leaflets of a self-expanding valve. There is a risk of valve embolization in the setting of pulsatility; it is crucial to ensure extremely reliable and rapid ventricular pacing.

If the TAVR height is unsatisfactory, careful assessment should be made as to whether the valve is able to be recaptured, or if an additional TAVR needs to be deployed at the correct height. This, of course, carries an increased risk of coronary obstruction given that the first TAVR leaflets will subsequently be stented open.

In some instances, significant post-deployment PVL is tolerated in the hybrid room, with the view to re-evaluation on TTE in the days to follow. In the case of self-expanding valves, the stent frame continues to expand over the hours and days following deployment. Thus, mild PVL has the potential to improve prior to discharge. If a focal leak with persistent symptoms is present, consideration should be given to use of a transcatheter occlusion device [21]. The final option for a large PVL is open surgery with removal of the TAVR and native leaflets, followed by surgical valve replacement.

8.6 Iatrogenic Ventricular Septal Defect

The 70-year-old TAVR patient has a mild paravalvular leak which typically would just be monitored. However, this patient's express wish for TAVR is to enable her to return to caring for her dependent husband more quickly. The surgeon is concerned about the long-term impact of leaving the small PVL alone. A balloon post-dilation is performed to fix the leak. Unfortunately, angiography shows that now the leak has worsened and that the left ventricular outflow tract has an additional jet across the septum.

8.6.1 Discussion

While a 2016 systematic review could only identify 18 case reports of iatrogenic ventricular septal defects after TAVR [22], another series reported an incidence of up to 1.5% [23]. This complication may be more common than previously believed. The basis of the anatomic etiology is focal annular rupture. This extends immediately into the right side or occurs as a result of delayed rupture of scar tissue from periprocedural trauma [24, 25].

A VSD occurs when calcium is directly compressed into the membranous septum [26]. This can manifest as AV block, hematoma of the septum, a fistula into the septal tissue, or a full-thickness VSD with left to right shunt (Table 8.1) [26].

A number of risk factors have been described, with subannular calcification in the setting of a large prosthesis as the most concerning (Table 8.2) [25]. There are similar risk factors for annular rupture, and described common characteristics of these patients help delineate those for whom extra care is required [26].

Management is conservative for small VSDs, and medical treatment is often successful even in the setting of initial heart failure symptoms [27]. Serial echo exams will be required to insure no extension. If increasing heart failure results, the VSD should be closed either surgically or percutaneously. Percutaneous closure should be the first choice if intervention is deemed necessary. The Amplatzer™ muscular VSD occluder (St. Jude Medical, St. Paul MN) closure devices have been effective for this indication [22].

An alternative to treat a very high VSD, which is especially useful in the setting of a paravalvular leak, is the deployment of a second valve-in-valve TAVR, deliberately placed deeper into the left ventricular outflow tract to cover the defect [28].

Table 8.1 Site and effects of subvalvular rupture after TAVR

Site of subvalvular rupture	Effects
Free myocardial wall	Acute myocardial rupture, pseudoaneurysm
Underneath the non-coronary sinus at the anterior mitral leaflet	Shunt from LVOT to LA
Interventricular septum	Hematoma, conduction disturbances (AV block), fistula formation

Table 8.2 Risk factors and common characteristics in subvalvular rupture

Risk factors for VSD	Common characteristics
Sub-annular calcification	Dialysis
Prosthesis oversizing	CKD
Narrow, calcified, oval-shaped annulus	Women
Post-dilation after TAVR >20%	Hx mitral valve surgery
Heavily calcified AV	Chest wall radiation

8.7 Valve Embolization

A straightforward transfemoral TAVR in a robust 75-year-old with excellent peripheral vessels and generous aortic root anatomy is in the hybrid suite. The balloon-expandable TAVR is placed across the annulus, and rapid pacing is commenced at 180 beats per minute via a femoral transvenous pacing wire in the right ventricle. Valve deployment is initiated, and at 75% deployment, the pacing wire loses capture—the heart returns to sinus rhythm with a thump, the blood pressure rises, and the valve is promptly ejected into the ascending aorta.

8.7.1 Discussion

Valve embolization occurs in <1% of TAVR cases and is becoming more infrequent with time [29, 30]. It remains a potential complication for both self-expanding and balloon-expandable valves, albeit with different mechanisms. Balloon-expandable valves tend to eject if there is loss of pacing capture during deployment, which was the cause of 60% of cases in one recent study [7]. The pacing threshold must be checked and the position verified, and the pacing setting should be a mandatory mode—VOO so that there is no risk of sensing leading to inhibition of pacing. The rate must also be high enough to abolish any pulsatility in the arterial waveform. The risk of pacemaker failure and ejection also occur with any post-deployment balloon aortic valvuloplasty in the setting of either balloon-expandable or self-inflating valve. At maximal inflation, the balloon becomes adherent to the prosthesis, so that if there is any ventricular ejection (such as with loss of pacing), the valve and balloon are carried out of the annulus and into the ascending aorta.

Embolization of self-expanding valves has been reported to occur more frequently and entails a slightly different mechanism, one of malpositioning [30]. In addition to valve undersizing, a horizontal aortic root makes it quite hard to line up the proximal portion of the stent with the annulus. This means that although the valve height appears satisfactory at partial deployment, it can be difficult to predict what the final alignment will be on release.

If a valve has ejected, the first step is to ensure hemodynamic stability. A self-expanding valve that gets lodged in the sinotubular junction often results in severe aortic insufficiency and possible coronary obstruction. Balloon-expandable valves that eject tend to "tumble" in the ascending aorta quite freely. Approximately 80% of valves embolize to the aorta, while the remainder are found in the ventricle [30]. The latter more commonly require surgical retrieval (40% vs <15%) [30]. Embolization was previously the leading cause of emergency cardiac surgery but now accounts for just 10–15% of cases [6].

The two steps to ensuring a good outcome are (1) snare the ejected valve and move it distally and (2) prepare a new valve for deployment. A large snare (20 mm) is usually required, inserted via the secondary arterial access. This usually means that a third access site (radial artery) will be required for contrast injection of the aortic root. Balloon-expandable valves are snared and pulled into the descending aorta where they will lodge, and self-expanding valves are pulled into the distal ascending aorta. Unfortunately, valve embolization results in a significantly higher all-cause mortality at both 30 days and 1 year [7, 30]. This event is best avoided through meticulous planning, accurate sizing, proper positioning, and verified pacing.

8.8 Aortic Root Rupture

A 70-year-old frail woman with severe aortic stenosis is referred for consideration of TAVR vs surgery. The planning CT shows a small aortic root and annulus with significant calcium in the aorto-mitral curtain. TAVR is certainly justified for the small root here present, as she would surgically fit only 19 mm valve without a root enlargement procedure. A small surgical valve would certainly result in patient-prosthesis mismatch. A balloon-expandable valve is selected to minimize a paravalvular leak in the setting of annular calcium. The case goes smoothly, the valve is deployed, and the height appears perfect. There are a trivial posterior jet and no gradient. Suddenly, the patient complains of pain, and the blood pressure begins to fall. Echocardiography shows a 10 mm pericardial effusion, and aortic root injection confirms contrast extravasation. A percutaneous pigtail drain is placed, but the patient is increasingly unstable and receiving epinephrine on her way to the operating theater.

8.8.1 Discussion

Aortic root rupture in TAVR is a rare but devastating complication with a low incidence of 0.5–1.1% and a high mortality of 50% [26, 31]. The time between the event and hemodynamic deterioration ranges from immediate to several days, depending on the injury. Small tears in the annulus are not always recognized, as they can be sealed by the skirt of the TAVR itself. These patients present either immediately in the hybrid suite or postoperatively with sudden hypotension and chest pain. Immediate echocardiography and aortic root angiography are indicated. Diagnostic findings vary. Annular rupture may also present as a hematoma of the root that results in the occlusion of a coronary artery. A pericardial effusion is often present.

Risk factors for this complication are similar to that of TAVR VSD and include the presence of a heavily calcified, small annulus. Oversizing, use of balloon-expandable valves, frailty, and a small sinotubular junction are also considered risk factors [26, 31, 32]. The anatomy of the root and location of calcium in the landing zone must be carefully reviewed (Table 8.3) [31].

Management of patients with annular rupture depends on hemodynamic stability [24, 26, 31–33] (Fig. 8.1). The stable patient with a small defect is generally managed with conservative therapy, including correction of coagulopathy and placement of a pericardial drain if an effusion is present. These patients will need close monitoring in the ICU and serial imaging with CTA.

Table 8.3 Anatomic findings based on sites of rupture

Supra-annular site	Annular site	Sub-annular site
Calcification of middle or upper aortic root	Ellipsoid annulus	Calcification in the LVOT or septum
Flat sinuses of Valsalva	Calcification at the base of cusps	Narrow LVOT
Calcified body or free edge of cusps	Calcified rim of annulus	

If the aortic root has a large rupture, pericardial drainage and urgent transfer to the OR are required. While there have been case reports of aortic rupture treated with a second TAVR deployed [24, 33], surgery is usually offered if the patient is a candidate. If the patient is hemodynamically unstable, the surgical team should be ready for femoral-femoral CPB and sternotomy. Peripheral bypass assists in regaining some control; however, many TAVR patients have poor peripheral vessels, mandating immediate sternotomy and central cannulation.

The type of repair needed is dependent on the site of injury. Repairing the aortic root starts with identifying the tear present in either the annulus or sinotubular junction. This information will determine whether repair can be achieved with a patch (usually bovine pericardium) or a full root replacement with reimplantation of the coronary arteries.

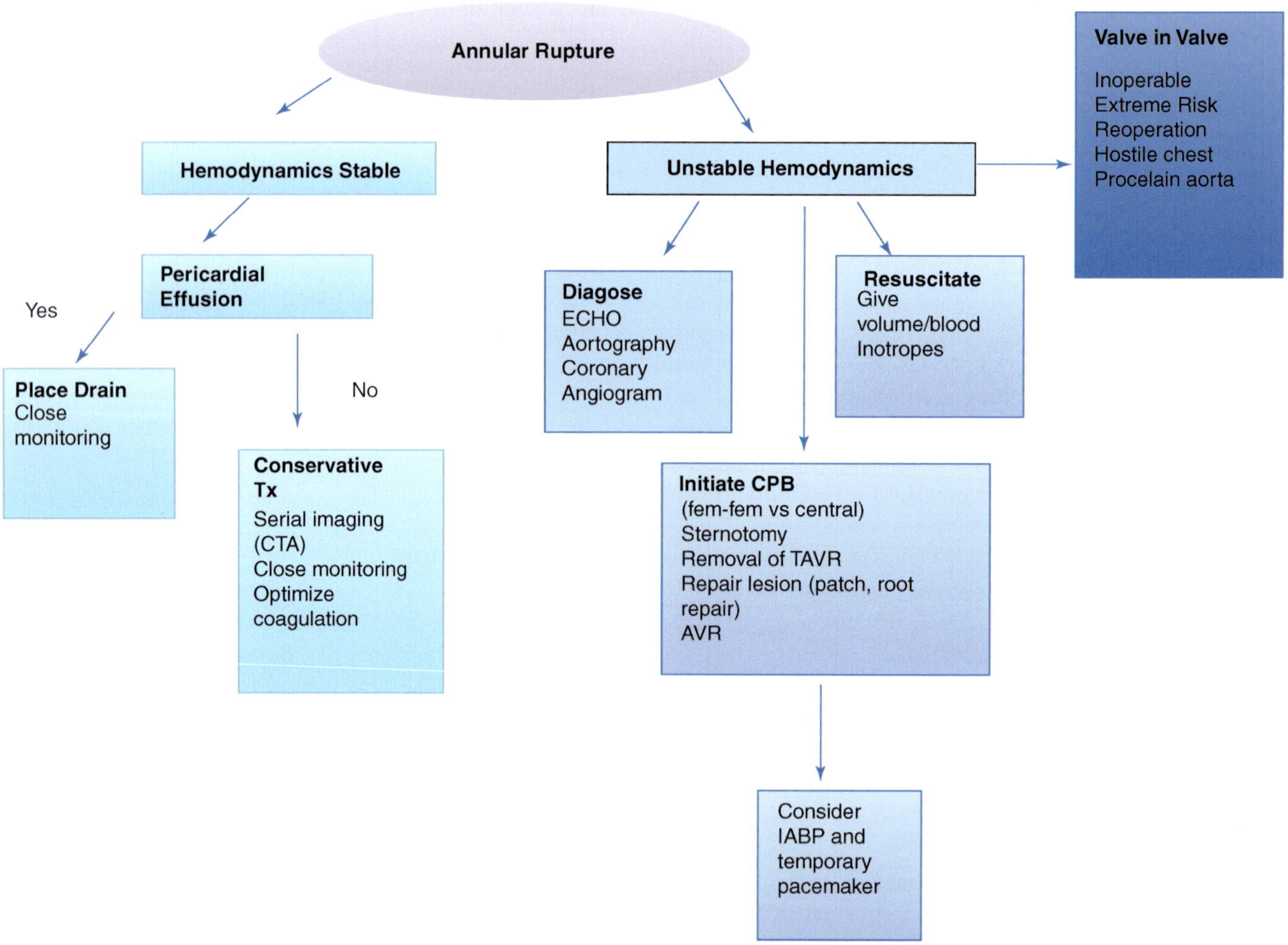

Fig. 8.1 Management of annular rupture post-TAVR

References

1. Généreux P, Webb JG, Svensson LG, Kodali SK, Satler LF, Fearon WF, Davidson CJ, Eisenhauer AC, Makkar RR, Bergman GW, Babaliaros V, Bavaria JE, Velazquez OC, Williams MR, Hueter I, Xu K, Leon MB, Trial PARTNER, Investigators. Vascular complications after transcatheter aortic valve replacement: insights from the PARTNER (Placement of AoRTic TraNscathetER Valve) trial. J Am Coll Cardiol. 2012;60(12):1043–52.
2. Ullery BW, Jin R, Kirker EB, Hayes G, Siwek L, Brevig J, et al. Trends in vascular complications and associated treatment strategies following transfemoral transcatheter aortic valve replacement. J Vasc Surg. 2020;72(4):1313–24.e5
3. Hayashida K, Lefèvre T, Chevalier B, Hovasse T, Romano M, Garot P, Mylotte D, Uribe J, Farge A, Donzeau-Gouge P, Bouvier E, Cormier B, Morice MC. Transfemoral aortic valve implantation new criteria to predict vascular complications. JACC Cardiovasc Interv. 2011;4(8):851–8.
4. Blakeslee-Carter J, Dexter D, Mahoney P, Ahanchi S, Steerman S, Larion S, et al. A novel iliac morphology score predicts procedural mortality and major vascular complications in transfemoral aortic valve replacement. Ann Vasc Surg. 2018;46:208–17.
5. Toggweiler S, Leipsic J, Binder RK, Freeman M, Barbanti M, Heijmen RH, et al. Management of vascular access in transcatheter aortic valve replacement: part 2: vascular complications. JACC Cardiovasc Interv. 2013;6(8):767–76.
6. Eggebrecht H, Vaquerizo B, Moris C, Bossone E, Lämmer J, Czerny M, et al. Incidence and outcomes of emergent cardiac surgery during transfemoral transcatheter aortic valve implantation (TAVI): insights from the European Registry on Emergent Cardiac Surgery during TAVI (EuRECS-TAVI). Eur Heart J. 2018;39(8):676–84.
7. Liang Y, Dhoble A, Pakanati A, Zhao Y, Kork F, Ruan W, Markham T, Smalling R, Balan P, Estrera A, Nguyen TC, Gregoric I, Kar B, Eltzschig H. Catastrophic cardiac events during transcatheter aortic valve replacement. Can J Cardiol. 2021; S0828-282X(21)00246-4
8. Owais T, El Garhy M, Fuchs J, Disha K, Elkaffas S, Breuer M, Lauer B, Kuntze T. Pathophysiological factors associated with left ventricular perforation in transcatheter aortic valve implantation by transfemoral approach. J Heart Valve Dis. 2017;26(4):430–6.
9. Akinseye OA, Jha SK, Ibebuogu UN. Clinical outcomes of coronary occlusion following transcatheter aortic valve replacement: a systematic review. Cardiovasc Revasc Med. 2018;19(2):229–36.
10. Ribeiro HB, Webb JG, Makkar RR, Cohen MG, Kapadia SR, et al. Predictive factors, management, and clinical outcomes of coronary obstruction following transcatheter aortic valve implantation: insights from a large multicenter registry. J Am Coll Cardiol. 2013;62(17):1552–62.
11. Mercanti F, Rosseel L, Neylon A, Bagur R, Sinning JM, et al. Chimney stenting for coronary occlusion during TAVR: insights from the chimney registry. JACC Cardiovasc Interv. 2020;13(6):751–61.
12. Khan JM, Greenbaum AB, Babaliaros VC, Rogers T, Eng MH, et al. The BASILICA Trial: prospective multicenter investigation of intentional leaflet laceration to prevent TAVR coronary obstruction. JACC Cardiovasc Interv. 2019;12(13):1240–52.
13. Palmerini T, Chakravarty T, Saia F, Bruno AG, Bacchi-Reggiani ML, et al. Coronary protection to prevent coronary obstruction during TAVR: a Multicenter international registry. JACC Cardiovasc Interv. 2020;13(6):739–47.
14. Khan JM, Babaliaros VC, Greenbaum AB, Spies C, Daniels D, et al. Preventing coronary obstruction during transcatheter aortic valve replacement: results from the multicenter international BASILICA registry. JACC Cardiovasc Interv. 2021;14(9):941–8.
15. Pibarot P, Hahn RT, Weissman NJ, Arsenault M, Beaudoin J, Bernier M, et al. Association of paravalvular regurgitation with 1-year outcomes after transcatheter aortic valve replacement with the SAPIEN 3 valve. JAMA Cardiol. 2017;2(11):1208–16.
16. Welle GA, El-Sabawi B, Thaden JJ, Greason KL, Klarich KW, Nkomo VT, et al. Effect of a fourth-generation transcatheter valve enhanced skirt on paravalvular leak. Catheter Cardiovasc Interv. 2021;97(5):895–902.
17. Rheude T, Pellegrini C, Lutz J, Alvarez-Covarrubias HA, Lahmann AL, Mayr NP, et al. Transcatheter aortic valve replacement with balloon-expandable valves: comparison of SAPIEN 3 ultra versus SAPIEN 3. JACC Cardiovasc Interv. 2020;13(22):2631–8.
18. Pollari F, Dell'Aquila AM, Söhn C, Marianowicz J, Wiehofsky P, Schwab J, et al. Risk factors for paravalvular leak after transcatheter aortic valve replacement. J Thorac Cardiovasc Surg. 2019;157(4):1406–15.e3.
19. Kim WK, Bhumimuang K, Renker M, Fischer-Rasokat U, Möllmann H, Walther T, et al. Determinants of paravalvular leakage following transcatheter aortic valve replacement in patients with bicuspid and tricuspid aortic stenosis. Eur Heart J Cardiovasc Imaging. 2021:jeab011.
20. Wang N, Lal S. Post-dilation in transcatheter aortic valve replacement: a systematic review and meta-analysis. J Interv Cardiol. 2017;30(3):204–11.
21. Waterbury TM, Reeder GS, Pislaru SV, Cabalka AK, Rihal CS, Eleid MF. Techniques and outcomes of paravalvular leak repair after transcatheter aortic valve replacement. Catheter Cardiovasc Interv. 2017;90(5):870–7.
22. Ando T, Holmes AA, Taub CC, Slovut DP, DeRose JJ. Iatrogenic ventricular septal defect following transcatheter aortic valve replacement: a systematic review. Heart Lung Circ. 2016;25(10): 968–74.
23. Zeniou V, Chen S, Gilon D, Segev A, Finkelstein A, Planer D, et al. Ventricular septal defect as a complication of TAVI: mechanism and incidence. Structural Heart. 2018;2(3):235–9.
24. Langer NB, Hamid NB, Nazif TM, Khalique OK, Vahl TP, White J, et al. Injuries to the aorta, aortic annulus, and left ventricle during transcatheter aortic valve replacement: management and outcomes. Circ Cardiovasc Interv. 2017;10(1):e004735.
25. Sardar MR, Khan ZA, Abraham A, Saeed W, Farhan M, Kolte D, Sharaf B, Gordon P, Sellke F, Sodha NR, Aronow HD, Ehsan A. Management strategies and possible risk factors for ventricular septal defects after transcatheter aortic valve replacement: case series from a single center and review of literature. Cardiovasc Revasc Med. 2017;18(6):462–70.
26. Pasic M, Unbehaun A, Buz S, Drews T, Hetzer R. Annular rupture during transcatheter aortic valve replacement: classification, pathophysiology, diagnostics, treatment approaches, and prevention. JACC Cardiovasc Interv. 2015;8(1 Pt A):1–9.
27. O'Gorman K, Smith A. A unique case of heart failure: iatrogenic ventricular septal defect following transcatheter aortic valve replacement. J Am Coll Cardiol. 2021;77(18 Supplement1):2411.
28. Hönemann KD, Hofmann S, Ritter F, Mönnig G. More in, better out? Successful valve-in-valve procedure of an iatrogenic ventricular septal defect following transcatheter aortic valve replacement: a case report. Eur Heart J Case Rep. 2021;5(5):ytab097.
29. Binder RK, Webb JG. Transcatheter heart valve migration and embolization: rare and preventable? Eur Heart J. 2019;40(38):3166–8.
30. Kim W-K, Schäfer U, Tchetche D, Nef H, Arnold M, Avanzas P, et al. Incidence and outcome of peri-procedural transcatheter heart valve embolization and migration: the TRAVEL registry (TranscatheteR HeArt Valve EmboLization and Migration). Eur Heart J. 2019;40(38):3156–65.

31. Schymik G, Heimeshoff M, Bramlage P, et al. Ruptures of the device landing zone in patients undergoing transcatheter aortic valve implantation: an analysis of TAVI Karlsruhe (TAVIK) patients. Clin Res Cardiol. 2014;103:912–20.
32. Barbanti M, Yang TH, Rodès Cabau J, Tamburino C, Wood DA, Jilaihawi H, et al. Anatomical and procedural features associated with aortic root rupture during balloon-expandable transcatheter aortic valve replacement. Circulation. 2013;128(3):244–53.
33. Yu Y, Vallely M, Ng MK. Valve-in-valve implantation for aortic annular rupture complicating transcatheter aortic valve replacement (TAVR). J Invasive Cardiol. 2013;25(8):409–10.

Index

S. K. Balaram, L. Bassin, *Cardiac Surgical Complications*, https://doi.org/10.1007/978-3-030-71563-2

Printed by Printforce, the Netherlands